Post-Transcriptional Gene Regulation

METHODS IN MOLECULAR BIOLOGY™

John M. Walker, SERIES EDITOR

METHODS IN MOLECULAR BIOLOGY™

Post-Transcriptional Gene Regulation

Edited by

Jeffrey Wilusz

*Department of Microbiology, Immunology and Pathology,
Colorado State University, Fort Collins, CO, USA*

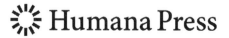 Humana Press

Editor
Jeffrey Wilusz
Department of Microbiology
Immunology and Pathology
Colorado State University
Fort Collins, CO, USA

Series Editor
John M. Walker
University of Hertfordshire
Hatfield, Herts., UK

ISBN: 978-1-58829-783-9   e-ISBN: 978-1-59745-033-1

Library of Congress Control Number: 2007930310

Cover Illustration: Figure 1, Chapter 13, "Monitoring the Temporal and Spatial Distribution of RNA in Living Yeast Cells," by Roy M. Long and Carl R. Urbinati.

Printed on acid-free paper

9 8 7 6 5 4 3 2 1

springer.com

Preface

This volume in the *Methods in Molecular Biology*™ series is organized into three sections. First, a series of bioinformatic approaches are presented in Chapters 1–3 to address the use of RNA databases and algorithms in the study of post-transcriptional regulation involving untranslated regions of transcripts. In the second section, a series of methods applicable to fundamental issues in mRNA biology are presented. These include RNA structure/function (Chapter 4), mRNP analysis (Chapters 5–8), and novel methods for mRNA labeling and isolation (Chapters 9–10). The third section of this volume presents 11 chapters that outline methodologies to study particular aspects of post-transcriptional control. This section includes methods for the study of alternative splicing and 3'-end processing (Chapters 11 and 12), mRNA localization (Chapters 13 and 14), mRNA translation (Chapters 15 and 16), mRNA stability (Chapters 17–19), and si/miRNA regulation (Chapters 20 and 21). Collectively, therefore, this volume strives to present current technical approaches to most aspects of post-transcriptional control and provide the reader with a useful and versatile laboratory bench resource.

Section I: Bioinformatics

Given the plethora of data available to the molecular biologist from available databases and high-throughput experimental techniques, bioinformatics and computational biology approaches are vital to the analysis of post-transcriptional control. This volume, therefore, contains three chapters that deal with aspects of bioinformatic analysis of mRNA processes. First, Bagga presents a chapter containing a very useful overview of the software tools and databases currently available that can be effectively applied to study the function of untranslated regions of mRNAs. As the untranslated regions (UTRs) of mRNAs undoubtedly contain numerous regulatory motifs that influence gene expression—many of which remain to be characterized—this chapter should prove to be a useful starting point for such experiments. Second, Lee and colleagues present a chapter detailing their computational approach for assigning mRNA polyadenylation sites. Given the high occurrence of alternative polyadenylation in mRNAs, this methodology is important to obtain a clear picture of the complexity of the ribonome. The methods and approaches outlined in this chapter are also readily adaptable to other areas of molecular biology as well. Finally, Doyle and colleagues present a chapter that outlines the University of Albany Training UTR database. This collection of validated and well-defined regulatory motifs represents a very practical resource that will allow the evaluation and fine tuning of new software programs to facilitate the development of tools to discover RNA regulatory motifs.

Section II: Fundamental Aspects of the Study of RNA Biology

RNA structure plays an important role in a variety of biological processes. Regulski and Breaker present a chapter on a technique called in-line probing of RNA structure. Although their assay is presented as a powerful way to evaluate the ligand-binding characteristics of riboswitches, the technique is readily adaptable to the study of structural changes in other RNAs as well.

Messenger RNAs clearly function as dynamic ribonucleoprotein (RNP) complexes rather than naked nucleic acids. RNPs, for example, are hypothesized to play a key role in defining post-transcriptional operons. However, our understanding of the dynamics of protein–RNA interactions in the cell is still rather rudimentary. Four chapters in this volume provide exciting methodologies to address this problem from both the RNA and RNA-binding protein perspectives. Beach and Keene describe a ribo-trap method to specifically purify targeted native RNPs through immunoaffinity technology and identify associated RNA-binding proteins. Baroni and colleagues present a chapter on ribonomic profiling using RNA-binding protein Immunoprecipitation-microarray/Chip (RIP-Chip) to identify mRNAs associated with a selected RNA-binding protein. These complementary chapters provide the latest and most effective methodologies currently available to unravel the dynamics of mRNP structure on a global scale. The study of reconstituted RNA–protein interactions also provides a great deal of insight into structure–function and biochemical relationships. The chapter by Tombelli and coworkers describes current methodologies to study the interaction of RNA aptamers and RNA-binding proteins using surface plasmon resonance. Finally, artificially tethering proteins to transcripts has become a powerful way to assess the influence of proteins on post-transcriptional processes. Clement and Lykke-Andersen present a chapter containing their detailed methods for tethering proteins to mRNAs through the MS2 coat protein. Although the chapter focuses on an application of protein tethering to the study of mRNA decay, the reader should note that the tethering technology is fully applicable to the study of a variety of post-transcriptional processes.

Two chapters in this volume deal with methodologies for labeling and isolating mRNAs in cells. First, the ability to label RNAs in vivo without using radioactivity and without significant biological consequences is a potentially very powerful addition to the toolbox to analyze gene expression. In the chapter by Zeiner et al., a methodology using cell-specific or pulsed expression of a Toxoplasma UPRT enzyme (that is usually missing in mammals) is outlined. This enzyme will specifically incorporate 4-thiouridine into RNA. The ability to biosynthetically tag RNA in a controlled fashion in vivo using this technology has numerous applications, including measurements of RNA decay rates and purification of tagged species. Second, poly(A) tail length is clearly regulated in cells. mRNAs that contain short poly(A) tails are often difficult to isolate using conventional oligo(dT) methods. To circumvent this problem, Bajak and Hagedorn present a chapter outlining a technique where mRNAs can be effectively purified on the basis of their 5' cap structure. This system should also prove very useful in the analysis of RNAs that are capped but are not naturally polyadenylated.

Section III: Techniques for Specific Aspects of RNA Biology

As the number of protein-encoding genes in an organism does not correlate with its cellular complexity, processes such as alternative splicing and polyadenylation must play a large role in defining the complexity of the proteome of a specific cell. Analyses of Expressed Sequence Tag (EST) and microarray data suggest that over two thirds of human genes contain one or more alternative exons. Identifying the full array of alternative splicing in a given gene is often difficult to infer from EST databases (as they are often incomplete), and computer-aided reconstructions of the possible complement of alternatively spliced species from a given gene do not necessarily reflect the array of spliced products that are produced in vivo. To address this problem, a PCR-based methodology to identify alternatively spliced forms from a given gene of interest is presented in the chapter by Venables. This methodology should prove invaluable to any laboratory interested in identifying the full range of alternatively spliced products generated by their gene of interest in a variety of cell types. Recent studies have demonstrated that almost 50% of genes contain more than one polyadenylation site. To effectively address questions of alternative poly(A) site usage, Hague and colleagues present a chapter outlining in vivo methodologies to assess relative usage of polyadenylation signals.

Collectively, these two chapters provide a good foundation for the reader to experimentally address questions of alternative mRNA processing.

mRNA localization in cells is another important and regulated aspect of gene expression. Two chapters in this volume describe interesting methodologies to address this issue. First, Long and Urbinati present a chapter discussing the use of tethered green fluorescent protein to a selected mRNA. In combination with microscopy, this technology allows for the visualization of the dynamics of mRNA localization in live cells. Next, Stephens and coworkers present a chapter discussing the analysis of mRNA partitioning between the cytosol and the endoplasmic reticulum. Given that recent data demonstrate that mRNAs lacking an encoded signal sequence are translated on the endoplasmic reticulum, this technology may prove very useful to obtain a full picture of the localization and translation of numerous mRNAs.

Regulated translation also makes a significant contribution to gene expression. Two chapters in this volume directly address methods for the assessment of relative translation efficiency and localization in cells. First, Peng and colleagues present a step-by-step guide for in vivo and in vitro polysome analysis using sucrose density gradients with a particular focus on the effect of poly(A) tail length on translation. These approaches should prove very useful in the identification of regulatory elements and factors. Second, Eldad and Arava present their methods for assessing the association of ribosomes with specific regions of mRNAs. This technology can be very helpful in assessing regulation at individual stages of the translation process and gives a complete picture of ribosome interactions with a transcript of interest.

Array-based analysis of changes in gene expression in response to changing environmental stimuli suggests that more than 40% are due to alterations in mRNA decay rates. The elucidation of the underlying mechanisms responsible for these mRNA-specific changes in stability will be very important, therefore, to obtain a full understanding of regulated gene expression. This volume, therefore, contains several chapters containing methodologies to assess mRNA stability. Olivas presents a step-by-step guide to assessing mRNA decay rates by northern blotting and real-time PCR. Murray and Schoenberg take the analysis of decay rates one step further and present a detailed discussion of Invader technology and its application to the study of mRNA stability. This highly sensitive signal amplification method appears to be an effective way to assess the relative turnover rates of different portions of an mRNA, allowing for the elucidation of the pathway(s) involved in the decay of selected mRNAs. Finally, the chapter by Sokoloski et al. presents an adaptation of an in vitro mRNA decay system that was first derived in HeLa cells to alternative cell types. The apparent versatility of this system to reproduce pathways of mRNA decay, often in a regulated fashion, should prove invaluable in understanding tissue-specific and organism-specific nuances in the process of regulated mRNA stability.

Studies involving the role of non-coding RNAs in the regulation of gene expression are revolutionizing our understanding of cell biology. To address this, please note that an entire volume of methodologies for the study of micro-RNAs was published in this series in April 2006. In this volume, two chapters are presented by Ford and colleagues that address key areas of this field. First, Ford and Cheng present a state-of-the-art approach to studying miRNA function in mammalian cells. Second, applications of siRNAs to assess the biological consequences of knocking down the expression of targeted genes are becoming a routine tool for the molecular biologist. Therefore, in this second chapter, Cheng, Johnson, and Ford present a comprehensive step-by-step protocol for performing such siRNA knockdown experiments.

Finally, I personally thank all of the authors for their contributions to this volume. Science advances very effectively when we share not only our data, insights, and reagents, but also our technical expertise. I applaud the contributors for their willingness to disseminate the tricks of the trade that their laboratories have perfected.

Jeffrey Wilusz

Contents

Contributors

JOHN R. ANDERSON • *Department of Microbiology, Immunology and Pathology, Colorado State University, Fort Collins, CO, USA*

YOAV ARAVA • *Department of Biology, Technion – Israel Institute of Technology, Haifa, Israel*

PARAMJEET S. BAGGA • *Bioinformatics Department, School of Theoretical and Applied Science, Ramapo College of New Jersey, Mahwah, NJ, USA*

EDYTA Z. BAJAK • *Department of Medicine and Pharmacology, University of Kansas Medical Center, Kansas City, KS, USA*

TIMOTHY E. BARONI • *Gen*NY*Sis Center for Excellence in Cancer Genomics, University at Albany-SUNY, Department of Biomedical Sciences, School of Public Health, Rensselaer, NY, USA*

DALE L. BEACH • *University of North Carolina at Chapel Hill and Department of Molecular Genetics and Microbiology, Duke University Medical Center, Durham, NC, USA*

JOHN C. BOOTHROYD • *Department of Microbiology and Immunology, Stanford University School of Medicine, Stanford, CA, USA*

RONALD R. BREAKER • *Department of Molecular, Cellular, and Developmental Biology, Department of Molecular Biophysics and Biochemistry, Yale University and Howard Hughes Medical Institute, New Haven, CT, USA*

ANGIE CHENG • *Ambion, Applied Biosystems Inc., Austin, TX, USA*

SRIDAR V. CHITTUR • *Gen*NY*Sis Center for Excellence in Cancer Genomics, University at Albany-SUNY, Department of Biomedical Sciences, School of Public Health, Rensselaer, NY, USA*

MICHAEL D. CLEARY • *Department of Microbiology and Immunology, Stanford University School of Medicine, Stanford, CA*

SANDRA L. CLEMENT • *Department of Molecular, Cellular and Developmental Biology, University of Colorado, Boulder, CO, USA*

REBECCA D. DODD • *Department of Cell Biology, Duke University Medical Center, Durham, NC, USA*

FRANCIS DOYLE • *Gen*NY*Sis Center for Excellence in Cancer Genomics, University at Albany-SUNY, Department of Biomedical Sciences, School of Public Health, Rensselaer, NY, USA*

NAAMA ELDAD • *Department of Biology, Technion – Israel Institute of Technology, Haifa, Israel*

LANCE P. FORD • *Bio Scientific Corporation, Austin, TX, USA*

ASHLEY E. FOUTS • *Department of Microbiology and Immunology, Stanford University School of Medicine, Stanford, CA, USA*

AJISH D. GEORGE • *Gen*NY*Sis Center for Excellence in Cancer Genomics, University at Albany-SUNY, Department of Biomedical Sciences, School of Public Health, Rensselaer, NY, USA*

CURT H. HAGEDORN • *Department of Medicine and Pharmacology, University of Kansas Medical Center, Kansas City, KS, USA*

LISA K. HAGUE • *Department of Biochemistry and Molecular Biology, UMDNJ-New Jersey Medical School and the Graduate School of Biomedical Sciences, Newark, NJ, USA*

TYRA HALL-POGAR • *Department of Biochemistry and Molecular Biology, UMDNJ-New Jersey Medical School and the Graduate School of Biomedical Sciences, Newark, NJ, USA*

CHARLES L. JOHNSON • *Ambion, Applied Biosystems Inc., Austin, TX, USA*

JACK D. KEENE • *Department of Molecular Genetics and Microbiology, Duke University Medical Center, Durham, NC, USA*

JU YOUN LEE • *Department of Biochemistry and Molecular Biology, New Jersey Medical School, University of Medicine and Dentistry of New Jersey, Newark, NJ, USA*

RACHEL S. LERNER • *Department of Biochemistry and Biophysics, University of North Carolina at Chapel Hill, Chapel Hill, NC, USA*

ROY M. LONG • *Department of Microbiology and Molecular Genetics, Medical College of Wisconsin, Milwaukee, WI, USA*

CAROL S. LUTZ • *Department of Biochemistry and Molecular Biology, UMDNJ-New Jersey Medical School and the Graduate School of Biomedical Sciences, Newark, NJ, USA*

JENS LYKKE-ANDERSEN • *Department of Molecular, Cellular, and Developmental Biology, University of Colorado, Boulder, CO, USA*

MARCO MASCINI • Università degli Studi di Firenze, *Dipartimento di Chimica, Sesto Fiorentino, Italy*

CHRISTOPHER D. MEIRING • *Department of Microbiology and Immunology, Stanford University School of Medicine, Stanford, CA, USA*

MARIA MINUNNI ● *Università degli Studi di Firenze, Dipartimento di Chimica, Sesto Fiorentino, Italy*

EDWARD S. MOCARSKI ● *Department of Microbiology and Immunology, Stanford University School of Medicine, Stanford, CA, USA*

ELIZABETH L. MURRAY ● *Department of Molecular and Cellular Biochemistry, The Ohio State University, Columbus, OH, USA*

CHRISTOPHER V. NICCHITTA ● *Department of Cell Biology, Duke University Medical Center, Durham, NC, USA*

WENDY M. OLIVAS ● *Department of Biology, University of Missouri-St. Louis, St. Louis, MO, USA*

JI YEON PARK ● *Department of Biochemistry and Molecular Biology, New Jersey Medical School, University of Medicine and Dentistry of New Jersey, Newark, NJ, USA*

JING PENG ● *Department of Molecular and Cellular Biochemistry, The Ohio State University, Columbus, OH, USA*

BROOK M. PYHTILA ● *Department of Cell Biology, Duke University Medical Center, Durham, NC, USA*

ELIZABETH E. REGULSKI ● *Department of Molecular, Cellular, and Developmental Biology, Yale University, New Haven, CT, USA*

ADELE RICCIARDI ● *Gen*NY*Sis Center for Excellence in Cancer Genomics, University at Albany-SUNY, Department of Biomedical Sciences, School of Public Health, Rensselaer, NY, USA*

DANIEL R. SCHOENBERG ● *Department of Molecular and Cellular Biochemistry, The Ohio State University, Columbus, OH, USA*

KEVIN SOKOLOSKI ● *Department of Microbiology, Immunology and Pathology, Colorado State University, Fort Collins, CO, USA*

ERIN K. STENSON● *Gen*NY*Sis Center for Excellence in Cancer Genomics, University at Albany-SUNY, Department of Biomedical Sciences, School of Public Health, Rensselaer, NY, USA*

SAMUEL B. STEPHENS ● *Department of Cell Biology, Duke University Medical Center, Durham, NC, USA*

SCOTT A. TENENBAUM ● *Gen*NY*Sis Center for Excellence in Cancer Genomics, University at Albany-SUNY, Department of Biomedical Sciences, School of Public Health, Rensselaer, NY, USA*

BIN TIAN ● *Department of Biochemistry and Molecular Biology, New Jersey Medical School, University of Medicine and Dentistry of New Jersey, Newark, NJ, USA*

SARA TOMBELLI ● *Università degli Studi di Firenze, Dipartimento di Chimica, Sesto Fiorentino, Italy*

CARL R. URBINATI • *Department of Biology, Loyola Marymount University, Los Angeles, CA, USA*

JULIAN P. VENABLES • *Laboratoire de génomique fonctionnelle de l'Université de Sherbrooke Centre de développement des biotechologies (CDB) de Sherbrooke, Québec, Canada*

JEFFREY WILUSZ • *Department of Microbiology, Immunology and Pathology, Colorado State University, Fort Collins, CO, USA*

CHRISTOPHER ZALESKI • *Gen*NY*Sis Center for Excellence in Cancer Genomics, University at Albany-SUNY, Department of Biomedical Sciences, School of Public Health, Rensselaer, NY, USA*

GUSTI M. ZEINER • *Department of Microbiology and Immunology, Stanford University School of Medicine, Stanford, CA, USA*

I

BIOINFORMATICS

1

Bioinformatics Approaches for Studying Untranslated Regions of mRNAs

Paramjeet S. Bagga

Summary

Interactions between RNA-binding proteins and *cis*-acting elements in the 5′- and 3′-untranslated regions (UTRs) of transcripts are responsible for regulating essential biological activities, such as mRNA localization, mRNA turnover, and translation efficiency. This chapter introduces some of the publicly available free bioinformatics resources, including software tools and databases, which can be used for predicting, mapping, and characterizing regulatory motifs found in the eukaryotic mRNA-untranslated regions.

Key Words: UTR; bioinformatics of UTR; RNA bioinformatics; untranslated regions; mRNA *cis*-acting elements; RNA regulatory motifs; RNA motif prediction; RNA motif databases.

1. Introduction

Untranslated regions of eukaryotic mRNAs contain motifs that are vital for regulation of gene expression at the post-transcriptional level. Specific interactions between RNA-binding proteins and *cis*-acting elements in 5′- and 3′-untranslated regions (UTRs) are responsible for regulating essential biological activities, such as mRNA localization, mRNA turnover, and translation efficiency. Much attention has been paid to study the composition of regulatory RNA motifs and the mechanism of their interactions with the cellular machinery [*see (1)* for a review].

The availability of sequenced human and other eukaryotic genomes makes it possible to envision bioinformatics-based approaches for identifying and

From: *Methods in Molecular Biology, Vol. 419: Post-Transcriptional Gene Regulation*
Edited by: J. Wilusz © Humana Press, Totowa, NJ

characterizing regulatory motifs. This will then allow for studying large number of genes to look at a bigger picture of the role that the mRNA-untranslated regions play in regulating gene expression. Development of smart algorithms for studying blocks of aligned UTR sequences can be very useful in identifying new regulatory motifs. The activity of some RNA motifs could be dependent on their structure. Therefore, better methods for predicting motifs must consider conservation at the level of structure in addition to sequence. Searching whole genomes with characterized motifs can help identify their relatives and also aid in the annotation of the genomic sequences. Use of transcriptomics for investigating RNA protein interactions can provide invaluable insights into the mechanism of these complex reactions. Studying the role of UTRs in regulating post-transcriptional gene expression in the post-genomic era will likely employ more of an integrated bioinformatics approach that utilizes comparative, structural as well as functional genomics.

This chapter introduces some of the publicly available free bioinformatics resources, including software tools and databases, which can be used for predicting, mapping, and characterizing *cis*-regulatory motifs found in the eukaryotic mRNA-untranslated regions.

2. Materials

All of the programs and databases described in this chapter either run directly on the web or are downloadable from a website/FTPsite and are free of charge for academic use (please cite relevant references when using these resources). You would need a PC or a Macintosh computer that is connected to the Internet and runs at least one Internet browser (*see* **Table 1** for free browser downloads). Netscape 7.1, Firefox 1.5, and Microsoft Internet Explorer (MSIE) 6.0 browsers were used for evaluation of the bioinformatics resources covered in the text. Most programs and database interfaces worked well with MSIE. However, some software worked better with Netscape. (See the **Heading 4**. section for useful tips when relevant.) It is important to enable "cookies" and turn on the Java plug-in by following specific instructions for your browser, generally through "Preferences" or "Internet Options." Some of the programs require specific plug-ins (helper programs for Internet browsers). *See* **Table 1** for free plug-in downloads.

A dial-up Internet connection will usually work. However, for better efficiency, especially when downloading large stand-alone program files, a high-speed connection would be highly desirable. Some downloads will require an FTP type of connection. Netscape browser can usually handle downloads from most FTP sites. However, some sites would require software like "Fetch" (for a Mac)

Table 1
Internet Browsers and Helper Software

Software program	Download URL
Netscape	http://browser.netscape.com/ns8/
Microsoft Internet Explorer (MSIE)	http://www.microsoft.com/windows/ie/default.mspx
Mozilla Firefox	http://www.mozilla.com/firefox/
Java plug-in	http://www.java.com/en/download/index.jsp
SVG plug-in	http://www.adobe.com/svg/viewer/install/main.html
Shockwave Flash Plugin	http://www.adobe.com/shockwave/download/download.cgi
Fetch (FTP for Mac)	http://www.dartmouth.edu/netsoftware
MacSSH & MacSFTP	http://pros.orange.fr/chombier/
SSH (secure file transfer clients for downloading software files)	http://www.chiark.greenend.org.uk/~sgtatham/putty/ http://www.openssh.org/ http://www.ssh.com/

and SSH (Secure SHell) clients (*see* **Table 1** for free downloads). Most of the downloaded files are "zipped." Windows-based utilities, like WinZip or analogous Unix programs, can be used to unzip the files before installation. Many of the downloadable programs would require a Unix or compatible environment.

Almost all sequence analysis programs described in the text require input in the FASTA format. This is a simplified format that starts with a one-line description identified with a ">" sign in the beginning. The actual nucleotide sequence, without any spaces, starts with a fresh line under the description. Most sequence databases, including GenBank/RefSeq and EMBL, provide options to view or download the sequence records in the FASTA format.

Some of the results of bioinformatics analysis are emailed to the user in the HTML format. It would be helpful to procure an email client software that can handle HTML. Netscape comes with such a program. Microsoft Outlook can also be used for this purpose. Many of the free webmail servers can handle HTML. Turning on HTML for email messages is important for correct formatting and functional cross-links.

3. Methods

Several algorithms have been developed for studying structure and function of the RNA motifs. Most of the methods that deal with RNA structure are not very dependable due to constraints inherent for studying structure. However, the quality of sequences analysis software has been relatively more acceptable.

The majority of the published programs have been developed for expert in-house laboratory use only. Limited numbers of computational tools have been made available to the public. A list of selected bioinformatics software and databases, including those covered in the text, is given in **Tables 2** and **3** along with URLs and relevant links to free full text articles that describe the applications in detail. A brief description of the applications is given here to provide a working knowledge of these resources.

Table 2
Web-Accessible Public Databases

ARED
A database of large number of human mRNAs containing AU-Rich Elements (AREs) (*see* **Section 3.1.1.**)
Web home: http://rc.kfshrc.edu.sa/ared
Reference: (3)
Full text: http://nar.oxfordjournals.org/cgi/content/full/34/suppl_1/D111

IRESite
IRESite database provides comprehensive coverage on information related to the IRES (Internal Ribosome Entry Site) elements found in viral and eukaryotic mRNAs (*see* **Section 3.2.1.**)
Web home: http://www.iresite.org
Reference: (5)
Full text: http://nar.oxfordjournals.org/cgi/content/full/34/suppl_1/D125

IRESdb
Provides basic information and some details for viral and cellular IRES elements. It also offers data on biotechnology based applications of IRES (*see* **Section 3.2.2.**)
Web home: http://ifr31w3.toulouse.inserm.fr/IRESdatabase/
Reference: (6)
Full text: http://nar.oxfordjournals.org/cgi/content/full/31/1/427

PolyA_DB
Database of poly(A) sites and PAS (polyA Signals) mapped in a large number (>25,000) of human and mouse genes (*see* **Section 3.3.1.**)
Web Home: http://polya.umdnj.edu/
Reference: (9)
Full Text: http://nar.oxfordjournals.org/cgi/content/full/33/suppl_1/D116

Rfam
A comprehensive database of RNA families that contains information about selected structural *cis*-regulatory elements found in UTRs of mRNAs
(*see* **Section 3.4.**)

Web home: http://www.sanger.ac.uk/Software/Rfam/ and http://rfam.janelia.org/
Reference: (10)
Full text: http://nar.oxfordjournals.org/cgi/content/full/33/suppl_1/D121

Transterm
Database of translation control RNA motifs and other biologically relevant regions of mRNAs (*see* **Section 3.5.**)
Web home: http://uther.otago.ac.nz/Transterm.html
Reference: (11)
Full Text: http://nar.oxfordjournals.org/cgi/content/full/34/suppl_1/D37

UTResource/UTRdb/UTRsite
A web server that provides a large collection of 5′- and 3′-untranslated region sequences of the eukaryotic mRNAs and a wide variety of regulatory motifs found in them. It consists of several individual databases and software utilities integrated with each other (*see* **Subheading 3.6.**)
Web home: http://www.ba.itb.cnr.it/BIG/UTRHome/
Reference: (12)
Full text: http://nar.oxfordjournals.org/cgi/content/full/33/suppl_1/D141

A list of selected web-accessible databases those are available freely to the public. The databases and accompanying utilities can be used to study *cis*-regulatory motifs of mRNA-untranslated regions in a variety of ways. See appropriate sections in the text for more details.

3.1. AU-Rich Element Database

AU-rich elements (AREs) are found in the 3′-untranslated regions of many mRNAs that are known to undergo rapid degradation through the deadenylation pathway. Roughly 5–8% of the human genes with diverse functions are known to code for ARE-containing mRNAs. The list is dominated by the genes that are involved in transient processes and therefore require strict expression control. Loss of regulated stability in ARE-mRNAs has been associated with disease. *See (2)* for a review on the nature of AREs and their role in biological processes.

ARE sequences can range up to approximately 150 bases in length and can be categorized into three classes. Bioinformatics and traditional techniques have helped detect multiple copies of a pentamer, AUUUA, and its extensions at the core of the first two classes which differ on the basis of the repetition pattern of this sequence motif. Class II has been further subdivided into clusters based on the number of pentamer repeats. The third class of AREs, although U rich, is not known to contain a well-defined sequence motif and has evaded computational analysis.

3.1.1. ARED

ARED 3.0 (3) is a third-generation database of a large non-redundant collection of human mRNAs containing class I and II AREs (*see* **Table 2**). The stored data have been derived by computational analysis of sequences retrieved from the NCBI/TIGR EST databases and mRNA sequences obtained from RefSeq/GenBank databases. The refined set of mRNAs were then searched for the ARE motif, WWWU(AUUUA)UUUW, and/or its variants in the 3′-UTR.

The data in ARED 3.0 have been classified into five searchable clusters. Clusters 1–4 contain AREs from class II, whereas the fifth cluster represents class I. The information provided by each record includes ARE classification, gene name, relevant information about Unigene and RefSeq, and a variety of cross-links (*see* **Note 1**). In addition, each ARE-mRNA has been functionally classified by Gene Ontology.

The search engine provided with the interface works in three modes. *Simple Search* allows the user to search all the clusters together or individually. In the *Advanced* mode, the database can be queried by a combination of NCBI IDs including Gene Name, Gene ID, Builtup Unigene, GenBank, and RefSeq in addition to the clusters. *Gene Ontology* terms can also be used as one of the search criteria in this mode to pull out ARE-mRNA records by the functional categories. ARED can also be searched by a collective *list* of several Accession Numbers, Gene Names, or Gene IDs separated by hard returns. All the search results can be downloaded in the form of tab-delimited simple text tables for further computational analysis.

3.2. Internal Ribosome Entry Site Databases

Internal ribosome entry sites (IRESs), *cis*-acting RNA elements found in the 5′-UTR, are capable of mediating cap-independent initiation of protein synthesis in viruses as well as eukaryotic mRNAs. IRESs length may vary between a few nucleotides to hundreds of bases. They may form a secondary structure and function by interacting with *trans*-acting cellular factors. The proteins synthesized through IRES pathway are involved in important biological processes like cell cycle, oncogensis, apoptosis, transcriptional control, and development [reviewed in *(1,4)*]. The list of genes containing IRES is continuously growing. However, owing to the complex and highly variable nature of these elements, reliable detection and characterization methodologies have been difficult to develop. Consequently, there are few known computational methods for studying the IRESs. Recently, there have been some efforts to curate the published information and experimental data about IRES elements (*see* **Table 2**).

3.2.1. IRESite

IRESite database *(5)* provides comprehensive coverage on a variety of non-redundant information related to the IRES elements found in viral and eukaryotic mRNAs (*see* **Table 2**). All of the data stored in this databank have been manually annotated from published experimental work. Each record provides detailed information about the experimental setup and methodology used for studying or utilizing the IRES elements. The database, therefore, stores an impressively large number of experimental characteristics. IRESite can be used to study primary as well as secondary structures associated with IRESs, activity under different experimental conditions, interactions with translation factors, and IRES-specific *trans* factors (ITAFs).

There are two types of records in the IRESite database. *Natural* records contain data associated with naturally occurring IRESs in the mRNA and have many important attributes including a description of the RNA, detailed information about ORFs (Open Reading Frames), functional/non-functional status of IRES, rRNA complementarity, data for RNA–protein interactions, and RNA secondary structure. *Engineered* IRES records cover information about artificial RNA constructs and include information about the relevant plasmid vector and promoter, proof of transcript integrity, and several other experimental characteristics.

The web-based interface is user friendly and provides a wide variety of options for querying the database (*see* **Note 2**). For example, in addition to the IRES name and sequence size, it is possible to search the database by organism name, gene name, function, name of the ITAF involved, or any combination of these parameters. The user can also use many of the experimental features as search criteria. The authors have provided helpful tips in the form of user-activated pop up "balloons" and convenient cross-links to outside databases like GenBank, PDB, and PubMed.

3.2.2. IRSdb

IRSdb (6) database provides basic information and some details for viral and cellular IRES elements (*see* **Table 2**). The data stored in IRSdb were originally derived from GenBank and EMBL sequence databases and has been classified into four categories: Viral IRES, Cellular IRES, Regulatory function, and ITAF interactions. In addition, a special section is devoted to biotechnology-based application of IRES elements.

A simple search form is available on the website for querying the database. IRES data can also be retrieved by visiting appropriate links to the four major classes. Each record consists of a gene name (or virus name) containing IRES, cellular conditions, information about relevant *trans*-acting factors, and a link to the IRES sequence. Appropriate cross-links to outside databases like GenBank and PubMed are provided as well. At the time this chapter was written, the database had not been updated since December 2002.

3.3. Poly(A) Database

The polyA signal or PAS, a *cis*-acting element in the 3′-UTR, serves as a binding site for cleavage polyadenylation specificity factor (CPSF) and plays an essential role in the cleavage-polyadenylation of mammalian pre-mRNAs. For a review on 3′-end processing, *see (7)*. The poly(A) tail serves a variety of important biological functions associated with mRNA metabolism, including its export from the nucleus, mRNA stability, and translation initiation. PAS, along with cytoplasmic polyadenylation element (CPE), has been found to regulate translation initiation during oocyte maturation in *Xenopus*. PAS-CPE-dependent cytoplasmic polyadenylation helps regulate cell cycle in *Schizosaccharomyces pombe* and may also be involved in mammalian somatic cell division. *See (8)* for a review.

A key element of the mammalian PAS is an AAUAAA located in the 3′-UTR, 10–30 nt upstream of the cleavage site. Although this is the most common PAS, several functional variants of the hexamer have been found. Studying PAS and associated *cis*-acting regulatory RNA elements at the genomic level can provide valuable insights into the mechanism of biological processes in which they are intimately involved. Only a fraction of the nucleic acid sequence entries in the primary databases provide PAS information, which is usually generated by wet-laboratory experiments and, therefore, is limited. Although some computational methods for PAS mapping have been reported, few have been made easily accessible to the public. A significant proportion of mammalian genes are known to have alternative poly(A) sites which can result in mRNAs with variable 3′-ends and hence different 3′-UTRs. This complexity presents challenges to the bioinformaticists for developing methods that can accurately predict PAS in the mRNAs. A recently published PolyA_DB database of several thousand poly(A) sites in the human and mouse genomes has attempted to address some of these computational challenges.

3.3.1. PolyA_DB

PolyA_DB *(9)* is a database of poly(A) sites and PAS mapped in a large number (>25,000) of human and mouse genes (*see* **Table 2**). A computational approach was used to retrieve and analyze sequence data from NCBI databases. The cDNA/EST sequence alignments were further refined and then analyzed with the help of custom computer programs to map poly(A) sites and predict PAS. The results of these experiments are stored in the database along with relevant information, such as genomic locations of the mapped sites, evidence of cDNA/EST alignments, PAS and its variants mapped in the vicinity of poly(A) sites, and tissue/organ information related to the sites.

PolyA_DB can be queried by Entrez Gene ID (*see* **Note 3** and **Table 4**). The search results can be viewed in five different interlinked modes. *Gene View* provides a summary of basic information about the gene itself and poly(A) sites in the form of a table and a graphic (*see* **Note 4**). *Evidence View* provides alignment evidence from cDNA/ESTs. *Ortholog View* can be used to compare human/mouse ortholog transcripts. The *Signal View* shows predicted PAS in the vicinity of the poly(A) sites and is very useful for comparison of alternatively polyadenylated mRNA products of a gene. The tissue/organ information for poly(A) sites can be obtained through the *Body View*. It is also possible to compare any two genes of the database solely in the graphic mode. The entire database can be downloaded as flat files for further computations.

3.4. Rfam Database

Rfam (10) is a comprehensive database of RNA families (*see* **Table 2**). In addition to a large number of non-coding genes, it contains information about selected structural *cis*-regulatory elements found in UTRs of mRNAs. The RNA families are represented by multiple sequence alignments and profile stochastic context-free grammars (SCFGs). This is somewhat similar to profile-hidden Markov models (HMMs) of Pfam.

Rfam contains data from all three major taxonomic domains, Eukaryota, Bacteria and Archaea, divided into three main sections. *Gene* represents a large collection of non-coding genes, *cis*-reg contains the UTR regulatory elements, and *Intron* stores data about self-splicing introns. The website provides an interface for querying Rfam in a variety of ways. Additionally, the website hosts a utility for searching homologs of the RNA elements in the user-provided query sequence or even a complete genome. The latter can be very useful for annotating nucleotide sequence entries. The sequence search is computationally intense and can take several minutes to complete even for short sequences. The entire database can also be browsed directly at the website or downloaded through FTP for further computations locally.

3.5. Transterm

Transterm (11) is a database of translation control RNA motifs and other biologically relevant regions of mRNAs (*see* **Table 2** and **Note 5**). The information in this database (e.g., UTR and CDS regions) has been extracted from GenBank/RefSeq with the help of computer programs. Each Transterm-defined motif has been linked to biological description with references. This information was extracted from literature and databases like UTRdb (*see* **Subheading 3.6.**) and Rfam (*see* **Subheading 3.5.**). Several useful utility programs associated with the database allow a variety of bioinformatics analysis, including identification of RNA motifs in the user-provided sequences or specific sections (e.g., 5′- or 3′-UTR) of the database itself. It is also possible for the users to define their own motifs for analysis. In addition to the "regular expression-" based patterns and matrices, BLAST may be used for database searches. Because extensive computations may take a few minutes, the results can be emailed to the user-provided address.

3.6. UTResource

UTResource (12,13) is a web server that provides a large collection of 5′- and 3′-untranslated region sequences of the eukaryotic mRNAs and a wide variety of regulatory motifs found in them (*see* **Table 2**). This highly valuable and comprehensive resource consists of several individual databases and software utilities. The databases were generated by a combination of computational and manual approaches and have been integrated among each other. The resource uses SRS retrieval system for querying all of its databases. As SRS also services the primary databases like EMBL, RefSeq, Swissprot, Uniprot, OMIM, and others, it is easy to combine the UTR search with them. Some of the databases and software accessible from this website are briefly described below:

3.6.1. UTRdb

UTRdb (12) is a databank of 5′- and 3′-UTR sequences collected from eukaryotic mRNAs. A custom computer program was used to mine UTR sequences and associated data from the primary databases (e.g., EMBL and GenBank). The collected data were subjected to further computations and refinements before curation into the UTRdb. Some of the information has been pulled out from published literature. The UTRdb has been subdivided into nine eukaryotic subclasses of EMBL nucleotide sequence database. In addition to this main section, UTRdb has three specialized divisions: *UTRef*, UTR sequences extracted from RefSeq; *UTRait*, a collection of UTRs of muscle-specific transcripts from TRAIT database, and *UTRexp* , consisting of UTR sequences in which functional motifs have been experimentally studied.

In addition to 5′- and 3′-UTR sequences, UTRdb provides information about mapped sequence repeats and regulatory motifs stored in the UTRSite (*see* **Subheading 3.6.2.**). Human UTR sequences have been mapped to genomic locations allowing useful experiments with sophisticated features of the Ensembl server. Some of the UTRs have also been associated to their specific proteins. The UTRdb records provide many useful cross-links to UTRSite and EMBL databases. Selected records are also cross-linked to International Protein Index (IPI) that stores data representing the selected mammalian proteomes (*see* **Table 4** for accessing IPI).

UTRdb can be queried through the SRS system with a large variety of data characteristics including motif names, UTR, repeat sequence name/type, organism, sequence length, genomic position, and protein IDs when available (*see* **Note 6**). The entire UTRdb can be downloaded for local computations.

3.6.2. UTRSite

UTRSite (12) is a repository of functional regulatory motifs mapped to 5′- or 3′-UTRs of eukaryotic mRNAs (*see* **Table 2**). A custom computational tool was used to collect information from a variety of databases. Only experimentally determined and published regulatory elements have been included in the UTRSite. The entries are manually curated by experts. Some examples of motifs covered in the UTRSite include PAS, CPE, IRE, IRES, ARE2, K-Box, SXL-binding site, and a variety of uORFs.

UTRSite can be searched by two methods: SRS and a simple search module available at the UTResource website. It is also possible to browse the complete list of UTRSite entries (currently >50) (*see* **Note 7**). In addition to the primary and, when applicable, secondary structure information, a typical UTRSite record includes the pattern syntax specific to the functional motif. The format of the syntax is compatible with PatSearch program *(14)* for pattern analysis (*see* **Table 3** and **Subheading 3.8.**). A detailed description of the regulatory motif is provided along with graphic representation if available. The records present a variety of other important information if known, including lists of genes/transcripts in which the regulatory element has been mapped and proteins that interact with it. Appropriate references and cross-links, both inside as well as outside the UTResource, are also provided.

Table 3
Free Software Tools for RNA Motif Search and Prediction

MEME (W, D)
Multiple Expectation Maximization for Motif Elicitations is a valuable software tool for searching novel motifs in user-provided set of nucleotide or protein sequences (*see* **Section 3.7.1.**)

MAST (W, D)
Motif Alignment and Search Tool is a tool that uses the motifs predicted by MEME and other motif programs to search primary sequence databases for matching motifs (*see* **Section 3.7.2.**)
Web home: http://meme.sdsc.edu/meme/intro.html
Reference: (15)
Full text: http://nar.oxfordjournals.org/cgi/content/full/34/suppl_2/W369

PatMatch (W, D)
PatMatch is a pattern-matching tool that can be used to predict short (<20 nt) *cis*-regulatory elements against *Arabidopsis thaliana* and other plant UTR datasets with a "regular expression" query
Web home: http://www.arabidopsis.org/cgi-bin/patmatch/nph-patmatch.pl
Reference: (25)
FTP site: ftp://ftp.arabidopsis.org/home/tair/Software/Patmatch/
Full text: http://nar.oxfordjournals.org/cgi/content/full/33/suppl_2/W262

PatSearch (W, D)
PatSearch is a software program that is able to search for specific combinations of consensus sequence patterns, secondary structure motifs, and position-weight matrices (*see* **Section 3.8.**)
Web home: http://bighost.area.ba.cnr.it/BIG/PatSearch/
Reference: (14)
Full text: http://nar.oxfordjournals.org/cgi/content/full/31/13/3608

MOST (W)
A web-based software tool that can be used through a multi-step semi-automatic process for the identification of seeded motifs in a set of nucleotide sequences. The program is based on a method that extracts overrepresented patterns from the input sequences to produce motif cores
Web home: http://telethon.bio.unipd.it/bioinfo/MOST/
Reference: (26)
Full text: http://nar.oxfordjournals.org/cgi/content/full/33/15/e135

QGRS Mapper (W)
Quadruplex forming G-Rich Sequences Mapper searches nucleotide sequences for the presence of G-quartet structural motifs (*see* **Section 3.9.**)
Web home: http://bioinformatics.ramapo.edu/QGRS/

Reference: (18)
Full text: http://nar.oxfordjournals.org/cgi/content/full/34/suppl_2/W676

RegRNA (W)

Regulatory RNA is a server that employs an integrated approach for identifying regulatory RNA sequence and structural motifs located in the 5´- and 3´-UTR (*see* **Section 3.10.**)

Web home: http://regrna.mbc.nctu.edu.tw/
Reference: (23)
Full text: http://nar.oxfordjournals.org/cgi/content/full/34/suppl_2/W429

RNAMotif (D)

Allows the user to define RNA structural motifs and search the genomic databases for sequences that are potentially capable of forming similar structure (*see* **Section 3.11.**)

Access: Downloadable stand-alone program
Web home: http://www.scripps.edu/mb/case/casegr-sh-3.5.html (not a web tool)
Downloadable: http://www.scripps.edu/case/rnamotif-3.0.4.tar.gz
Reference: (24)
Full text: http://nar.oxfordjournals.org/cgi/content/full/29/22/4724

RNAMST (W)

RNA Motif Search Tool searches a large, pre-processed, database of RNA sequences for homologs of pre-defined RNA structural motifs
Web home: http://bioinfo.csie.ncu.edu.tw/~rnamst/
Reference: (27)
Full text: http://nar.oxfordjournals.org/cgi/content/full/34/suppl_2/W423

UTRscan (W)

Searches user-provided sequence for patterns defined in the UTRSite (*see* **Subheadings 3.6.2.** and **3.6.5.**)
Web home: http://www.ba.itb.cnr.it/BIG/UTRScan/
Reference: (13)
Full text: http://www.ba.itb.cnr.it/BIG/UTRScan/TIGUTR.pdf

A selection of free bioinformatics software tools available to the public. These can be used to predict and map a variety of *cis*-regulatory motifs in the user-provided nucleotide sequences. Some of the programs also allow searching eukaryotic genomes for elements of interest in the untranslated regions of mRNAs. Selected tools are briefly described in the text. D, downloadable for running as stand-alone program; W, program runs on the web server.

3.6.3. UTRgenome Browser

UTRgenome Browser (12) is a web-based facility that allows search and retrieval of UTR subsets by using accession numbers of a variety of outside data sources like IPI, PDB, Swissprot, RefSeq, EMBL, MIM, and so on. For

Table 4
Selected Databases and Other Bioinformatics Resources on the Internet

Resource	URL
RefSeq NCBI Reference Sequence Database	http://www.ncbi.nlm.nih.gov/RefSeq/
EMBL Nucleotide Sequence Database	http://www.ebi.ac.uk/embl/
Ensembl A Joint Project for Annotation of Eukaryotic Genomes	http://www.ensembl.org/index.html
Entrez Gene Database of Genes from RefSeq Genomes	http://www.ncbi.nlm.nih.gov/entrez/query.fcgi?db=gene
OMIM Database of Human Genes and Disorders	http://www.ncbi.nlm.nih.gov/entrez/query.fcgi?db=OMIM
Gene Ontology Database	http://www.geneontology.org/
IPI International Protein Index	http://www.ebi.ac.uk/IPI
EMBOSS Open Source, Free Software for Molecular Biology	http://emboss.sourceforge.net/
NAR Database Database of Published Databases	http://www3.oup.co.uk/nar/database/c/
NAR Methods Archives of Published Bioinformatics Methods	http://nar.oxfordjournals.org/collections/index.shtml
Bioinformatics Links Directory	http://bioinformatics.ubc.ca/resources/links_directory/
Bioinformatics Resource Site	http://phobos.ramapo.edu/~pbagga/binf/binf_int _res.htm

example, it is possible to use a Gene Ontology (*see* **Table 4**) term to search for UTRs that are associated with the desired gene function category. The resulting UTRgenome records can either be downloaded in the FASTA format or viewed as interactive web pages that contain information and links for several genomic features related to the UTR, such as coordinates on the chromosome, corresponding mRNAs, ESTs, and proteins. A variety of important cross-links are also provided.

3.6.4. UTRblast

UTRblast (12) is an online utility that can be used to query the entire UTRdb or its sections with user-input sequences in the FASTA format. Many of the standard BLAST options including choices of matrices, expect values and data filtration, are provided. The utility requires free online registration prior to use. The results are emailed to the registered users.

3.6.5. UTRscan

UTRscan (13) is a software program that searches user-provided sequence for the patterns defined in the UTRSite (*see* **Subheading 3.6.2.**) (*see* **Table 3**). This tool can be useful for predicting UTR regulatory motifs in the nucleotide sequences that have not been studied before. The program requires free online registration ahead of time. The query sequence, in the FASTA format, can be directly pasted into the space provided on the web page or uploaded in the form of a text file (*see* **Note 8**). The results, in the user-opted plain text or HTML format, are emailed to the registered address. It is useful to request the results in the HTML format for direct links to the relevant UTRSite records (*see* **Note 9**).

3.7. Multiple Expectation Maximization for Motif Elicitations and Motif Alignment and Search Tool

Multiple Expectation Maximization for Motif Elicitations (MEME) and *Motif Alignment and Search Tool* (MAST) *(15)*, a pair of computer programs, can together help identify regulatory elements in the user-provided sequences and primary sequence databases (*see* **Table 3**).

3.7.1. MEME

MEME *(15,16)* is a software tool for searching novel motifs in user-provided set of nucleotide sequences (*see* **Table 3**). This program can be useful in a situation where one believes the presence of a regulatory motif in the same

UTR region of a set of genes, which, for example, may share a common mechanism of post-transcriptional regulation. MEME uses a method that is similar to creation of ungapped multiple sequence alignments to search for statistically significant sequence motifs in the set of sequences provided by the user in the FASTA format. The input set should be carefully chosen to avoid sequences that are less likely to contain a motif. MEME prefers short (<1000 nt) sequences from which low-information segments and sequence repeats, that are not expected to be part of the motif of interest, have been removed.

The detailed program output, which is emailed to the user in the HTML format, shows identified motifs as local multiple alignments and several other formats. Up to three different motifs can be identified in a set. Several useful links are provided in the output, which can directly forward the motifs for further analysis by other web servers like MAST (*see* **Subheading 3.7.2.**). MEME can be downloaded and installed on local computer servers like Linux and Solaris. It is also possible to install MEME on Linux clusters to accommodate high traffic and extensive computing. Locally running MEME has many features that are not supported at the website.

3.7.2. MAST

MAST *(15,17)* is a software tool that uses the motifs predicted by MEME and other programs to search primary sequence databases for matching motifs. This program can be useful in identifying motifs of interest in other genes or genomes. MAST accepts only a special format of queries that are generated by MEME outputs, other motif programs and databases. Currently, a limited number of target databases are supported for MAST searches. However, it is possible to upload a small (up to 1,000,000 characters maximum size) sequence database in the FASTA format for motif searches. Like MEME, MAST can also be downloaded for local installations.

3.8. PatSearch

PatSearch *(14)* is a software program that is able to match patterns in the nucleotide sequences (*see* **Table 3**). It is different from other pattern-finding bioinformatics tools in that it supports a variety of queries that can be combined for the same search, thereby reducing the chances of false positive results. PatSearch can search UTRdb or user-provided nucleotide sequence(s) for any combinations of consensus sequence patterns, secondary structure motifs, and position-weight matrices (PWMs). The latter are formed by assigning weight to each nucleotide in every position of the putative motif. The PWM calculations are based on the observed occurrence of that nucleotide in the known motifs.

PatSearch requires registration prior to use. It accepts sequences to be searched in several standard formats, including FASTA, GenBank, and EMBL. Queries need to be built with a custom syntax *(14)* that supports a variety of patterns including PWMs. The same syntax is used for representing functional motifs in the UTRSite (*see* **Subheading 3.6.2.**). The output results are returned to the registered email address.

3.9. Quadruplex forming G-Rich Sequences Mapper

Quadruplex forming G-Rich Sequences (QGRS) *Mapper (18)* searches nucleotide sequences for the presence of G-quartet structural motifs (*see* **Table 3**). The G-quartet structure, also known as a G-quadruplex, is formed by repeated folding of either the single polynucleotide molecule or by association of two or four molecules. The structure consists of stacked G-tetrads, which are square co-planar arrays of four guanine bases each. G-quadruplex structure has been implicated in a variety of regulatory processes in the cytoplasm, including mRNA turnover through exoribonuclease action *(19)*, interaction with FMRP (Fragile X Mental Retardation Protein) *(20)*, cap-independent translation initiation *(21)*, and translation repression *(22)*. A G-quadruplex structure was found to determine IRES function in human fibroblast growth factor 2 (FGF-2) mRNA *(21)*. QGRS Mapper is a useful bioinformatics tool for exploring the role of G-quadruplex motifs in regulation of gene expression.

Although designed primarily to map G-quadruplexes in genomic entries, QGRS Mapper provides a utility for analyzing any user-provided nucleotide sequence in the FASTA or raw format. The program offers many search options and allows the user to define the size and composition of G-quadruplex motifs. The method for motif prediction follows a scoring system that evaluates a QGRS for its likelihood to form a stable G-quadruplex. QGRS Mapper output can be accessed in three major modes. *Data View* provides the mapped positions as well as the sequence of the predicted G-quadruplex motifs. *Sequence View* shows mapped G-quadruplexes in the context of input sequence itself. The output can also be accessed in a highly interactive *Graphic View* that provides a variety of utilities, including a zoom tool which is helpful in navigation of a large sequence (*see* **Note 10**).

3.10. RegRNA

RegRNA *(23)* is a server that uses an integrated approach for identifying RegRNA sequence and structural motifs in mRNA sequences, including specific motifs located in the 5´- and 3´-UTR (*see* **Table 3** and **Note 11**). The program also supports a variety of other RNA motifs. The server employs several published software tools such as RNAMotif (*see* **Subheading 3.11.**) and PatSearch (*see* **Subheading 3.8.**) to look for regulatory motif homologs of interest in the user-provided nucleotide sequence by searching against RegRNA resource of motifs. The latter was built by collecting information from literature surveys and RNA motif databases like UTRdb/UTRSite (*see* **Subheading 3.6.**). The integrated approach allows for identification of sequence as well as structural homologs of regulatory RNA motifs. RegRNA also provides an option for users to define their own RNA structural motif for analysis. Owing to extensive computations, the queries may take 30 s or more to run. The results, containing a detailed account of the identified RNA motifs, are presented in text as well as graphic formats. The latter includes appropriate links to detailed description of the motifs. Secondary structures of the structural motifs, generated by *mfold*, are also provided.

3.11. RNAMotif

RNAMotif (24) allows the user to define RNA structural motifs and search the genomic databases for sequences that are potentially capable of forming similar structure (*see* **Table 3**). The program represents a major enhancement over the earlier RNAMOT tool and permits searching for almost any kind of structures that can be defined, including helices, hairpins, internal loops, multistem loops, and complicated motifs like E-loops and tetraloops. In addition, the user can define scoring rules for evaluation of the matches, which is very useful in combining the structural elements for enhanced capabilities. RNAMotif program can be used to identify structural motifs in the untranslated regions of mRNAs.

RNAMotif is not a web tool. It is provided as a downloadable "zipped" package along with detailed instructions and help manual (*see* **Table 3**). It can be "unzipped" by most Unix servers. The program would need to be installed on a Unix machine and run from a command line. Some expertise in Unix operating system may be required to properly install and configure the program.

4. Notes

1. *ARED:* The exact location of AREs is not provided in the ARED records. However, UTRdb and UTRscan (*see* **Subheading 3.6**) can be used to map class II AREs.
2. *IRESite:* Although Netscape and Firefox browsers work well for accessing and searching the database, some of the IRESite features are active only through MS Internet Explorer.
3. *PolyA_DB* was originally designed to be searched by Locus Link ID. Although that option is still available, NCBI Locus Link has been superseded by Entrez Gene. *See* **Table 4** to access Entrez Gene database.

4. *PolyA_DB:* You would need to download and install SVG plug-in (*see* **Table 1**) for accessing part of the results in the interactive mode. The plug-in works best with MSIE. If you do not want to install the plug-in, you will still be able to view the data in static graphics.

5. *Transterm* website worked best with Netscape in my experience. It did not work well with MSIE.

6. *UTRdb:* The lists of selectable motif names and repeats are accessible through "Extended Query Form" of the SRS system.

7. *UTRSite:* If you see an error message while using the search option at the UTResource website, you would need to exit the browser, close all the browser windows, and restart the search.

8. *UTRscan:* Make sure to check the correct input mode. Pasting the sequence does not automatically select the right format choice.

9. *UTRscan:* Write down the codes provided on the screen after job submission, especially if you are running multiple searches. The results do not identify sequence names submitted as queries.

10. *QGRS Mapper:* Java plug-in for the browser needs to be turned on for Graphics View.

11. *RegRNA* does not collect information for IRES elements and hence will not be able to analyze the sequences for this motif.

References

1. Mignone, F., Gissi, C., Liuni, S., and Pesole, G. (2002) Untranslated regions of mRNAs. *Genome Biol* **3,** REVIEWS0004.

2. Khabar, K. S. (2005) The AU-rich transcriptome: more than interferons and cytokines, and its role in disease. *J Interferon Cytokine Res* **25,** 1–10.

3. Bakheet, T., Williams, B. R., and Khabar, K. S. (2006) ARED 3.0: the large and diverse AU-rich transcriptome. *Nucleic Acids Res* **34,** D111–D114.

4. Lopez-Lastra, M., Rivas, A., and Barria, M. I. (2005) Protein synthesis in eukaryotes: the growing biological relevance of cap-independent translation initiation. *Biol Res* **38,** 121–146.

5. Mokrejs, M., Vopalensky, V., Kolenaty, O., Masek, T., Feketova, Z., Sekyrova, P., Skaloudova, B., Kriz, V., and Pospisek, M. (2006) IRESite: the database of experimentally verified IRES structures (www.iresite.org). *Nucleic Acids Res* **34,** D125–D130.

6. Bonnal, S., Boutonnet, C., Prado-Lourenco, L., and Vagner, S. (2003) IRESdb: the Internal Ribosome Entry Site database. *Nucleic Acids Res* **31,** 427–428.

7. Gilmartin, G. M. (2005) Eukaryotic mRNA 3' processing: a common means to different ends. *Genes Dev* **19,** 2517–2521.

8. de Moor, C. H., Meijer, H., and Lissenden, S. (2005) Mechanisms of translational control by the 3' UTR in development and differentiation. *Semin Cell Dev Biol* **16,** 49–58.

9. Zhang, H., Hu, J., Recce, M., and Tian, B. (2005) PolyA_DB: a database for mammalian mRNA polyadenylation. *Nucleic Acids Res* **33**, D116–D120.
10. Griffiths-Jones, S., Moxon, S., Marshall, M., Khanna, A., Eddy, S. R., and Bateman, A. (2005) Rfam: annotating non-coding RNAs in complete genomes. *Nucleic Acids Res* **33**, D121–D124.
11. Jacobs, G. H., Stockwell, P. A., Tate, W. P., and Brown, C. M. (2006) Transterm–extended search facilities and improved integration with other databases. *Nucleic Acids Res* **34**, D37–D40.
12. Mignone, F., Grillo, G., Licciulli, F., Iacono, M., Liuni, S., Kersey, P. J., Duarte, J., Saccone, C., and Pesole, G. (2005) UTRdb and UTRsite: a collection of sequences and regulatory motifs of the untranslated regions of eukaryotic mRNAs. *Nucleic Acids Res* **33**, D141–D146.
13. Pesole, G., and Liuni, S. (1999) Internet resources for the functional analysis of 5' and 3' untranslated regions of eukaryotic mRNAs. *Trends Genet* **15**, 378.
14. Grillo, G., Licciulli, F., Liuni, S., Sbisa, E., and Pesole, G. (2003) PatSearch: a program for the detection of patterns and structural motifs in nucleotide sequences. *Nucleic Acids Res* **31**, 3608–3612.
15. Bailey, T. L., Williams, N., Misleh, C., and Li, W. W. (2006) MEME: discovering and analyzing DNA and protein sequence motifs. *Nucleic Acids Res* **34**, W369–W373.
16. Bailey, T. L., and Elkan, C. (1994) Fitting a mixture model by expectation maximization to discover motifs in biopolymers. *Proc Int Conf Intell Syst Mol Biol* **2**, 28–36.
17. Bailey, T. L., and Gribskov, M. (1998) Combining evidence using p-values: application to sequence homology searches. *Bioinformatics* **14**, 48–54.
18. Kikin, O., D'Antonio, L., and Bagga, P. S. (2006) QGRS Mapper: a web-based server for predicting G-quadruplexes in nucleotide sequences. *Nucleic Acids Res* **34**, W676–W682.
19. Bashkirov, V. I., Scherthan, H., Solinger, J. A., Buerstedde, J. M., and Heyer, W. D. (1997) A mouse cytoplasmic exoribonuclease (mXRN1p) with preference for G4 tetraplex substrates. *J Cell Biol* **136**, 761–773.
20. Darnell, J. C., Jensen, K. B., Jin, P., Brown, V., Warren, S. T., and Darnell, R. B. (2001) Fragile X mental retardation protein targets G quartet mRNAs important for neuronal function. *Cell* **107**, 489–499.
21. Bonnal, S., Schaeffer, C., Creancier, L., Clamens, S., Moine, H., Prats, A. C., and Vagner, S. (2003) A single internal ribosome entry site containing a G quartet RNA structure drives fibroblast growth factor 2 gene expression at four alternative translation initiation codons. *J Biol Chem* **278**, 39330–39336.
22. Oliver, A. W., Bogdarina, I., Schroeder, E., Taylor, I. A., and Kneale, G. G. (2000) Preferential binding of fd gene 5 protein to tetraplex nucleic acid structures. *J Mol Biol* **301**, 575–584.

23. Huang, H. Y., Chien, C. H., Jen, K. H., and Huang, H. D. (2006) RegRNA: an integrated web server for identifying regulatory RNA motifs and elements. *Nucleic Acids Res* **34,** W429–W434.
24. Macke, T. J., Ecker, D. J., Gutell, R. R., Gautheret, D., Case, D. A., and Sampath, R. (2001) RNAMotif, an RNA secondary structure definition and search algorithm. *Nucleic Acids Res* **29,** 4724–4735.
25. Yan, T., Yoo, D., Berardini, T. Z., Mueller, L. A., Weems, D. C., Weng, S., Cherry, J. M., and Rhee, S. Y. (2005) PatMatch: a program for finding patterns in peptide and nucleotide sequences. *Nucleic Acids Res* **33,** W262–W266.
26. Pizzi, C., Bortoluzzi, S., Bisognin, A., Coppe, A., and Danieli, G. A. (2005) Detecting seeded motifs in DNA sequences. *Nucleic Acids Res* **33,** e135.
27. Chang, T. H., Huang, H. D., Chuang, T. N., Shien, D. M., and Horng, J. T. (2006) RNAMST: efficient and flexible approach for identifying RNA structural homologs. *Nucleic Acids Res* **34,** W423–W428.

2

Identification of mRNA Polyadenylation Sites in Genomes Using cDNA Sequences, Expressed Sequence Tags, and Trace

Ju Youn Lee, Ji Yeon Park, and Bin Tian

Summary

Polyadenylation of nascent transcripts is an essential step for most mRNAs in eukaryotic cells. It is directly involved in the termination of transcription and is coupled with other steps of pre-mRNA processing. Recent studies have shown that transcript variants resulting from alternative polyadenylation are widespread for human and mouse genes, contributing to the complexity of mRNA pool in the cell. In addition to 3′-most exons, alternative polyadenylation sites (or poly(A) sites) can be located in internal exons and introns. Identification of poly(A) sites in genomes is critical for understanding the occurrence and significance of alternative polyadenylation events. Bioinformatic methods using cDNA sequences, Expressed Sequence Tags (ESTs), and Trace offer a sensitive and systematic approach to detect poly(A) sites in genomes. Various criteria can be employed to enhance the specificity of the detection, including identifying sequences derived from internal priming of mRNA and polyadenylated RNAs during degradation.

Key Words: Polyadenylation; EST; Trace; genome; intron; exon; 3′-UTR; internal priming.

1. Introduction

The poly(A) tail, which is located at the 3′-end of all mature mRNAs except some histone genes (*1,2*), is critical for many aspects of mRNA metabolism, including mRNA stability, translation, and transport (*3,4*). The cellular process of making poly(A) tails, called polyadenylation, is a two-step reaction (*5,6*), involving an endonucleolytic cleavage at a site determined by adjacent *cis*-elements and their binding factors, followed by polymerization of an adenosine

From: *Methods in Molecular Biology, Vol. 419: Post-Transcriptional Gene Regulation*
Edited by: J. Wilusz © Humana Press, Totowa, NJ

tail to a length specific to the species, for example, 200–250 nucleotides (nt) in mammals and 70–90 nt in yeasts. Polyadenylation is coupled with transcription and other steps of pre-mRNA processing, such as splicing and termination *(7, 8)*. For each polyadenylation site (or poly(A) site), the cleavage position usually is not precisely defined and can take place anywhere within a window of approximately 24 nt *(9, 10)*. However, the nucleotide preference A > U > C ≫ G has been observed in biochemical assays *(11)*. More than half of the human genes have alternative poly(A) sites, with locations in 3′-most exons, internal exons, and introns *(9, 12)*, leading to transcript variants with different open reading frames (ORFs) and/or variable 3′-untranslated regions (UTRs). Some alternative polyadenylation events have been shown to be tissue specific *(13, 14)*. Recently, polyadenylation has been implicated in the degradation of some nuclear RNAs in eukaryotic cells by the exosome *(15)*. The polyadenylation reaction in this process involves a distinct set of factors than those responsible for polyadenylation of nascent transcripts, and the resulting poly(A) tail is usually short and contains variable nucleotides *(16)*.

Poly(A) sites can be identified using various experimental approaches, including examining the transcript size using the northern blot and determining the 3′-end of a transcript using techniques such as RNase protection assays *(17)* and 3′ rapid amplification of cDNA 3′-ends (3′-RACE). In addition, the polyadenylation activity of a given sequence can be examined in vivo by reporter assays *(17, 18)*. Bioinformatic techniques using cDNA sequences, Expressed Sequence Tags (ESTs), and Trace sequences offer a high-throughput approach for identification of poly(A) sites *(9, 12, 19–21)*. cDNAs are sequenced mRNAs that are usually assembled from sequences of multiple sequencing reactions. Most cDNAs contain a full-length ORF, and some even contain complete UTR regions. In contrast, ESTs are single-pass 5′ or 3′ sequences of cDNA clones from large sequencing projects, and thus, they are only partial sequences of mRNA. Trace sequences are the original sequence reads for ESTs and cDNAs. As of now, there are many more EST Trace records than cDNA ones in public databases. It is noteworthy that other functional genomic techniques are beginning to be utilized for poly(A) site studies, such as DNA microarrays *(22)* and serial analysis of gene expression (SAGE) tags *(10, 18)*.

Here, we present a detailed bioinformatic approach for identifying poly(A) sites in genomes using cDNA, EST, and Trace sequences. cDNA/ESTs often contain a consecutive adenosine region, or poly(A) region, at the 3′-end or a consecutive thymidine region, or poly(T) region, at the 5′-end that are derived from mRNA poly(A) tails, as illustrated in **Fig. 1A.** As the poly(A/T) regions are not present in the genome, they can be used to identify poly(A) sites through

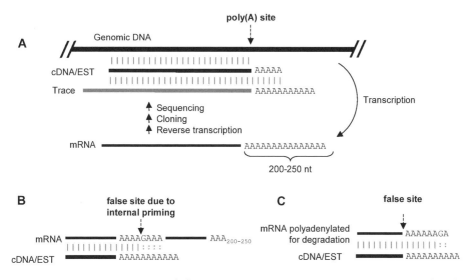

Fig. 1. Identification of poly(A) sites using cDNA, Expressed Sequence Tag (EST), and Trace. (**A**) Transcribed mRNAs are reverse transcribed, cloned, and sequenced to give rise to cDNA/EST/Trace sequences, as indicated in the figure. Sequences are aligned with the genome sequence for identification of poly(A) sites. Vertical lines indicate nucleotide matches. (**B**) A-rich mRNA internal sequences can lead to internal priming, resulting in false identification of poly(A) sites. Dotted vertical lines are those nucleotides that can hybridize with oligo(dT) primers but cannot align with poly(A/T) tail sequences of cDNA/ESTs by sequence alignment tools. (**C**) RNAs to be degraded by the exosome can be polyadenylated by the TRAMP complex, resulting in false identification of poly(A) sites. Their poly(A) tails are usually shorter than canonical poly(A) tails, and may contain other nucleotides in addition to adenosines.

the alignment of cDNA/EST sequences with the genome. Trace sequences can contain additional poly(A/T) regions that are absent in cDNA/ESTs. These regions are removed from sequences submitted to public cDNA/EST databases because of their low sequencing quality or low complexity (being consecutive As or Ts). This information can be recovered by the alignment of Trace sequences with their corresponding cDNA/ESTs. Although false negatives arise in various situations as discussed in the Notes (4.1) below, the primary concern for bioinformatic detection of poly(A) sites is false positives, that is, genomic locations that are not genuine sites for polyadenylation of nascent transcripts. Major sources include internal priming of A-rich sequences (*see* **Fig. 1B**) *(23)* and mRNA sequences that are polyadenylated for degradation by the exosome (*see* **Fig. 1C**) *(15,16)*. Examination of genome sequences and poly(A) tail length can enhance the quality of poly(A) site identification. Other criteria can also be employed to minimize the false identification of poly(A) sites, as discussed in the Notes section. Although the methods presented here are intended for research on poly(A) sites, most bioinformatic databases and tools described are broadly applicable in other areas of molecular biology. We provide a rationale for most steps, and our comments are based on our experience.

2. Materials

2.1. Computers

Almost all working computers can be used, including desktops, laptops, and servers. However, computers with large memory are preferred, as some steps, such as running basic local alignment search tool (BLAST)-like alignment tool (BLAT), require storage of a large amount of data in the computer memory. We routinely use a computer with two Intel Xeon CPUs (2.4 GHz) and 3 GB memory. The choice of operating system (OS) is up to each individual's preference and the availability of the required programs (see 2.2) for the OS. We use the Linux OS on our computers. Ample hard disk space is also recommended, as various sequence databases and result files consume a large amount of storage. We use a 480 GB disk array for data storage.

2.2. Computer programs

2.2.1. Practical Extraction and Reporting Language (Perl)

Perl is a programming language widely used in bioinformatics. Its simple syntax and extensive functions for manipulation of text make it particularly useful for handling sequences and parsing text-based results from other programs, such as BLAST and BLAT (see 2.2.2 and 2.2.3). Debugging and running Perl programs are also straightforward. All that is required is a Perl interpreter program that converts text-based scripts to commands understandable to computers. Linux/UNIX systems usually have it by default. One can use the ActivePerl program (http://www.activestate.com) for Windows and Mac systems.

2.2.2. Basic Local Alignment Search Tool

BLAST is a commonly used sequence alignment program *(24)*. Its popularity lies in its efficiency, which is attributable to the two computing steps used for aligning sequences: exact matching of short words followed by extending the words by dynamic programming, an accurate but not-so-efficient computational algorithm for matching two sequences. Two versions of the BLAST program are available, that is, NCBI BLAST and WU-BLAST, which are based on an identical central algorithm but have slightly different running options. We use NCBI BLAST here. The stand-alone command line-based suite of NCBI BLAST can be downloaded from http://www.ncbi.nlm.nih.gov/BLAST/download.shtml, which also contains a number of utility tools for formatting sequence databases and retrieving sequence from databases.

2.2.3. BLAST-Like Alignment Tool

BLAT is a sequence alignment program that is particularly suitable for aligning mRNAs with genome sequences *(25)*. Compared with other similar programs, such as Sim4 *(26)* and Spidey *(27)*, BLAT is highly efficient and accurate, which is largely attributable to its unique algorithm to index genome sequences in the computer memory. Thus, a large computer memory is required for running BLAT. The program can be obtained from http://www.soe.ucsc.edu/~kent/exe/.

2.3. Databases

2.3.1. cDNA Sequences

cDNA sequences can be downloaded from the NCBI nucleotide (nt) database (ftp://ftp.ncbi.nih.gov/blast/db/FASTA/nt.gz), which contains both GenBank cDNA sequences and RefSeq sequences. Sequences are in FASTA format.

2.3.2. EST Sequences

EST sequences can be obtained from the NCBI dbEST database (ftp://ftp.ncbi.nih.gov/repository/dbEST/). Files for FASTA format sequences can be downloaded from ftp://ftp.ncbi.nih.gov/blast/db/FASTA/, which often contain the clone ID for EST. Some clones, particularly those from the I.M.A.G.E. consortium (http://image.llnl.gov), can be purchased from several vendors, such as American Type Culture Collection (ATCC, Manassas, VA), Open Biosystems (Huntsville, AL), and Research Genetics/Invitrogen (Carlsbad, CA).

2.3.3. Trace Sequences

Trace sequences can be downloaded from the NCBI Trace Archive (http://www.ncbi.nlm.nih.gov/Traces/), which also stores original sequencing chromatograms and quality scores. A Perl script named "query_tracedb" is available from the site, which allows batch download of Trace sequences based on user queries. For example, the query "SPECIES_CODE = 'HOMO SAPIENS' and TRACE_TYPE_CODE = 'EST' and ACCESSION!=NULL" specifies that Trace sequences for human ESTs that contain Accession numbers are selected for download.

2.3.4. NCBI UniGene Database

The UniGene database from NCBI contains entries representing transcripts that appear to come from the same transcription locus, together with information on protein similarities, gene expression, cDNA clones, genomic locations, and so on. It is available for most model species. The database is constructed by sequence clustering *(28)*. Each entry is called a UniGene Bin or Cluster. Some entries contain anti-sense transcripts, and expressed pseudogenes are also included. The UniGene database can be downloaded from ftp://ftp.ncbi.nlm.nih.gov/repository/UniGene/.

2.3.5. NCBI Gene Database

The Gene database from NCBI has a variety of information about genes, including genome locus, RefSeq sequences, protein structures, homology data, and so on. Each gene has a unique identifier called Gene ID. The database was previously known as the LocusLink database. It can be downloaded from NCBI (ftp://ftp.ncbi.nlm.nih.gov/gene/).

2.3.6. Genome Databases

Assembled genome sequences can be downloaded from various websites. We download them from the UCSC Genome Bioinformatics site (http://genome.ucsc.edu).

3. Methods

Our purpose is to identify all poly(A) sites for any given gene in the genome. To this end, we use all available cDNA/ESTs to align with their corresponding genomes (*see* **Fig. 2**). For efficiency, we use the sequence cluster information from the UniGene database to find all sequences for a gene (*see* **Fig. 2A**). To ensure quality, we apply a sequence clean-up step (*see* **Fig. 2B**) to eliminate erroneous entries in UniGene. For simplicity, we use the human genome as an example for the methods described below, and only Trace sequences for ESTs are described. We also provide simple commands, when necessary, to illustrate the methods.

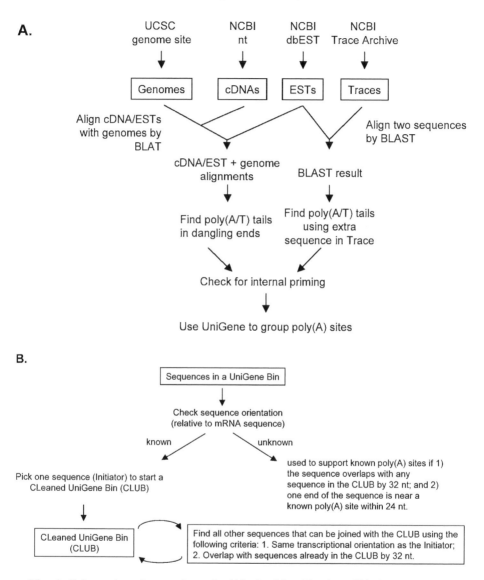

Fig. 2. Schematics of steps for poly (A) site identification. (**A**) Overall approach of identifying poly(A) sites using cDNA/Expressed Sequence Tag (EST)/Trace and genome sequences. (**B**) A strategy to clean up sequences in UniGene Bins. The initiator for a Cleaned UniGene Bin (CLUB) is selected based on the sequence type according to the following order: RefSeqs > other cDNAs > ESTs, where other cDNAs are cDNAs that are not RefSeq sequences. When multiple sequences of the same type are encountered, the longest sequence is chosen.

3.1. Identify Poly(A) Tails by Aligning cDNA/ESTs with Genome Sequences

1. cDNA/ESTs are grouped according to their corresponding chromosomes using the UniGene database, so that a sequence is only compared with the chromosome on which it is located. This step can significantly shorten the computer running time. However, incorrect chromosome assignment by UniGene can lead to missing alignments.

2. cDNA/EST sequences corresponding to a chromosome are aligned with the chromosome sequence using BLAT, as shown below:

```
>blat genome_file mrna_file blat_output -
minIdentity=95 -noTrimA -out=blast8 -ooc=11.ooc
```

In the command, genome_file contains genome sequences, mrna_file contains cDNA/EST sequences, and blat_output is the output file. The minimum sequence identity option "-minIdentity" is set to 95 (in percent), and the "-noTrimA" option is chosen to avoid trimming of poly(A) tail. The "11.ooc" file is used to mask short sequences (11-mers) frequently occurring in the genome. It can be downloaded from UCSC genome web site or can be made by the BLAT program use the "-makeOoc" option. Many output formats are available for BLAT. We use NCBI BLAST format "-out=blast8."

3. Use Perl to parse the BLAT results. For each sequence, BLAT returns a number of blocks or high scoring pairs (HSPs), representing alignments with the genome. Some HSPs are exons, and others are just homologous regions. To piece together correct exons for a sequence, we use a simple dynamic programming strategy based on the maximum matching score. We have also used the "-out=sim4" option to get ordered exons from BLAT directly.

4. Examine splicing signals for each intron at the 5′ and 3′ splice sites to obtain its type, that is, GT-AG (5′ splicing site-3′ splicing site), GC-AG, or AT-AC. Both sense and anti-sense orientations (cDNA/EST sequence orientation relative to mRNA) are examined. The result can determine which chromosome strand encodes the mRNA (with the same direction), that is, "+" or "−" strand, and the orientation of the cDNA/EST sequence relative to its mRNA, that is, sense or anti-sense.

5. For cDNA/EST sequences in the anti-sense orientation, examine the 5′ unaligned sequence (called dangling end sequence) for a stretch of Ts. For sequences in the sense orientation, examine the 3′ dangling end sequence for a stretch of As. If the orientation of a sequence is not available based on its splicing pattern (single exon or no reliable splicing signals being identified), both ends are examined. In this case, poly(A) tail information can assist in the determination of the sequence orientation, as a poly(A) sequence at the 3′-end indicates sense orientation, whereas a poly(T) sequence at the 5′-end indicates anti-sense orientation. For quality purposes, we discard sequences that contain both poly(T) sequence at the 5′-end and poly(A)

sequence at the 3′-end, which result from A- or T-rich internal sequences in the mRNA. Sequences in the dangling end that contain eight or more consecutive As or Ts next to the end of cDNA/EST versus genome alignment are considered to be derived from mRNA poly(A) tails. We also allow one non-A nucleotide in the sequence to account for sequencing errors. A Perl script for parsing poly(A) tails is provided below, where "$seq" is a dangling end sequence.

```
if($seq=~/^(A{8,})/i){
    # >= 8 consecutive As
}elsif($seq=~/^(A{0,}[GTCN]A{8,})/i){
    # one mismatch allowed, but still need >= 8 As
}else{
    # no poly(A) tail
}
```

3.2. Compare EST with Trace to Obtain Additional Poly(A) Tail Information

1. Format EST sequences by the "formatdb" function in the BLAST suite, so that they can be retrieved one sequence at a time.
2. Use Perl to read Trace sequences downloaded from the Trace Archive. A useful sequence should have a Trace ID or Ti and an Accession number for EST.
3. Use the Accession and the "fastacmd" command in the BLAST suite to retrieve the corresponding EST sequence.
4. Use function "bl2seq" in the BLAST suite to compare the Trace sequence with the EST sequence.

```
>bl2seq -i trace_file -j est_file -p blastn -D 1 -e
0.001 -F "m D"
```

In the command, `trace_file` contains a Trace sequence and `est_file` contains an EST sequence. The "-D 1" option specifies that only one strand is used for BLAST search, as Trace sequences are in the same orientation as ESTs. The "-F 'm D'" option specifies that low complexity regions, such as poly(A/T) sequences, are masked by the program DUST at the word-matching stage but not in the final alignment.

5. Use Perl to parse the BLAST result. As Trace sequences contain low-quality regions, particularly at the 3′-end, we allow 1 and 3 nucleotide mismatches at 5′ and 3′ regions, respectively, between a Trace sequence and an EST sequence. This is carried out by examining the start position and end position of the EST in the alignment. A Trace sequence is discarded if its corresponding EST has the start position in the alignment >2 nucleotides from the 5′-end or has the end position in the alignment >3 nucleotides away from the 3′-end.

6. Retrieve the regions in the Trace sequence that are not present in the EST sequence by Perl and use the same strategy described above to find poly(A/T) sequences. To ensure quality, we discarded Trace sequences for which poly(A/T) tail information conflicts with that of their corresponding ESTs, for example, poly(T) sequence in Trace sequence but poly(A) sequence in EST. Poly(A) tail information is then added to the EST for poly(A) site identification.

3.3. Internal Priming Identification

1. For a cDNA/EST sequence that contains a poly(A/T) tail sequence, including that derived from Trace, its putative poly(A) cleavage site on the genome is considered to be immediately downstream to the end of the alignment with the genome. Use Perl to retrieve the −10 to +10 nt genomic region surrounding the putative poly(A) site.
2. Use Perl to examine whether there exist six or more consecutive As or seven or more As in a 10 nt window in the region. If yes, the cleavage site is an internal priming candidate. The reason to examine both upstream and downstream sequences is that some nucleotides of the A-rich genomic sequence align with the poly(A/T) tail sequence and some do not, as illustrated in **Fig. 1B**.

3.4. Gene-Based Grouping of Poly(A) Sites

1. Use Perl to group together poly(A) sites belonging to the same gene. For efficiency, we first use the UniGene database to group cDNA/ESTs according to UniGene Bins. The UniGene database has comprehensive coverage of all nucleotide sequences but often contains transcripts generated from both strands of the genome. In addition, erroneous sequences can lead to incorrect merging of genes. Thus, it is necessary to clean up sequences in UniGene Bins. On the other hand, the situation where one gene is divided into multiple UniGene Bins occurs much less frequently, and thus is less a concern.
2. Use Perl to clean up UniGene Bins according to the method shown in **Fig. 2B**. The cleaned UniGene Bin is called CLeaned UniGene Bin (CLUB), which only contains transcripts with known orientation. One UniGene Bin may have more than one CLUB. Each CLUB has a representative sequence called initiator. We use the order RefSeqs > other cDNAs > ESTs to select initiators.
3. To maximize the number of supporting sequences for a poly(A) site, the 3′-ends of sequences without poly(A/T) tails are compared with identified poly(A) sites. An cDNA/EST is considered to support a poly(A) site if its 3′-end is near the poly(A) site within 24 nt. Transcripts with unknown orientation can also be used by this method (*see* **Fig. 2B**), but both ends are examined.
4. Use Perl to merge poly(A) sites that are located within 24 nt from one another. Clustering is carried out in the 5′ to 3′ direction.
5. The genome location for each poly(A) site can be compared with the gene structure of the initiator sequence of the CLUB. When an initiator is a RefSeq entry, use the

RefSeq-to-Gene mapping information from the Gene database to find corresponding Gene ID and related gene information for the CLUB.

4. Notes

4.1. Sources for False Negatives

1. Many sequences do not contain usable poly(A) tail information: some EST sequences are generated by 5′ sequencing of cDNA clones; almost all cDNA sequences and a large fraction of ESTs do not have Trace data; some sequences have poor quality in the poly(A/T) tail sequence.

2. Some poly(A) sites are refractory to the detection by using the cDNA/EST/Trace approach due to biological reasons: some poly(A) sites, and/or their corresponding genes, are rarely used/expressed or only used/expressed under certain conditions; usage of some poly(A) sites can lead to rapid degradation of transcript and thus is hard to detect.

3. Poly(A) sites located in large introns or intergenic regions are difficult to detect using the approach described here. This is because their supporting sequences usually do not overlap with other sequences for the same gene. Thus, they are deleted during the UniGene cleaning up stage.

4.2. Concerns for False Positives

4. The length of the poly(A/T) tail sequence in a cDNA/EST provides important information as to the quality of poly(A) site identification. This is because poly(A/T) sequences derived from internal A-rich sequence and the poly(A) tail used to mark mRNA for degradation are usually short, and thus their resulting poly(A/T) sequences in cDNA/ESTs are expected to be shorter than those for real poly(A) tails. The upper bound for their length is approximately 15–20 nt, which is the length of most oligo(dT) primers. In this regard, poly(A) tail information from Trace is highly valuable for both the sensitivity and selectivity of poly(A) site identification, as some poly(A) sites are otherwise elusive due to lack of poly(A/T)-containing cDNA/ESTs, and longer poly(A/T) sequences suggest higher quality for the identification. As shown in **Fig. 3**, when Trace sequences are used, the maximum poly(A/T) sequence length derived from cDNA/ESTs can be significantly extended.

5. In addition to poly(A/T) tail length, several other characteristics of a poly(A) site can indicate its high quality: (1) large number of cDNA/ESTs; (2) several cleavage sites for a poly(A) site; (3) presence of canonical polyadenylation signal elements, AAUAAA or AUUAAA; (4) presence of other regulatory elements, such as UGUA element *(29)* and other putative elements reported in *(30)*; (4) high-sequence conservation in orthologous regions in other species; (5) high probability of having a poly(A) site by prediction tools *(31,32)*. However, it is noteworthy that specificity is at odds with sensitivity, that is, applying stringent parameters will leave out some genuine poly(A) sites.

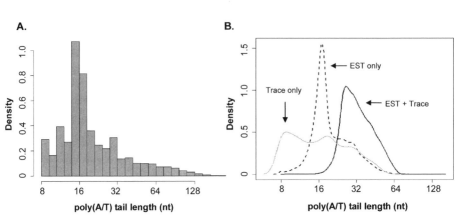

Fig. 3. Statistics of poly(A/T) tail length identified by cDNA, Expressed Sequence Tag (EST), and Trace. (**A**) Histogram of poly(A) tail length for all poly(A) sites identified in the human genome using cDNA/EST sequences. Only the maximum value for each site is plotted. (**B**) Distribution of poly(A/T) tail length for sites when Trace sequences are used. Only poly(A) sites whose maximum poly(A) tail length can be extended by Trace are plotted. Trace only, poly(A) sites which do not have any poly(A/T)-tailed ESTs; EST only, poly(A) sites without using Trace; and EST + Trace, poly(A) sites with Trace sequences used.

4.3. Other Issues

6. When a poly(A) site is supported by ESTs, it is often possible to obtain information as to which tissues utilize the poly(A) site based on the source of cDNA libraries from the UniGene database. The mapping between ESTs and cDNA libraries is available in the UniGene data file. Tissue annotations for cDNA libraries can be obtained from the "Hs.lib.info" file in the UniGene database. Currently, tissue source, developmental stage, and cancer source are annotated for most cDNA libraries. Additional information about cDNA libraries, including detailed library description, their original source, experimental procedure, such as normalization or subtraction information, are available through the Library Browser web site at NCBI. For example, for human cDNA library no. 1, the query URL is http://www.ncbi.nlm.nih.gov/UniGene/library.cgi?ORG=Hs&LID=1.

7. Phylogenetic information can be obtained from UCSC genome browser (http://genome.ucsc.edu), which contains multiple genome alignments. The sequence surrounding a poly(A) site can be examined for its conservation across species. High conservation of sequence suggests conservation of poly(A) sites. This information can also provide validation for poly(A) site identification from a phylogenetic viewpoint.

Acknowledgments

We thank Carol S. Lutz and members of B.T. laboratory for helpful discussions. This work was supported by The Foundation of the University of Medicine and Dentistry of New Jersey.

References

1. Edmonds, M. (2002) A history of poly A sequences: from formation to factors to function. *Prog Nucleic Acid Res Mol Biol*, **71**, 285–389.
2. Marzluff, W.F. (2005) Metazoan replication-dependent histone mRNAs: a distinct set of RNA polymerase II transcripts. *Curr Opin Cell Biol*, **17**, 274–280.
3. Mangus, D.A., Evans, M.C. and Jacobson, A. (2003) Poly(A)-binding proteins: multifunctional scaffolds for the post-transcriptional control of gene expression. *Genome Biol*, **4**, 223.
4. Wickens, M., Anderson, P. and Jackson, R.J. (1997) Life and death in the cytoplasm: messages from the 3´ end. *Curr Opin Genet Dev*, **7**, 220–232.
5. Colgan, D.F. and Manley, J.L. (1997) Mechanism and regulation of mRNA polyadenylation. *Genes Dev*, **11**, 2755–2766.
6. Zhao, J., Hyman, L. and Moore, C. (1999) Formation of mRNA 3´ ends in eukaryotes: mechanism, regulation, and interrelationships with other steps in mRNA synthesis. *Microbiol Mol Biol Rev*, **63**, 405–445.
7. Minvielle-Sebastia, L. and Keller, W. (1999) mRNA polyadenylation and its coupling to other RNA processing reactions and to transcription. *Curr Opin Cell Biol*, **11**, 352–357.
8. Proudfoot, N. (2004) New perspectives on connecting messenger RNA 3´ end formation to transcription. *Curr Opin Cell Biol*, **16**, 272–278.
9. Tian, B., Hu, J., Zhang, H. and Lutz, C.S. (2005) A large-scale analysis of mRNA polyadenylation of human and mouse genes. *Nucleic Acids Res*, **33**, 201–212.
10. Pauws, E., van Kampen, A.H., van de Graaf, S.A., de Vijlder, J.J. and Ris-Stalpers, C. (2001) Heterogeneity in polyadenylation cleavage sites in mammalian mRNA sequences: implications for SAGE analysis. *Nucleic Acids Res*, **29**, 1690–1694.
11. Chen, F., MacDonald, C.C. and Wilusz, J. (1995) Cleavage site determinants in the mammalian polyadenylation signal. *Nucleic Acids Res*, **23**, 2614–2620.
12. Yan, J. and Marr, T.G. (2005) Computational analysis of 3´-ends of ESTs shows four classes of alternative polyadenylation in human, mouse, and rat. *Genome Res*, **15**, 369–375.
13. Zhang, H., Lee, J.Y. and Tian, B. (2005) Biased alternative polyadenylation in human tissues. *Genome Biol*, **6**, R100.
14. Edwalds-Gilbert, G., Veraldi, K.L. and Milcarek, C. (1997) Alternative poly(A) site selection in complex transcription units: means to an end? *Nucleic Acids Res*, **25**, 2547–2561.

15. Houseley, J., LaCava, J. and Tollervey, D. (2006) RNA-quality control by the exosome. *Nat Rev Mol Cell Biol*, **7**, 529–539.

16. West, S., Gromak, N., Norbury, C.J. and Proudfoot, N.J. (2006) Adenylation and exosome-mediated degradation of cotranscriptionally cleaved pre-messenger RNA in human cells. *Mol Cell*, **21**, 437–443.

17. Hall-Pogar, T., Zhang, H., Tian, B. and Lutz, C.S. (2005) Alternative polyadenylation of cyclooxygenase-2. *Nucleic Acids Res*, **33**, 2565–2579.

18. Pan, Z., Zhang, H., Hague, L.K., Lee, J.Y., Lutz, C.S. and Tian, B. (2006) An intronic polyadenylation site in human and mouse CstF-77 genes suggests an evolutionarily conserved regulatory mechanism. *Gene*, **366**, 325–334.

19. Gautheret, D., Poirot, O., Lopez, F., Audic, S. and Claverie, J.M. (1998) Alternate polyadenylation in human mRNAs: a large-scale analysis by EST clustering. *Genome Res*, **8**, 524–530.

20. Graber, J.H., Cantor, C.R., Mohr, S.C. and Smith, T.F. (1999) In silico detection of control signals: mRNA 3′-end-processing sequences in diverse species. *Proc Natl Acad Sci USA*, **96**, 14055–14060.

21. Iseli, C., Stevenson, B.J., de Souza, S.J., Samaia, H.B., Camargo, A.A., Buetow, K.H., Strausberg, R.L., Simpson, A.J., Bucher, P. and Jongeneel, C.V. (2002) Long-range heterogeneity at the 3′ ends of human mRNAs. *Genome Res*, **12**, 1068–1074.

22. Cheng, J., Kapranov, P., Drenkow, J., Dike, S., Brubaker, S., Patel, S., Long, J., Stern, D., Tammana, H., Helt, G., et al. (2005) Transcriptional maps of 10 human chromosomes at 5-nucleotide resolution. *Science*, **308**, 1149–1154.

23. Chen, J., Sun, M., Lee, S., Zhou, G., Rowley, J.D. and Wang, S.M. (2002) Identifying novel transcripts and novel genes in the human genome by using novel SAGE tags. *Proc Natl Acad Sci USA*, **99**, 12257–12262.

24. Altschul, S.F., Gish, W., Miller, W., Myers, E.W. and Lipman, D.J. (1990) Basic local alignment search tool. *J Mol Biol*, **215**, 403–410.

25. Kent, W.J. (2002) BLAT–the BLAST-like alignment tool. *Genome Res*, **12**, 656–664.

26. Florea, L., Hartzell, G., Zhang, Z., Rubin, G.M. and Miller, W. (1998) A computer program for aligning a cDNA sequence with a genomic DNA sequence. *Genome Res*, **8**, 967–974.

27. Wheelan, S.J., Church, D.M. and Ostell, J.M. (2001) Spidey: a tool for mRNA-to-genomic alignments. *Genome Res*, **11**, 1952–1957.

28. Wheeler, D.L., Barrett, T., Benson, D.A., Bryant, S.H., Canese, K., Chetvernin, V., Church, D.M., DiCuccio, M., Edgar, R., Federhen, S., et al. (2006) Database resources of the National Center for Biotechnology Information. *Nucleic Acids Res*, **34**, D173–D180.

29. Venkataraman, K., Brown, K.M. and Gilmartin, G.M. (2005) Analysis of a noncanonical poly(A) site reveals a tripartite mechanism for vertebrate poly(A) site recognition. *Genes Dev*, **19**, 1315–1327.

30. Hu, J., Lutz, C.S., Wilusz, J. and Tian, B. (2005) Bioinformatic identification of candidate cis-regulatory elements involved in human mRNA polyadenylation. *RNA*, **11,** 1485–1493.
31. Cheng, Y., Miura, R.M. and Tian, B. (2006) Prediction of mRNA polyadenylation sites by support vector machine. *Bioinformatics*, **22,** 2320–2335
32. Tabaska, J.E. and Zhang, M.Q. (1999) Detection of polyadenylation signals in human DNA sequences. *Gene*, **231,** 77–86.

3

Bioinformatic Tools for Studying Post-Transcriptional Gene Regulation

The UAlbany TUTR Collection and Other Informatic Resources

Francis Doyle*, Christopher Zaleski*, Ajish D. George, Erin K. Stenson, Adele Ricciardi, and Scott A. Tenenbaum

Summary

The untranslated regions (UTRs) of many mRNAs contain sequence and structural motifs that are used to regulate the stability, localization, and translatability of the mRNA. It should be possible to discover previously unidentified RNA regulatory motifs by examining many related nucleotide sequences, which are assumed to contain a common motif. This is a general practice for discovery of DNA-based sequence-based patterns, in which alignment tools are heavily exploited. However, because of the complexity of sequential and structural components of RNA-based motifs, simple-alignment tools are frequently inadequate. The consensus sequences they find frequently have the potential for significant variability at any given position and are only loosely characterized. The development of RNA-motif discovery tools that infer and integrate structural information into motif discovery is both necessary and expedient. Here, we provide a selected list of *existing* web-accessible algorithms for the discovery of RNA motifs, which, although not exhaustive, represents the current state of the art. To facilitate the development, evaluation, and training of new software programs that identify RNA motifs, we created the UAlbany training UTR (TUTR) database, which is a collection of validated sets of sequences containing experimentally defined regulatory motifs. Presently, eleven training sets have been generated with associated indexes and "answer sets" provided that identify where the previously characterized RNA motif [the iron responsive element (IRE), AU-rich class-2 element (ARE),

---

*These authors contributed equally.

From: *Methods in Molecular Biology, Vol. 419: Post-Transcriptional Gene Regulation*
Edited by: J. Wilusz © Humana Press, Totowa, NJ

selenocysteine insertion sequence (SECIS), etc.] resides in each sequence. The UAlbany TUTR collection is a shared resource that is available to researchers for software development and as a research aid.

Key Words: mRNA; RNA motif; RNA binding; RNA consensus; RNA folding; post-transcription; bioinformatics; informatics.

1. Introduction

Post-transcriptional regulation of genes and transcripts is a vital aspect of cellular processes and unlike transcriptional regulation remains a largely unexplored domain. One of the most obvious and most important questions to explore is the discovery of new functional motifs in RNA. Unfortunately, very few RNA motifs have presently been well characterized, and it is believed that many hundreds (possibly thousands) more exist. The characterization, organization, and analysis of these motifs very likely will hold the key to understanding the post-transcriptional regulatory code.

One practical approach for an algorithm that identifies RNA regulatory elements is to search for a recurring pattern among a set of sequences, where one assumes that the sequences share a common motif. For this reason, we developed the UAlbany training untranslated region (TUTR) Collection—to act as a standard set of data, upon which motif discovery algorithms can be tested and trained (*see* **Table 1**). Additionally, these training sets and their accompanying control sequences can be used to empirically benchmark an algorithm over sets with varying signal to noise ratios.

We have also compiled a list of web-based bioinformatic tools and resources that represent the current state of the art for RNA motif discovery in post-transcriptionally regulated sets of mRNAs (*see* **Table 2**). Other tools and algorithms exist for similar analyses; however, they may be limited in their ability to be used for motif discovery.

There are a variety of ways to generate sets of related sequences that contain a shared motif, such as ribonomic profiling *(1)* and ChIP-Chip. Determination of the consensus sequence motifs requires new high-throughput analysis methods. Where applicable, these methods must account for potentially high deviation in the actual nucleotide sequence composition forming a particular structural motif. Consideration must also be given to the possibility of variant forms of the structure (*see* **Fig. 1**).

Software tools based solely on alignment techniques are unlikely to provide suitable analyses of similarities among related sequences when those similarities are due to secondary structures whose stems are not highly conserved. New generation computer software tools that utilize adaptive techniques

Table 1
RNA Motifs Presently Contained in the UAlbany training untranslated region (TUTR) Collection

RNA motif	Acronym	Reference
Histone 3´-UTR stem-loop structure	HSL3	*(5)*
Iron responsive element	IRE	*(6)*
Selenocysteine insertion sequence—type 1	SECIS—type 1	*(7–9)*
Selenocysteine insertion sequence—type 2	SECIS—type 2	*(7–9)*
Cytoplasmic polyadenylation element	CPE	*(10)*
TGE translational regulation element	TGE	*(11)*
15-Lipoxygenase differentiation control element	15-LOX-DICE	*(12)*
AU-rich class-2 element	ARE2	*(13)*
Terminal oligopyrimidine tract	TOP	*(14–16)*
Glusose transporter type-1 3´-UTR *cis*-acting element	GLUT1	*(17)*
Internal ribosome entry site	IRES	*(18)*

(such as natural language processing, artificial neural networks, and genetic programming) provide more sophisticated means of complex pattern matching that may prove capable of identifying common mRNA secondary structures.

Development of such advanced algorithms will require test data sets that mirror the expected experimental data. To provide valid benchmarking of any algorithm while avoiding introduction of artificial bias, these sets should be biologically relevant (i.e., not containing randomly generated portions or fabricated motifs). The current UAlbany TUTR collection is a first generation example of such sets, and its members have been compiled from UTR sequences catalogued in the UTRdb known to contain a given binding motif *(2)*. Each UAlbany TUTR collection set is composed of a number of human genome 3´- or 5´-UTR sequences containing a shared consensus binding site.

Currently, all sequences in a given set are positive for a common signal and consist solely of UTR regions. However, as noted, the sets are intended to eventually resemble experimental data. This will require the addition of noise in the form of some intronic, coding, or non-genic sequences and sequences which contain none or only part of the signal.

At present, we provide generated control sets that are formed from random concatenations of 3´- and 5´-UTR sequence, matched for length and GC content to their respective training set sequences (UTR genomic coordinates are generated from UCSC Known Genes annotation: http://genome.ucsc.edu/cgi-bin/hgTrackUi?g=knownGene based on NCBI

Table 2
Overview of Available Web-Based Motif Discovery Tools

1. Carnac
Carnac is a software tool for analyzing the hypothetical secondary structure of a family of homologous RNA. It aims at predicting whether the sequences actually share a common secondary structure. When this structure exists, Carnac is then able to correctly recover a large amount of the folded stems. The input is a set of single-stranded RNA sequences that need not to be aligned. The folding strategy relies on a thermodynamic model with energy minimization. It combines information coming from locally conserved elements of the primary structure and mutual information between sequences with covariations as well. (http://bioinfo.lifl.fr/RNA/carnac/index.php)

2. comRNA
comRNA predicts common RNA secondary structure motifs in a group of related sequences. The algorithm applies graph-theoretical approaches to automatically detect common RNA secondary structure motifs in a group of functionally or evolutionarily related RNA sequences. The advantages of this method are that it does not require the presence of global sequence similarities (but can take advantage of it), does not require prior structural alignment, and is able to detect pseudoknot structures. (http://ural.wustl.edu/~yji/comRNA/)

3. GPRM
GPRM is aimed at finding common secondary structure elements, not a global alignment, in a sufficiently large family (e.g., more than 15 members) of unaligned RNA sequences. (http://bioinfo.cis.nctu.edu.tw/service/gprm/)

4. CMFinder
CMfinder is an RNA-motif prediction tool. It is an expectation maximization algorithm using covariance models for motif description, carefully crafted heuristics for effective motif search, and a novel Bayesian framework for structure prediction combining folding energy and sequence covariation. (http://bio.cs.washington.edu/yzizhen/CMfinder/)

5. Gibbs Motif sampler
The Gibbs Motif Sampler is a software package for locating common elements in collections of biopolymer sequences. The Gibbs Motif Sampler will allow you to identify motifs, conserved regions, in DNA or protein sequences. The web version of the Gibbs Motif Sampler operates in three basic modes: site sampling, motif sampling, and recursive sampling (http://bayesweb.wadsworth.org/gibbs/gibbs.html)

6. *RSAT*

Regulatory Sequence Analysis Tools (RSAT) provides a series of modular computer programs specifically designed for the detection of regulatory signals in non-coding sequences. RSA provides a set of tools providing multiple approaches for pattern discovery across multiple sequences. Structural patterns are not accommodated (http://rsat.ulb.ac.be/rsat/)

7. *RNAalifold*

RNAalifold is a program for computing the consensus structure of a set-aligned RNA sequence taking into account both thermodynamic stability and sequence covariation. (http://rna.tbi.univie.ac.at/cgi-bin/alifold.cgi)

8. *FoldAlign*

FoldAlign is an algorithm for local simultaneous folding and aligning two or more RNA sequences. (http://foldalign.ku.dk/)

9. *SCARNA*

SCARNA is a fast, convenient tool for structural alignment of a pair of RNA sequences. It aligns two RNA sequences and calculates the similarities between them based on the estimated common secondary structures (http://www.scarna.org/scarna/)

10. *Pfold*

Pfold takes the alignment of RNA sequences as input and predicts a common structure for all sequences. It requires input sequences to be aligned. The server implementation is restricted to a maximum of 40 sequences with maximum length of 500 bases (http://www.daimi.au.dk/~compbio/rnafold/)

---

genome build 35). Although not proven negative for any sequence or structure, they should not be enriched for such motifs beyond what a random sampling of the UTR population provides.

We also intend to provide experimental "negative control sets." Many adaptive software tools for pattern recognition are aided by (or rely on) differential analysis with comparative sets known to have no (or little) representation of the signal found in the positive set. These sets will also be produced from actual genome data. Their use is only relevant if an analogous negative control set can be provided for the eventual experimental data sets. A possible means of accomplishing this, in the case of immunoprecipitation, is by collecting sequences represented by array probes that do not show signals across multiple experiments with the same protocol.

We are only just beginning to realize the role of UTRs in gene regulation. Quantifying and categorizing the various elements involved will first require

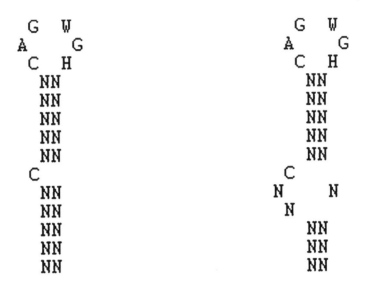

Iron Response Element Variants

Fig. 1. Two identified major structural variations of the iron responsive element (IRE) binding site *(6)*. Both IREs can have varied lower stem lengths and show high variability in acceptable bases throughout most stem positions, the only constraint being that these positions form pairs. Allowing for this degree of variability (including the possibility of non-Watson-Crick pairing) calls into question the utility of simple alignment methods for finding the motif.

that we identify them. The current state of the art in RNA motif detection leaves much to be desired in terms of applicability to set-based motif discovery, independence from seed sequence alignments, stability of implementation, and perplexity of usage. The development of realistic test sets for the desired applications should help focus new development away from spurious example cases and toward a more systematic look at de novo motif detection.

This first release of the TUTR collection provides researchers the ability to assess current software and experiment with new algorithms on relevant data sets. Although algorithm development continues toward robustly detecting common motifs in our training sets, the repertoire of sets themselves will be curated and expanded using newly available databases of functional RNA motifs (such as Rfam and RNABase), recasting this available motif data in terms of test and control sets. Future releases will also include test sets with varying signal to noise ratios that more closely reflect experimentally derived data.

2. Materials

2.1. Generation of the TUTR Collection

1. Each of the TUTR sets was generated by downloading all UTR sequence records from the UTRdb known to contain a specified motif. These records had already been purged for redundancy by the UTRdb through the CLEANUP program *(3)* using elimination threshold criteria of 95% similarity and 90% overlap.
2. We implemented a Java application to parse these records, filter for only those UTR's belonging to humans, and output three related files.
3. The first of these files is the actual training set, presented in modified FASTA format (no line breaks within sequences) with a non-descriptive unique header tag.
4. The second file is an index file that references the training set header tags to their original sequence record descriptor.
5. The third file is an "answer" file whose first line specifies the acronym of the common motif for a given training set and whose subsequent lines specify the start and end position of the motif region within each sequence as well as the actual subsequence representing the consensus.

2.2. TUTR Collection Contents

1. The collection consists of a parent folder ("TUTR_revn.n" where "n.n" is the revision number) with four sub-folders (*see* **Fig. 2.**): "blinded sets," "non-blinded sets," "control sets," and "additional_set_data."
2. The blinded_sets folder contains a number of text files with the naming convention set_Cnn.txt, where nn is a two-digit ID.
3. Each of these files contains multiple UTR sequence entries in FASTA format, and each of these sequences contains one or more instances of a motif common to that file (*see* **Fig. 3**).
4. The non-blinded sets folder contains a number of sub-folders. Each of these is named so that it conforms to "setCnn motif name," where Cnn corresponds to the name of a blinded sets file and motif name is an acronym for the motif contained by the sequences in that set.
5. Within each of these folders are four files (*see* **Figs. 2–5**):

   a. Named motif name.txt that describes the hidden motif and its consensus sequence.

   b. Named indexOfSet_Cnn.txt that maps the substituted FASTA header to what was originally associated to the sequence as part of its EMBL record.

   c. Named set_Cnn_Answers.txt, top line gives motif name, subsequent lines have a header from the training set file followed by a start and end position of the motif within that given sequence followed by the subsequence extracted to contain only the motif.

   d. A copy of the relevant training set so the user can work easily within this folder.

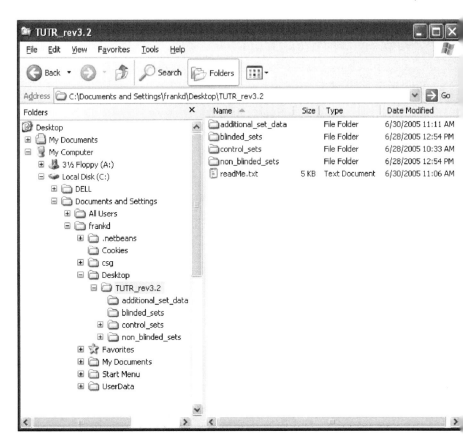

Fig. 2. The UAlbany training untranslated region (TUTR) file hierarchy. Top-level directory contents. "blinded_sets" contain only training set files with no additional data to betray contents. "non-blinded_sets" contain multiple sub-folders for each of the motifs These folders contain the training set file for that motif, a text file describing the motif, an index file mapping the training set entries to their original FASTA headers, and an answer file describing the location of the motif within each sequence. "control_sets" contains three control sequence files for each training set and a sub-folder, "controlGenerationLogfiles" with logs detailing the generation of each control file. "additional_set_data" contains files describing basic characteristics of the sequences in the training sets. The readme.txt file gives a general overview of the TUTR collection.

6. The control sets folder contains three control sequence sets for each of the training sets.

7. Controls were generated using a program we developed called cogent—control set generator for genomics (Doyle, Zaleski and Tenenbaum, manuscript i preparation).

a.)
>STLAB_C01_1
ggcgttcctcggcgtcctgaacccaaaggctcttttcagagccacccacacgatcaagaaagggtttcgaacacggtgaa
gggttatgtaacctaatgtgtctccatgcaccctagcggttgccgggtgcggccgaatctgagccgtgtgtgtgtgtgtg
tgtgtgtgtg...

b.)
>STLAB_C01_1   >utr|CC165450|3HSA054868 3'UTR in Homo sapiens cDNA FLJ33901 fis,
clone CTONG2008321, highly similar to HISTONE H2B F.

Fig. 3. An example of a training set entry. Example (**a**) is a training set entry
(partial) with non-descriptive header tag and (**b**) is the corresponding index file entry
that maps the trainings set sequence's header to its original header as given by the
UTRdb (2).

8. The control sequences were chosen by starting with a 50/50 choice of a random
   3′- or random 5′-UTR entry from the Known Genes annotation for NCBI human
   genome build 35. Then a randomly chosen starting point within its sequence
   is selected, and alternating random 3′- and 5′-UTRs are concatenated until the
   control is matched to a respective training set sequence by length. The prospective
   control was then checked for GC content (±2%) compared to its respective training
   set sequence. Matches are written to file, failures are rejected, and the process
   repeats until an acceptable match is found (*see* **Note 1**).
9. The naming convention for these control sets is "training set name" +
   "_Controln.txt," where n is 1, 2, or 3 (*see* **Note 2**).
10. The "additional set data" folder contains files that hold basic descriptive infor-
    mation about the sequences in the training sets whose name corresponds to
    their own. The files are comma separated value (csv) format with three fields
    per line:

    a. Training set sequence header
    b. Length of that sequence
    c. GC content of that sequence

11. Presently, the TUTR collection is composed of sets for 11 motifs (*see* **Table 1**).

MOTIF: HSL3
STLAB_C01_1   Start: 23   End: 45,   ccaaaggctcttttcagagccac

Fig. 4. An example of an answer set file. The first line names the common motif
acronym and second line describes the motif's location within the sequence shown
Fig. 3. The first field is the training set sequence's header. The second and third
fields are the start and end position, respectively, of the motif within that sequence.
The fourth field is the sub-sequence found at this location.

```
ID:              U0001;  date: 30-07-1997
Name:            Histone 3'UTR stem-loop structure (HSL3)
Pattern:         r1={au,ua,gc,cg,gu,ug}
                 0...1 mmmm p1=ggyyy u hhuh a r1~p1 mm 0...3
Taxon_Range:     UTR, 5' and 3' UTRs
Description:     Metazoan histone 3'-UTR mRNAs, lacking a polyA tail,
                 contain a highly conserved stem-loop structure with a
                 six base stem and a four base loop. This stem-loop
                 structure plays a different role in the nucleus and in
                 the cytoplasm. In the nucleus, it is involved in pre-mRNA
                 processing and nucleocytoplasmic transport, whereas in the
                 cytoplasm it enhances translation efficiency and regulates
                 histone mRNA stability. The trans-acting factor which
                 interacts with the 3'-UTR hairpin structure of histone
                 mRNAs is a 31 kDa stem-loop binding protein in mammals
                 (SLBP) present both in nuclei and polyribosomes.
                 In mammals in addition to SLBP histone mRNA processing requires
                 at least one additional factor: the U7 snRNP, which binds
                 a purine-rich element 10-20 nt downstream of the stem-loop
                 sequence (Histone Downstream Element, HDE).
                 The consensus stem-loop structure is shown below:

                          H U
                       H        H
                       U-A
                       Y-R
                       Y-R
                       Y-R
                       G-C
                       G-C
                     CCAAA  ACCCA

                 In all histone mRNAs analyzed so far  no G has been
                 observed in the four base loop. In all metazoan except
                 C. elegans , there are two invariant urydines in the first
                 and third base of the loop.
                 In C.elegans the first base of the loop is C. Either 5' and 3'
                 flanking sequences are necessary for high affinity binding
                 of SLBP. The 5' flanking sequence consensus is CCAAA and
                 the 3' flanking sequence consensus is ACCCA or ACCA with
                 cleavage occuring after the A.
                 The histone 3'-UTR hairpin structure is peculiar in that
                 the bases of the stem are conserved unlike most functional
                 hairpin motifs where conserved bases are found in single
                 stranded loop regions only. The sequence of the stem and
                 flanking sequences are critical for binding of the SLBP.
Bibliography:    [1]
   Authors       A.S. Williams and W.F.Marzluff (1995)
   Title         "The sequence of the stem and flanking sequences at the 3'end of histone
   Title         mRNA are critical determinants for the binding of the stemm-loop binding
   Title         protein"
   Journal       Nucleic Acid Research vol.23 n.4
//
```

Fig. 5. Example of a motif description retrieved from the UTRdb. The text gives a summary description of the motif. Note the "Pattern:" entry. This dictates the match criteria used for locating the motif within UTRs. This syntax is defined for use by the PatSearch program *(19)*.

3. Methods

3.1. Application of TUTR Sets to Select Algorithms

1. The TUTR collection is currently available to browse or download in zip file format: http://www.albany.edu/cancergenomics/faculty/stenenbaum/bioinfoMain/ TUTR/ (*see* **Fig. 2**).
2. Questions regarding the collection can be directed to czaleski@albany.edu.

3. We have found several web-based algorithms (*see* **Table 2**) that attempt to address the problem of new motif discovery in the same manner we have described.

4. There are many algorithms which address similar issues—such as searching sequences for a user-defined motif, generating Minimum Free Energy structures, sequence only alignments, and so on. However, it seems that few, if any, are able make a true de novo discovery of a motif containing both sequential and structural components (*see* **Notes 3** and **4**).

5. We used three motif identification tools (comRNA, cmFinder, and CARNAC) as test cases for acting upon the TUTR collection.

6. Our intent was to gauge the usability of the applications and to help put some perspective on the size and difficulty of the problem at hand—that of *de novo motif discovery*.

7. We sought to make an inexhaustive "first-pass" at testing these algorithms on a small number of the TUTR sets. In all cases, we attempted to use the tools "as is," utilizing provided web interfaces (*see* **Note 5**).

8. We attempted to use several sequence sets from the TUTR collection, and add variations such as the addition of noise (as described previously).

9. The histone stem loop (HSL) was intentionally used as a first test case with all three applications. We did this because the HSL is by far the most highly conserved motif—both in its simple stem-loop structure and in its sequence—in the collection and thus should be the "easiest" to find (*see* **Note 6**).

10. Both cmFinder and CARNAC were run through their web interfaces, and comRNA was downloaded and run locally on our test server (RedHat Linux—Dual Intel Xeon 3 GHz, 2 G memory). Default parameters were used in all cases, for the initial test of each program.

11. Each of the three applications failed to return results for the first test case—the unaltered histone stem-loop set. A submission was made multiple times to each web server for cmFinder and CARNAC separated by 2–3 days, and results were never returned to the user through email as stated. In addition, the local comRNA instance was run multiple times and would begin to formulate its output file but would simply exit pre-completion after approximately 4 h of run time. Given that all three applications were unable to return results for what we consider the "simplest" set, no further attempts were made to analyze any other sets.

12. Our current assumption is that the size of the data set we presented to these applications is beyond the scope of their computational resources.

13. We believe that the TUTR collection accurately represents common experimental results (UTR sequences), and in most cases, the size and number of sequences is certainly not extravagant. However, in almost all cases, these sequences are indeed many times larger than the test data sets presented by each application in their reference manuscript.

14. Certainly, there are likely to be extenuating circumstances for our experience with these applications, and given a greater amount of time and resources, we may have been able to garner results. However, it is apparent that there remains a great amount of work to be done before investigators have a *reasonable* solution to this problem.

4. Notes

1. Users may note that some logs show upward of 0.5 million rejects (where GC content is highly skewed). Portions of sequences in these sets vary by upper and lower case order to conserve the characteristics found in the NCBI genome downloaded from UCSC: http://hgdownload.cse.ucsc.edu/downloads.html. The case is imparted by RepeatMasker, http://repeatmasker.org, and Tandem Repeats Finder *4*.
2. In this folder is a sub-folder (only present in downloadable zip archive) "control-GenerationLogfiles," containing logs of the generation steps for the control sets so that interested parties can confirm validity and see their exact origination criteria.
3. Of the algorithms we found, all require that some minimum amount of information regarding the motif be known a priori. Although this may be the case in a few contrived circumstances, it is far from ideal.
4. It was not our intent to compare and contrast an algorithm's ability to successfully discover a motif in the TUTR collection but rather to provide these algorithms with testable data sets, which we believe most accurately represent true experimental results.
5. Exceptions-cases it was necessary to download and run a program locally, and cases where no results were ever produced from existing web interfaces.
6. It was even possible to use simple multiple alignment tools to footprint the HSL in this set.

Acknowledgments

We acknowledge the helpful input from Timothy Baroni, Jennifer Bean, David Tuck, Georgi Shablovski, and the rest of the Tenenbaum lab. This research was supported by National Institute of Health National Human Genome Research Institute grant R21 HG003679.

References

1. Tenenbaum SA, Lager PJ, Carson CC, Keene JD (2002) Ribonomics: identifying mRNA subsets in mRNP complexes using antibodies to RNA-binding proteins and genomic arrays. *Methods* 26:191–198.
2. Mignone F, Grillo G, Licciulli F, Iacono M, Liuni S, Kersey PJ, Duarte J, Saccone C, Pesole G (2005) UTRdb and UTRsite: a collection of sequences and regulatory motifs of the untranslated regions of eukaryotic mRNAs. *Nucleic Acids Res* Database Issue:D141–D146.

3. Grillo G, Attimonelli M, Liuni S, Pesole G (1996) CLEANUP: a fast computer program for removing redundancies from nucleotide sequence databases. *Comput Appl Biosci* 12:1–8.
4. Benson G (1999) Tandem repeats finder: a program to analyze DNA sequences. *Nucleic Acids Res* 2:573–580.
5. Williams AS, Marzluff WF (1995) The sequence of the stem and flanking sequences at the 3′ end of histone mRNA are critical determinants for the binding of the stem-loop binding protein. *Nucleic Acids Res* 23:654–662.
6. Hentze MW, Kuhn LC (1996) Molecular control of vertebrate iron metabolism: mRNA-based regulatory circuits operated by iron, nitric oxide, and oxidative stress. *Proc Natl Acad Sci USA* 93:8175–8182.
7. Hubert N, Walczak R, Sturchler C, Schuster C, Westhof E, Carbon P, Krol A (1996) RNAs mediating cotranslational insertion of selenocysteine in eukaryotic selenoproteins. *Biochimie* 78:590–596.
8. Walczak R, Westhof E, Carbon P, Krol A (1996) A novel RNA structural motif in the selenocysteine insertion element of eukaryotic selenoprotein mRNAs. *RNA* 2:367–379.
9. Fagegaltier D, Lescure A, Walczak R, Carbon P, Krol A (2000) Structural analysis of new local features in SECIS RNA hairpins. *Nucleic Acids Res* 28: 2679–2689.
10. Verrotti A, Thompson S, Wreden C, Strickland S, Wickens M (1996) Evolutionary conservation of sequence elements controlling cytoplasmic polyadenylylation. *Proc Natl Acad Sci USA* 93:9027–9032.
11. Goodwin EB, Okkema PG, Evans TC, Kimble J (1993) Translational regulation of tra-2 by its 3′-untranslated region controls sexual identity in *C. elegans. Cell* 75:329–339.
12. Ostareck-Lederer A, Ostareck D, Standart N, Thiele B (1994) Translation of 15-lipoxygenase mRNA is inhibited by a protein that binds to a repeated sequence in the 3' untranslated region. *EMBO J* 13:1476–1481.
13. Chen C, Shyu A (1995) AU-rich elements: characterization and importance in mRNA degradation. *Trends Biochem Sci* 20:465–470.
14. Amaldi F, Pierandrei-Amaldi P (1997) TOP genes: a translationally controlled class of genes including those coding for ribosomal proteins. *Prog Mol Subcell Biol* 18:1–17.
15. Kaspar RL, Kakegawa T, Cranston H, Morris DR, White MW (1992) A regulatory *cis* element and a specific binding factor involved in the mitogenic control of murine ribosomal protein L32 translation. *J Biol Chem* 267: 508–514.
16. Morris DR, Kakegawa T, Kaspar RL, White MW (1993) Polypyrimidine tracts and their binding proteins: regulatory sites for posttranscriptional modulation of gene expression. *Biochemistry* 32:2931–2937.
17. Boado RJ, Pardridge WM (1998) Ten nucleotide *cis* element in the 3′-untranslated region of the GLUT1 glucose transporter mRNA increases gene expression via mRNA stabilization. *Brain Res Mol Brain Res* 59:109–113.

18. Le SY, Maizel JV Jr (1997) A common RNA structural motif involved in the internal initiation of translation of cellular mRNAs. *Nucleic Acids Res* 25:362–369.
19. Grillo G, Licciulli F, Liuni S, Sbisa E, Pesole G (2003) PatSearch: a program for the detection of patterns and structural motifs in nucleotide sequences. *Nucleic Acids Res* 13:3608–3612.

II

FUNDAMENTAL ASPECTS OF THE STUDY OF RNA BIOLOGY

4

In-Line Probing Analysis of Riboswitches

Elizabeth E. Regulski and Ronald R. Breaker

Summary

Riboswitches are intricate, metabolite-binding RNA structures found in the non-coding regions of mRNA. Allosteric changes in the riboswitch that are induced by metabolite binding are harnessed to control the genes of a variety of essential metabolic pathways in eubacteria and in some eukaryotes. In this chapter, we describe an RNA structure analysis technique called "in-line probing." This assay makes use of the natural instability of RNA to elucidate secondary structure characteristics and ligand-binding capabilities of riboswitches in an entirely protein-free manner. Although this method has most frequently been used to examine riboswitches, the structures of any RNAs can be examined using this technique.

Key Words: Allosteric; aptamer; phosphoester transfer; RNA structure; spontaneous RNA cleavage.

1. Introduction

1.1. Description of Riboswitches

Riboswitches are structurally complex RNA genetic control elements that act as receptors for specific metabolites and that mediate gene expression in response to ligand binding *(1,2)*. These gene control elements are embedded in the non-coding portions of mRNAs in many bacteria although examples of the thiamine pyrophosphate riboswitches have been identified in some eukaryotes *(3,4)*. Numerous examples of riboswitches exist that selectively bind a variety of organic molecules critical for fundamental biochemical processes including coenzymes, purines, amino acids, and an aminosugar *(5)*. Riboswitches typically

From: *Methods in Molecular Biology, Vol. 419: Post-Transcriptional Gene Regulation*
Edited by: J. Wilusz © Humana Press, Totowa, NJ

consist of two functional components: a metabolite-binding aptamer and an expression platform. An aptamer is an RNA domain that forms a precise three-dimensional structure and selectively binds a target molecule. Allosteric changes in the aptamer, induced by metabolite binding, cause downstream structural changes that are translated into gene expression control by altering the folding patterns of the adjoining expression platform. The sizes of known aptamers range from 35 to approximately 200 nucleotides in length. In contrast to the highly conserved aptamer domain, expression platforms range widely in size, nucleotide sequence, and structural configuration. This great variety of expression platforms allows aptamers from the same riboswitch class to exert control over gene expression at several different levels. These include modulation of gene expression through control of transcription elongation of mRNA and efficiency of translation initiation. Riboswitches also have the ability to distinguish between closely related analogs and recognize their cognate effectors with values for apparent dissociation constant (K_d) ranging from picomolar to low micromolar concentrations *(2)*.

1.2. Methods for Identifying Riboswitch Candidates

Initially, a number of riboswitch classes were identified by searching the literature for instances of genetic regulation in which a protein factor could not be found even after extensive investigation. Currently, comparative genomic approaches are proving to be invaluable tools in the identification of new riboswitch candidates *(6, 7)*. The Breaker Lab Intergenic Sequence Server (BLISS) reports lists of homologous RNAs in intergenic regions that were identified using BLAST. Secondary-structure models are then built for promising RNA motifs. These are refined and extended by additional BLAST searches. By employing increasingly automated methods, it is likely that these comparative genomic approaches will continue to uncover interesting novel RNA structures.

1.3. Methods for Confirming the Existence of Riboswitches

Once a promising riboswitch candidate is discovered, it is important to demonstrate that it is indeed a genetic control element that binds a metabolite with high affinity and specificity in the complete absence of proteins. To determine whether an RNA motif is a regulator of gene expression in bacteria, the RNA motif is fused to a reporter construct such as *lacZ (8)*. The putative metabolite ligand is then added to the medium, and the effect it has on gene expression by binding to the RNA motif is monitored by measuring the production of the reporter. Mutations are then made in the predicted stem structures of the RNA motif to demonstrate that disruption and subsequent restoration of the RNA secondary structure affect gene regulation. Direct metabolite binding to RNA without the obligate involvement of proteins is a hallmark of riboswitches. Biochemical evidence for metabolite binding in the absence of accessory proteins can be obtained by several methods including in-line probing assays, equilibrium dialysis assays, and in vitro transcription termination assays *(9)*.

1.4. Background on the In-Line Probing Chemical Reaction

In-line probing assays have been used extensively to elucidate the secondary structure and binding capabilities of riboswitch aptamers. This technique exploits the natural tendency for RNA to differentially degrade according to its structure *(10)*. The phosphodiester bonds in the covalent backbone of RNA are subject to slow, non-enzymatic cleavage through the "in-line" nucleophilic attack of the 2′ oxygen on the adjacent phosphorus center (*see* **Fig. 1**). When the 2′ oxygen, the phosphorus, and adjacent 5′ oxygen enter an "in-line" conformation, the 2′ oxygen acts as a nucleophile in the intramolecular displacement of the 5′ oxygen in the adjacent phosphorus center and efficiently cleaves the RNA linkage. The relative speeds of spontaneous cleavage are dependent on local structural characteristics of each RNA linkage (*see* **Fig. 2**). Single-stranded RNA lacks secondary structure and is free to sample a range

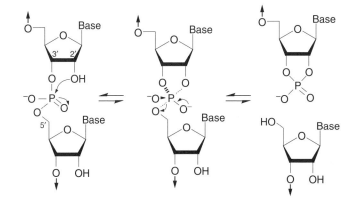

Fig. 1. "In-line" nucleophilic attack is necessary for efficient cleavage of an RNA phosphodiester linkage. When the 2′ oxygen enters a linear "in-line" arrangement with the phosphorus and the 5′ oxygen-leaving group of the linkage (dotted line), the 2′ oxygen executes a productive nucleophilic attack on the adjacent phosphorus center, cleaving the RNA linkage.

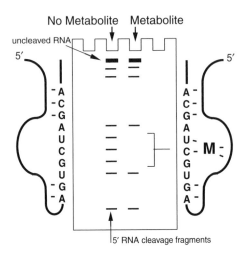

Fig. 2. Schematic of an in-line probing gel. Notice the difference in cleavage patterns (indicated by bracket) between reactions without and with metabolite (M) present. Changes occur in the secondary structure configuration of the RNA upon binding of the metabolite. Certain linkages are now locked into position making them less flexible and less likely to sample an in-line conformation, thus reducing cleavage.

of conformations, including in-line geometry. Consequently, these linkages are susceptible to spontaneous cleavage through in-line attack. In contrast to flexible single-stranded RNA domains, linkages found in the highly structured regions of a folded RNA are locked into position. The linkages in these structured RNA regions are less likely to sample an in-line geometry and therefore have a reduced chance of cleaving. Occasionally, RNA secondary structure actually locks the molecule into an in-line conformation resulting in accelerated cleavage of the linkage involved *(10)*. However, this is a rare occurrence and will only infrequently lead to misassignment of structured and unstructured RNA regions.

The initial step for an in-line probing experiment is to prepare the RNA for analysis. Usually, the RNAs of interest are produced by in vitro transcription using double-stranded DNA templates. These can be prepared in many ways, including DNA synthesis or PCR amplification as described in this protocol. The resulting DNA template is transcribed into RNA using T7 RNA polymerase and then is dephosphorylated and 5′-end labeled with [γ-^{32}P] ATP. An in-line probing reaction is assembled with radiolabeled RNA and the metabolite of interest. Denaturing (8 M urea) polyacrylamide gel electrophoresis (PAGE) is used to separate the products of spontaneous RNA cleavage. The products are visualized and quantitated by a phosphorimager. Finally, the pattern of spontaneous RNA cleavage is compared between RNAs incubated in the absence of metabolite versus RNAs incubated in the presence of the metabolite. Changes in the cleavage pattern indicate altered RNA structure due to metabolite binding. Further analysis of this pattern provides detailed information about the secondary structure characteristics of the RNA motif *(11)*.

2. Materials
2.1. Preparation of DNA Template

1. PCR buffer (10×): 15 mM $MgCl_2$, 500 mM KCl, 100 mM Tris–HCl (pH 8.3 at 20°C), and 0.1% gelatin in deionized water (dH_2O).
2. Deoxyribonucleoside 5′ triphosphate (dNTP) mix (10×): 2 mM each of dATP, dCTP, dGTP, and dTTP in dH_2O.
3. Appropriate 5′ and 3′ DNA oligonucleotide primers, 10 µM each in dH_2O.
4. About 5–100 ng double-stranded DNA template.
5. Thermostable DNA polymerase *(Taq)* (New England BioLabs).
6. 100% ethanol chilled to –20°C.

2.2. RNA Transcription

1. Transcription buffer (10×): 150 mM $MgCl_2$, 20 mM spermidine, 500 mM Tris–HCl (pH 7.5 at 25°C), and 50 mM dithiothreitol (DTT).
2. Nucleoside 5′ triphosphate (NTP) mix (10×): 20 mM each of ATP, CTP, GTP, and TTP in dH_2O.
3. Bacteriophage T7 RNA polymerase.
4. DNA template: approximately 50 pmol of PCR product.
5. Reagents and apparatus for PAGE.
6. Denaturing 8% polyacrylamide gel (1.5 mm thick).
7. Tris-Borate-EDTA (TBE) gel-running buffer (1×): 90 mM Tris, 90 mM borate, and 10 mM ethylenediaminetetraacetic acid (EDTA; pH 8.0 at 20°C).
8. Gel-loading buffer (2×): 18 M urea, 20% (w/v) sucrose, 0.1% (w/v) SDS, 0.05% (w/v) bromophenol blue sodium salt, 0.05% (w/v) xylene cyanol, 90 mM Tris–HCl, 90 mM borate, and 1 mM EDTA (pH 8.0 at 25°C).
9. Hand-held UV light, shortwave (254 nm).
10. Fluor-coated Thin Layer Chromatography (TLC) plate (Applied Biosystems).
11. Crush/soak buffer: 200 mM NaCl, 10 mM Tris–HCl (pH 7.5 at 25°C), and 1 mM EDTA (pH 8.0 at 25°C).
12. 100% ethanol chilled to –20°C.

2.3. 5′ ^{32}P End-Labeling of RNA

1. RNA: approximately 5 pmoles.
2. Phosphatase buffer (10× as supplied by manufacturer).

3. 500 mM Tris–HCl and 1 mM EDTA (pH 8.5 at 20°C) (Roche Diagnostics).
4. Calf intestinal alkaline phosphatase (CIP): 1 U/µL (Roche Diagnostics).
5. Phenol saturated with 10 mM Tris–HCl buffer (pH 7.9 at 20°C) with hydroxquinoline indicator (American Bioanalytical).
6. 24:1 chloroform: isoamyl alcohol (CHISAM).
7. T4 polynucleotide kinase (T4 PNK): 10 U/µL (New England BioLabs).
8. Kinase buffer (5×): 25 mM MgCl$_2$, 125 mM CHES (pH 9.0 at 20°C), and 15 mM DTT.
9. [γ-^{32}P] ATP: 6000 Ci/mmole total ATP on reference date (*see* **Note 1**).
10. Reagents and apparatus for PAGE.
11. Gel-loading buffer (2×): *See* **Subheading 2.2, item 8**.
12. Denaturing 8% polyacrylamide gel (1.5 mm thick).
13. TBE gel-running buffer (1×): *See* **Subheading 2.2, item 7**.
14. Stratagene Autorad markers.
15. X-ray film BioMax MR (Kodak).
16. Crush/soak buffer: *See* **Subheading 2.2, item 11**.
17. Glycogen 20 mg/mL (Roche Diagnostics).
18. 100% ethanol chilled to –20°C.

2.4. In-line Probing Reaction

1. In-line reaction buffer (2×): 100 mM Tris–HCl (pH 8.3 at 20°C), 40 mM MgCl$_2$, and 200 mM KCl.
2. 5' ^{32}P-labeled RNA: 30 kcpm/reaction.
3. Solutions of metabolites (10×).
4. Colorless gel-loading solution (2×): 10 M urea and 1.5 mM EDTA (pH 8.0 at 20°C).
5. Na$_2$CO$_3$ buffer (10×): 0.5 M Na$_2$CO$_3$ (pH 9.0 at 23°C) and 10 mM EDTA.
6. Sodium citrate buffer (10×): 0.25 M sodium citrate (pH 5.0 at 23°C).
7. RNase T1: 1 U/µL (Roche Diagnostics).

2.5. PAGE Analysis of In-line Probing Reaction

1. Reagents and apparatus for denaturing PAGE.
2. Denaturing 10% polyacrylamide gel, 0.75 mm thick.
3. Gel dryer model 583 (Bio-Rad).
4. Blotting paper 703 (VWR International).
5. PhosphorImager cassette (GE Healthcare).
6. PhosphorImager with ImageQuant software (GE Healthcare).

3. Methods

It is critical that the following procedures are carried out in an RNase-free environment including the use of sterile, RNase-free reagents, and materials. Frequent glove changes should be employed to protect the RNA sample from

RNases present on the hands or in the environment. RNA should always be kept on ice when preparing reactions to reduce unwanted spontaneous cleavage.

3.1. Preparation of the DNA Template

The region of interest is cloned by established methods into an appropriate vector and then sequenced to ensure that the correct construct is present, as even a single nucleotide error in sequence can alter RNA secondary structure (*see* **Note 2**). The purified plasmid is then used as the DNA template in the following PCR.

1. Design a 5′ DNA oligonucleotide primer beginning with the T7 RNA polymerase recognition sequence (TAATACGACTCACTATA) at the 5′ end. One to three G residues should immediately follow the T7 promoter, which favors robust transcription yields. These two sections should be followed by 15–20 nucleotides that are homologous with the 5′-end of the region of interest. Design a 3′ DNA oligonucleotide primer 15–20 nucleotides in length that is complementary to the 3′-end of the region of interest.
2. Combine the following in a 0.5-mL microfuge tube to perform a PCR to amplify the region of interest:

    a. 10 μL 10× PCR buffer
    b. 10 μL 10× dNTP mix
    c. 5–100 ng double-stranded DNA template

    7.5. μL 10 μM 5′ DNA oligonucleotide primer

    d. μL 10 μM 3′ DNA oligonucleotide primer
    e. *Taq* DNA polymerase 0.05 U/μL final concentration
    f. dH$_2$O to 100 μL

3. Using standard thermocycler conditions, conduct 25–30 cycles with appropriate annealing temperatures for the selected primers.
4. Precipitate the PCR product by adding 2.5 vol of 100% ethanol and mixing well. Pellet by centrifugation for 20 min at 14,000 *g* and 4°C.
5. Decant supernatant and dry pellet. Resuspend in 25 μL dH$_2$O.

3.2. RNA Transcription Reaction

1. Combine the following in a 1.5 mL microfuge tube to perform in vitro transcription:

    a. Approximately 50 pmol DNA template
    b. 2.5 μL 10× transcription buffer
    c. 2.5 μL 10× NTP mix
    d. 25 U/μL T7 RNA polymerase, final concentration
    e. dH$_2$O to 25 μL

2. Incubate for 2 h at 37°C (*see* **Note 3**).
3. Purify the RNA transcription product by PAGE. Add 25 μL of 2× gel-loading buffer and load the sample into two lanes of a denaturing 8% polyacrylamide gel (1.5 mm thick).
4. Run the gel until the band of interest is midway down the plate. Use the location of the bromophenol blue and xylene cyanol markers to estimate the product location according to molecular weight.
5. Separate the glass plates and transfer the gel to plastic wrap.
6. Use UV-shadowing to locate the band of interest. In a darkened room, place the plastic-wrapped gel on top of the fluor-coated TLC plate. Visualize the RNA bands by shining a hand-held ultraviolet light (shortwave 254 nm) on the gel. The RNA will appear as a dark band, whereas the TLC plate will fluoresce.
7. Excise the RNA band of interest with a razor blade. Use the razor blade to chop the isolated gel into small cubes and transfer to a microfuge tube.
8. Add 400 μL of crush/soak buffer to microfuge tube, covering all gel cubes. Incubate at room temperature for 30 min or incubate longer for greater yields.
9. Using a micropipette tip, remove the crush/soak buffer from the gel cubes and transfer buffer to a new microfuge tube.
10. Precipitate the RNA transcription product with ethanol and centrifuge as in **Subheading 3.1, step 4**.
11. Decant the supernatant and dry pellet. Resuspend in 30 μL dH$_2$O. This may be used immediately for the 5′-end-labeling reaction or stored at –20°C.

3.3. 5′ ^{32}P End-Labeling of RNA

1. Calculate the concentration of the RNA transcription product by measuring the absorbance at 260 nm (A$_{260}$) and entering this and the RNA transcript sequence into a program such as Oligonucleotide Properties Calculator (http://www.basic.northwestern.edu/biotools/oligocalc.html).
2. Combine the following in a 1.5-mL microfuge tube to perform the dephosphory-lation reaction, incubate at 50°C for 15 min:

   a. 10 pmol RNA
   b. 2 μL 10× dephosphorylation buffer
   c. 2 μL CIP
   d. dH$_2$O to 20 μL

3. After incubation, add 172 μL dH$_2$O and 8 μL 5 M NaCl.
4. Remove the CIP by phenol/chloroform extraction as follows. Combine equal parts of Tris-buffered phenol and CHISAM. Then add 200 μL of phenol/CHISAM to the dephosphorylation reaction and vortex. Separate the phases by centrifugation and transfer aqueous phase to a new microfuge tube. Add 200 μL of CHISAM to the aqueous layer and vortex. Separate the layers by centrifugation and transfer the aqueous phase to a new microfuge tube.

5. Add 1 µL glycogen to act as a carrier for the RNA during precipitation with ethanol.
6. Precipitate RNA from the dephosphorylation reaction with ethanol and pellet by centrifugation.
7. Decant the supernatant and dry pellet. Resuspend in 15 µL dH_2O. Employ this dephosphorylated RNA in the 5′ ^{32}P end-labeling reaction.
8. Combine the following in a 1.5 mL microfuge tube for the 5′ ^{32}P end-labeling reaction, and incubate at 37°C for 30–60 min:

   a. 10 pmol dephosphorylated RNA
   b. 4 µL 5× kinase buffer
   c. 6 µL [γ-^{32}P] ATP (1.6 µM)
   d. 2.5 µL T4 polynucleotide kinase
   e. dH_2O to 20 µL

9. Purify the 5′ ^{32}P end-labeled RNA by PAGE. Add 20 µL 2× gel-loading buffer and load the sample onto a denaturing 8% polyacrylamide gel (1.5 mm thick).
10. Run gel until the band of interest is midway down the plate. Use the location of the bromophenol blue and xylene cyanol markers to estimate the location of the 5′ ^{32}P end-labeled RNA according to the expected molecular weight of the product.
11. Separate the glass plates, removing the top plate and covering the gel with plastic wrap. Attach the Stratagene markers to the plastic wrap on either side of the lane that contains the sample.
12. In the darkroom, expose an X-ray film to the gel with the radiolabeled RNA and develop. The labeled RNA should be visible as a dark defined band on the film. Any unincorporated γ-^{32}P ATP may also be visible as a dark band of very small molecular weight at the bottom of the gel. The Stratagene marker images will also be visible on the film.
13. Using the Stratagene marker images to align the film to the gel, mark the location of the band of interest and excise with a razor blade. Use a razor blade to chop the band of interest into small cubes and transfer to a microfuge tube.
14. Recover the 5′ ^{32}P-labeled RNA from the gel by crush/soak and precipitate with ethanol as in **Subheading 3.2, steps 8–10**.
15. Decant supernatant and dry pellet. Resuspend in 100 µL dH_2O. This may be used immediately for the in-line probing reaction or stored at –20°C. Use labeled RNA before radioactive decay depletes the signal or causes excessive RNA damage.
16. Estimate the amount of RNA by using a scintillation counter.

3.4. In-line Probing Reaction: Binding Confirmation and K_d Determination

In-line probing assays provide information as to whether or not a particular metabolite is bound to the RNA of interest (*see* **Note 4**). In-line probing reactions containing a metabolite concentration of 1 mM usually provide sufficient metabolite to determine whether there are any changes in the structure of the RNA.

Metabolite concentrations above 1 mM can result in non-specific interactions, and the data from these reactions should be interpreted with caution.

In-line probing analysis also can be used to gather information about relative binding of the RNA and metabolite by establishing the apparent K_d. This is done by measuring changes in the level of cleavage of metabolite-sensitive regions of the RNA over a range of metabolite concentrations. To obtain an accurate plot of concentration-dependent ligand-mediated structure modulation, it is usually necessary to collect data at metabolite concentrations ranging from 1 nM to 1 mM with two to three data points per order of magnitude concentration change. During the in-line probing reaction, the metabolite should not be the limiting factor in the reaction and must be in excess over the amount of radiolabeled RNA. Therefore, the low range of metabolite concentration will depend on the efficiency of the RNA labeling. 5′ ^{32}P end-labeling of the RNA must be efficient enough so that only trace amounts of RNA are present in the in-line probing reaction while maintaining the 20,000–100,000 cpm per reaction necessary for high-quality visualization of the gel image. With any in-line probing gel, it is important to prepare a no-reaction sample of undigested precursor RNA, precursor RNA subjected to RNase T1 and partial alkaline digestion, and a control in-line probing reaction containing spontaneously degraded RNA with no metabolite.

1. Combine the following in 0.5-mL microfuge tubes for the in-line probing reaction and incubate at room temperature for approximately 40 h:

   a. 5 µL 2× in-line reaction buffer
   b. 1 µL 5′ ^{32}P-labeled RNA
   c. Metabolite solution to yield the final desired concentration
   d. dH$_2$O to 10 µL

2. Quench the reaction by adding 10 µL of 2× colorless gel-loading solution. Without excessive delay, the samples should be loaded onto a denaturing polyacrylamide gel to avoid unwanted RNA cleavage that will complicate data interpretation.

3. The following three reactions, the RNase T1 and partial alkaline-digested RNA and the undigested precursor RNA, should be prepared when the in-line probing reactions are approaching completion. Without excessive delay, the samples should be loaded immediately onto a denaturing polyacrylamide gel.

4. Combine the following in a 0.5-mL microfuge tube to prepare RNase T1-digested RNA and incubate at 55°C for 5–15 min depending on the length of the RNA:

   a. 1 µL 10× sodium citrate buffer
   b. 1 µL RNase T1
   c. 1 µL 5′ ^{32}P-labeled RNA
   d. 7 µL 2× colorless gel-loading solution

5. Quench the reaction by adding 3 μL of 2× colorless gel-loading solution and 7 μL dH$_2$O (*see* **Subheading 3.4., step 3**).
6. Combine the following in a 0.5-mL microfuge tube to prepare a partial alkaline digest of the precursor RNA and incubate at 90°C for 5–10 min depending on the length of the RNA:

  a. 1 μL 10× Na$_2$CO$_3$ buffer
  b. 1 μL 5′ ^{32}P-labeled RNA
  c. dH$_2$O to 10 μL

7. Quench the reaction by adding 10 μL of 2× colorless gel-loading solution (*see* **Subheading 3.4., step 3**).
8. Combine the following in a 0.5-mL microfuge tube to prepare the uncleaved precursor (no reaction) sample (*see* **Subheading 3.4., step 3**):

  a. 1 μL 5′ ^{32}P-labeled RNA
  b. 9 μL dH$_2$O
  c. 10 μL 2× colorless gel-loading solution

3.5. PAGE Analysis of In-line Probing Reaction

1. Prepare a 10% denaturing polyacrylamide gel (40 cm long, width depends on number of samples, approximately 1 cm wide wells, and 0.75 mm thick) (*see* **Note 5**).
2. Load 10 μL of each sample on the gel in the following order (*see* **Fig. 3**):

  a. 1× gel-loading buffer with dyes to monitor electrophoresis progress
  b. Uncleaved precursor RNA (no reaction)
  c. RNase T1-digested precursor RNA
  d. Partial alkaline digestion of precursor RNA
  e. In-line probing reaction samples with appropriate ligand
  f. 1× gel-loading buffer with dyes

3. Run the gel at 45 Watts for approximately 3 h until the dyes in the 1× gel-loading buffer are at the desired position on the gel.
4. Separate the glass plates and transfer the gel to blotting paper. Cover with plastic wrap.
5. Dry the gel for 2 h and then expose in a PhosphorImager cassette overnight.
6. Image the screen on a PhosphorImager and collect the data using software such as ImageQuant.

3.6. Analysis of Data

The extent of metabolite binding by a riboswitch can be determined by establishing the apparent K_d value. The apparent K_d of a riboswitch represents the concentration of metabolite that is required to convert half of the RNA

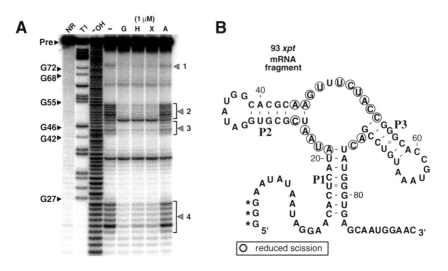

Fig. 3. (**A**) In-line probing gel image of the guanine sensing riboswitch *(12)*. Lanes 1–3 contain the precursor (Pre) RNA undigested (NR), partially digested by T1 RNase (T1) and subjected to partial alkaline digestion ($^-$OH). Lane 4 shows the cleavage pattern of the riboswitch in the absence (–) of any metabolites. Lanes 5–7 contain guanine (G) and the related compounds, hypoxanthine (H) and xanthine (X). Their cleavage patterns show reduced cleavage, indicating that they bind to the guanine riboswitch. While the cleavage pattern of adenine (A) closely resembles that of the metabolite-free reaction, indicating that it does not bind to the guanine riboswitch. Arrows 1–4 indicate regions of cleavage modulation in response to metabolite binding. (**B**) Secondary structure model of the guanine-sensing riboswitch that corresponds to the in-line probing gel image. The circles indicate linkages of the RNA that demonstrate reduced scission in the presence of the metabolite.

aptamers that are present in a mixture to their metabolite-bound form. This is calculated by measuring spontaneous cleavage at different metabolite-sensitive positions in the RNA over a range of metabolite concentrations (*see* **Note 6**).

1. Check to make sure that there are no breakdown bands in the undigested precursor RNA (no reaction) sample as this may indicate degraded RNA that will interfere with interpretation of the gel (*see* **Note 7**).
2. Use the lane with precursor RNA treated with RNase T1, producing cleavage products at each guanine residue, and the precursor RNA subjected to partial alkaline digestion, producing cleavage products at each nucleotide position, to provide markers for mapping the sites of spontaneous RNA cleavage.

3. Examine the in-line probing gel for bands that increase or decrease in intensity as the concentration of metabolite changes. This indicates a nucleotide position that is sensitive to metabolite binding.
4. Record the intensity of this band over the range of metabolite concentrations. Repeat at each modulating band within the RNA. Subtract background intensity from values.
5. The apparent K_d value is the concentration at which RNA cleavage is at half maximum. To find this, plot the fraction of RNA cleaved versus the logarithm of the concentration of metabolite, excluding the zero metabolite sample. The fraction of RNA cleaved is calculated as:

$$\text{Fraction RNA cleaved} = \frac{\text{sample value-min.value}}{\text{max.value-min.value}}$$

Max. and min. refer to the highest and lowest cleavage values measured for each modulating band. The concentration of metabolite needed to yield half maximal modulation of cleavage provides an estimate of the K_d for the complex.

4. Notes

1. Standard precautions should be observed when using radioactive isotopes.
2. This sensitivity to mutations can also be a useful tool. Incorporating disruptive and compensatory mutations in the pairing regions of the RNA and then observing how this affects binding affinity will allow you to pinpoint regions that are critical to the maintenance of RNA secondary structure and metabolite binding.
3. A white precipitate sometimes develops as the transcription reaction proceeds. This is the result of an insoluble complex forming between Mg^{2+} ions and inorganic pyrophosphate and usually indicates a robust transcription reaction. Incubation beyond this point is unnecessary as the lack of free Mg^{2+} ions will prevent any further significant production of RNA.
4. Riboswitches are usually *cis*-acting genetic control elements. As a result, the genomic context of the RNA of interest may be useful in determining which metabolites to begin testing as possible ligands. The functions of proteins encoded downstream might indicate compounds that are important to test. The molecular recognition capabilities of the RNA can be explored using panels of chemical analogs that are derivatives of the metabolite but that contain modifications of various functional groups.
5. To reliably obtain high-quality gel images, we recommend using PAGE gel solutions made "in-house" starting with acrylamide powder rather than commercially available gel solutions, which we find are inconsistent in their propensity to yield high-resolution separation of bands (particularly when small amounts of

RNA are used). Use care in handling powdered acrylamide as it is extremely neurotoxic until polymerized. Wear a dust mask, gloves, and eye protection and use a spatula to weigh powdered acrylamide. Make a 29:1 (w/w) [37% w/v] solution of acrylamide/bis-acrylamide and filter through #1 Whatman paper using a Buchner funnel. Combine 270 mL 37% acrylamide/bis, 480 g urea, 100 mL 10× TBE and 270 mL dH_2O to make denaturing 10% polyacrylamide gel solution.

6. The values for RNA–metabolite interactions obtained in vitro may be different than those actually found under physiological conditions within the cell due to changes in Mg^{2+} ion concentrations and other factors affecting RNA structure.

7. Unwanted RNA degradation will result in the presence of a constant band that may mask metabolite-induced modulation.

References

1. Winkler, W. C., and Breaker, R. R. (2005) Regulation of bacterial gene expression by riboswitches. *Annu Rev Microbiol* **59**, 487–517.

2. Mandal, M., and Breaker, R. R. (2004) Gene regulation by riboswitches. *Nat Rev Mol Cell Biol* **5**, 451–63.

3. Winkler, W., Nahvi, A., and Breaker, R. R. (2002) Thiamine derivatives bind messenger RNAs directly to regulate bacterial gene expression. *Nature* **419**, 952–6.

4. Sudarsan, N., Barrick, J. E., and Breaker, R. R. (2003) Metabolite-binding RNA domains are present in the genes of eukaryotes. *RNA* **9**, 644–7.

5. Breaker, R. R. (2006) Riboswitches and the RNA world, in *The RNA World* (Gesteland, R. F., Cech, T.R., and Atkins J. F., eds), Cold Spring Harbor Laboratory Press, Cold Spring Harbor, NY, pp. 89–107.

6. Corbino, K. A., Barrick, J. E., Lim, J., Welz, R., Tucker, B. J., Puskarz, I., Mandal, M., Rudnick, N. D., and Breaker, R. R. (2005) Evidence for a second class of S-adenosylmethionine riboswitches and other regulatory RNA motifs in alpha-proteobacteria. *Genome Biol* **6**, R70.

7. Barrick, J. E., Corbino, K. A., Winkler, W. C., Nahvi, A., Mandal, M., Collins, J., Lee, M., Roth, A., Sudarsan, N., Jona, I., Wickiser, J. K., and Breaker, R. R. (2004) New RNA motifs suggest an expanded scope for riboswitches in bacterial genetic control. *Proc Natl Acad Sci USA* **101**, 6421–6.

8. Miller, J. H. (1992) *A Short Course in Bacterial Genetics: A Laboratory Manual and Handbook for Escherichia coli and Related Bacteria*, Cold Spring Harbor Laboratory Press, Plainview, NY.

9. Nahvi, A., Sudarsan, N., Ebert, M. S., Zou, X., Brown, K. L., and Breaker, R. R. (2002) Genetic control by a metabolite binding mRNA. *Chem Biol* **9**, 1043.

10. Soukup, G. A., and Breaker, R. R. (1999) Relationship between internucleotide linkage geometry and the stability of RNA. *RNA* **5**, 1308–25.

11. Soukup, G. A., DeRose, E. C., Koizumi, M., and Breaker, R. R. (2001) Generating new ligand-binding RNAs by affinity maturation and disintegration of allosteric ribozymes. *RNA* **7,** 524–36.

12. Mandal, M., Boese, B., Barrick, J. E., Winkler, W. C., and Breaker, R. R. (2003) Riboswitches control fundamental biochemical pathways in *Bacillus subtilis* and other bacteria. *Cell* **113,** 577–86.

5

Ribotrap

Targeted Purification of RNA-Specific RNPs from Cell Lysates
Through Immunoaffinity Precipitation to Identify Regulatory
Proteins and RNAs

Dale L. Beach and Jack D. Keene

Summary

Many elegant methodologies have been devised to explore RNA-protein as well as RNA–RNA interactions. Although the characterization of messages targeted by a specific RNA-binding protein (RBP) has been accelerated by the application of microarray technologies, reliable methods to describe the endogenous assembly of ribonucleoproteins (RNPs) are needed. However, this approach requires the targeted purification of a select mRNA under conditions favorable for the copurification of associated factors including RNA and protein components of the RNP. This chapter describes previous methods used to characterize RNPs in the context of in vitro approaches and presents the Ribotrap methodology, an in vivo protocol for message-specific purification of a target RNP. The method was developed in a yeast model system, yet is amenable to other in vivo cell systems including mammalian cell culture.

Key Words: RNP; ribonomics; RNA-binding protein; post-transcriptional regulation; MS2; aptamer.

1. Introduction

The fundamental unit of post-transcriptional regulation is the ribonucle-oprotein complex (RNP complex) (or RNA protein particle), a complex composed of an RNA harboring distinct *cis*-acting elements decorated by corresponding *trans*-acting factors. Component factors of the RNP include proteins and noncoding RNAs associated with the mRNA. The exact composition of

From: *Methods in Molecular Biology, Vol. 419: Post-Transcriptional Gene Regulation*
Edited by: J. Wilusz © Humana Press, Totowa, NJ

the RNP dictates the regulatory state of the message within a given cellular environment and is responsive to changing cellular requirements. Several properties of RNA-binding proteins (RBPs) contribute to the dynamic nature of the RNP: multiple RBPs may associate with a given RNA, each RBP may associate with a number of RNAs, and the RNA–RBP interactions are responsive to regulatory signals [reviewed in *(1–4)*]. Therefore, the composition of any RNP is combinatorial at both the RNA and protein levels and subject to biological perturbations, so that the characterization of *trans*-acting factors forming the RNP requires a reliable, in vivo, approach.

The dynamic nature of the RNP contributes directly to the Post-Transcriptional Operon model *(3,5)*, which predicts a network of functionally related messages, and regulators exists to integrate regulatory signals with post-transcriptional alterations in RNA metabolism. Dissection of post-transcriptional regulatory pathways essentially requires two complementary approaches: identification of the mRNA populations bound by each RBP and characterization of the RNP assembled on each mRNA. *R*NA-binding protein *I*mmuno*P*recipitation-microarray/*C*hip (RIP-Chip) methodologies use microarray analysis to identify messages that coimmunoprecipitate with a given RBP *(6,7)*. Using these methods, the RNA targets of a number of RBPs from different organisms have been identified and indicate a post-transcriptional role for the regulation of a range of cellular pathways (*see* **Table I** of ref. *7*). The converse approach, the identification of factors associated with a message of interest, will allow researchers to identify the full complement of RBPs and noncoding RNAs for an endogenously assembled RNP. Isolation of a specific mRNA of interest from a cell requires molecular targeting techniques to specifically identify and biochemically purify the message along with associated factors to characterize the endogenous RNP. The iterative application of these discovery tools leads to a ribonomic discovery cycle (*see* **Fig. 1**) for which networks of messages and regulators assembling into discrete RNPs can be studied.

Several in vitro approaches have been described, extensively targeting the assembly of RNPs from cell lysates (*see* **Fig. 2**). Synthetic oligonucleotides or in vitro transcribed RNAs can function as templates for the assembly of RNPs. Incorporation of labeled nucleotides generates in vitro transcripts with a range of functions from live cell imaging of RNP transport *(8)* to the capture of RNPs assembled from cell lysates *(9)*. RNP complexes assembled in vitro remain valuable for the analysis of a variety of static complexes and biochemical systems such as the spliceosome; however, these constructs are not amenable to the dynamic characterization of post-transcriptional regulatory systems. In vivo

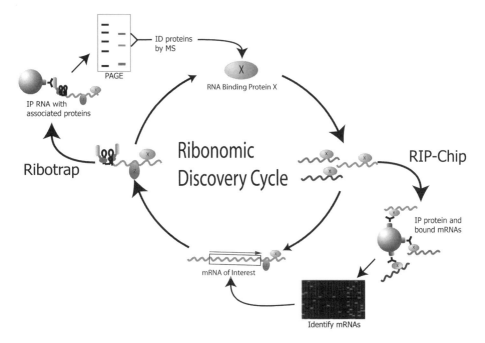

Fig. 1. Ribonomic discovery cycle. The cyclic discovery of RNA-binding proteins (RBPs) and their targets is described by the ribonomic discovery cycle. Starting at the top and moving clockwise: Given any RBP (protein X), a subset of the transcriptome will be bound under a specific cellular environment. Bound RNAs (right of cycle) can be copurified with the RBP by immunoprecipitation (IP) and then identified using microarray and related techniques [RIP-Chip method, *see (7)*]. Any of the identified RNAs can then become the target of Ribotrap analysis to determine bound proteins. The full length of a mRNA (bottom of cycle) including the 5′-untranslated region (UTR), coding region (boxed with arrow), and 3′-UTR will endogenously form an RNP, presumably with regulatory elements bound to the 3′-UTR as depicted by factors X and Z. A reporter construct including the 3′-UTR serves as a template for the endogenous assembly of the RNP when expressed in the cell (left of cycle). Including specific sites for a known RBP (i.e., MS2 CP) facilitates the affinity capture of the coexpressed RBP along with the RNP. Analysis on 1D PAGE (right lane in figure) with suitable controls (left lane) distinguishes novel RBPs that can be identified with standard mass spectrometric analysis. These novel RBPs can then re-enter the cycle, and through iterative analytical cycles, post-transcriptional networks can be mapped.

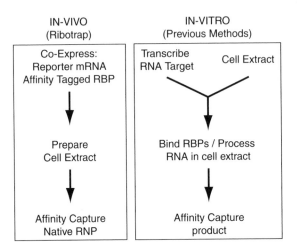

Fig. 2. Comparison of in vivo and in vitro methods. A number of methods have been used to describe RNA–protein complexes in terms of paired binding partners as well as functional complexes. The vast majority of the methods determine these interactions using an in vitro approach (right column) combining synthetic RNAs with cellular extracts. As a result, proteins decorating the transcript can be purified and/or RNA processing reactions can ensue. This approach has found great success for in vitro splicing reactions [as examples *see (21,45)*]. In contrast, the goal to purify endogenously assembled RNPs under a variety of dynamic biological perturbations requires an in vivo approach. Although complicated in vivo approaches using transfection of synthetic oligonucleotides to capture target messages have been described *(11)*, the Ribotrap method (left column) is facilitated through the coexpression of capture protein and target mRNA. Following cell lysis and extract preparation, a completely in vivo assembled ribonucleoprotein (RNP) is affinity purified. Thus, Ribotrap is intended to be an in vivo methodology for RNP characterization.

application of synthetic transcripts exposed to cell lysates *(10)* or injected directly into the cell *(11)* target endogenous messages by hybridization to complementary sequences. Subsequent affinity precipitation of these in vitro transcripts from cell lysates then is copurified with the target messages and their associated RNPs (*see* **Fig. 2**). These methods have the benefit of targeting endogenous mRNAs directly; however, the necessary step of hybridization between the target RNA and the solid phase probe can be complicated by high stringency buffers that disrupt the RNP, obstruction of probe-binding sites within the RNP, and nonspecific hybridization to other RNAs.

In vivo expression of reporter transcripts provides an improved approach where a recombinant mRNA facilitates in vivo RNP assembly and biochemical purification of the complex (*see* **Fig. 2**). Messages including the RNA sequence of interest are engineered to include elements useful for mRNA purification, such as aptamers or specific binding sites for well characterized RBPs that can attach to a solid support. The incorporation of defined RNA elements, such as RNA aptamers, provides affinity motifs encoded by the transcript that support affinity capture of the message. Aptamers are RNA or DNA structures developed for high-affinity binding to a range of molecular targets including antibiotic agents like streptomycin *(12)* and tobramycin *(13)*, the cellulose matrix, sephadex *(14)*, and many proteins including streptavidin *(15)* and antibodies *(16)*. Although an aptamer-containing message is appealing for its small size, direct binding to ligands on bead matrixes, and elution with small molecules, their dependence upon the folding of complicated three-dimensional structures may interfere with RNP formation or conversely RNP assembly may disrupt aptamer folding.

Another successful approach utilizes a site-specific RBP to capture a reporter construct engineered with the target-binding site. Short RNA sequences within a reporter mRNA can mediate high-affinity binding to endogenous transcripts and the associated native RNPs (*see* **Fig. 3**). Site-specific RBPs can circumvent the need for complicated secondary structures, and the RBP itself can be expressed as a fusion with a variety of affinity ligands, providing a broad choice of affinity matrixes for purification. Coexpression of the RBP fusion protein and reporter transcript produces a single product comprising an affinity-capture site and endogenously assembled RNP (*see* **Fig. 2**). The coat protein (CP) of the bacteriophage MS2 and the spliceosome component, U1-A protein *(17,18)*, has been used recently to tag mRNAs in vivo. The bacteriophage MS2 CP has been used extensively to tag RNA transcribed in vitro *(19–22)* and in vivo as a genetic method to determine protein–RNA interactions through a "Three-Hybrid" system *(23–25)*. Fusion with green fluorescent protein (CP-GFP) facilitates imaging of RNA dynamics in living yeast and mammalian cells [examples include *(26–28)*]. The CP binds a target stem loop structure with a K_d of 3–300 × 10^{-9} M depending on the CP isoform *(29)* or stem loom sequence variant *(30)*, sufficient to block translation in eukaryotic cells *(31)*.

Herein, we describe a method, termed as "Ribotrap," to isolate an in vivo-expressed RNA and identify copurifying *trans*-acting factors (*see* **Fig. 3**). We chose to use the MS2 CP to develop the Ribotrap method in a *Saccharomyces cerevisiae* model system. Our previous experience of constructing a in vivo-labeling system for live cell imaging of dynamic mRNA movements utilized the MS2 CP *(26,32)* and targeted stem loop-binding sites.

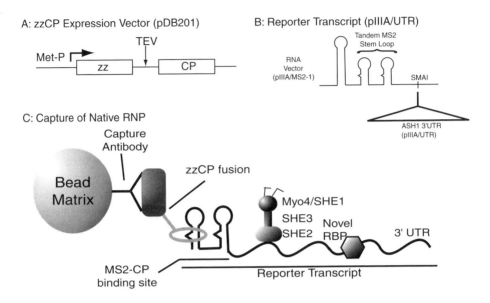

Fig. 3. Components of the Ribotrap ribonucleoprotein (RNP) capture. The Ribotrap methodology relies on the coexpression of a reporter RNA and a recombinant RNA-binding protein (RBP) to form an in vivo RNP amenable for affinity capture. (**A**) We utilize the MS2 coat protein (CP) fused to the IgG-binding domains of protein-A (zz), with a TEV cleavage site between the two domains for elution from the bead matrix (pDB201). Because the MS2 CP can encapsulate RNAs containing the target stem loop at high intracellular concentrations, a nonoligomeric allele (dlFG) of the MS2 CP is used *(32,50)*. The fusion protein is expressed from the MET25 promoter that allows attenuation of transcriptional activity to modulate the final protein levels *(51)*. (**B**) As a reporter mRNA, the ASH1 3′-untranslated region (UTR) is inserted downstream of the tandem stem loop-binding sites of the CP and an RNAseP leader sequence (pIIIA/UTR) *(26,32)*. The reporter RNA is ubiquitously expressed through the RNA polymerase III promoter, and the parent vector (pIIIA/MS2-1) *(23)* provides a negative control for the analysis of IP eluates. (**C**) The captured native RNP complex is associated with the bead matrix during the immunoprecipitation reaction. The reporter RNA and associated proteins are depicted as predicted for the yeast ASH1 3′-UTR including members of the localization complex (see text: She2p, She3p, and Myo4p/She1p) along with a novel RBP. The zzCP fusion protein forms the link between the affinity beads and captured RNP. Note that the MS2 CP binding in sites are oriented 5′ to the ASH1 3′-UTR to ensure survival of the reporter message following post-transcriptional processing. Elution of the *trans*-factors bound to the RNA is facilitated by TEV-dependant cleavage of the zzCP fusion protein or RNAse digestion of the RNA template.

The yeast protein, She2p, facilitates the asymmetric localization of a set of 24 mRNAs in yeast, including the Ash1 message *(17,33,34)*, as part of a localization complex [*see* **Fig. 3C**; reviewed in *(35)*]. Elements within the 3′-untranslated region (UTR) of the Ash1 mRNA are sufficient for localization of a reporter transcript to the bud tip *(26,32)*. Coexpression of the reporter transcript with a green fluorescent protein fusion of the CP (CP-GFP) yielded a GFP-labeled RNA (termed "gRNA") that could be imaged using standard fluorescent microscopes *(26,32)*. The reporter transcript containing tandem target stem loop structures and the complete Ash1 3′-UTR (*see* **Fig. 3B**) was localized to the bud tip as predicted from in situ hybridization results *(26,33,36)*. Further, live cell imaging established the dynamic localization of the Ash1 message to sites of polarized growth during different phases of cell growth, including the bud tip, incipient bud site, and tip of the mating projection. Following the successful demonstration that our reporter RNA acted as predicted for the localization complex, we adapted the imaging system for use in devising Ribotrap.

The Ribotrap method utilizes the complete Ash1 3′-UTR for the reporter construct [as described in *(26)*] and a modification of the CP fusion protein. The anticipated RNP complex purified with the system includes She2p as well as other components of the transport complex (She3p and Myo4p), translation inhibitor, Puf6p *(37)*, and potentially unknown proteins. Briefly, a reporter transcript including tandem MS2 CP-binding sites followed by an RNA sequence of interest (*see* **Fig. 3B**) is coexpressed with a fusion protein (*see* **Fig. 3A**) including the MS2 CP and the immunoglobin-binding domain of the *Staphylococcus aureus* protein A (zz domain from pEZZ18; Amersham Biosciences). The resulting native RNP is purified from a cell lysate using immunoglobin-conjugated beads, Sepharose™ beads (*see* **Fig. 3C**), and material eluted from the immunoprecipitation (IP) can be analyzed for protein content using PAGE, Western, or Mass Spectrometric (MS) techniques or RNA content (including noncoding, regulatory RNAs) using agarose gel, Northern, RT–PCR, or microarray-based methods.

Application of the Ribotrap method produced specific and reproducible IP of the localization complex members (*see* **Fig. 3C**). To determine the integrity of the Ribotrap method as well as the extent of known components within the immunoprecipitate RNP, the endogenous proteins, She2p and She3p, were tagged with the Myc epitope for Western analysis. As shown in **Fig. 4A**, She2p and She3p were present in the IP eluate following IP using the IgG-conjugated beads. Further, the presence of She2p-Myc and She3p-Myc is dependent on the inclusion of the ASH1 3′-UTR in the reporter mRNA, as both proteins were absent in samples expressing the empty vector.

Beach and Keene

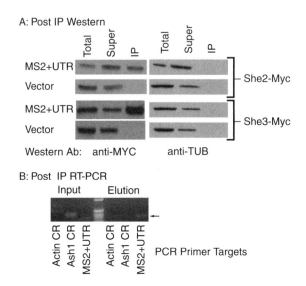

Fig. 4. Integrity of the Ribotrap immunoprecipitation (IT). The Ribotrap methodology aims to specifically purify a single target RNA from the cell extract along with the assembled ribonucleoprotein (RNP). Post-IP analyses were used to determine the integrity of the Ribotrap protocol. (**A**) Western analysis demonstrated the specific enrichment of She2p and She3p in the IP eluate. Separate yeast strains were generated harboring She2-Myc (BDY11) and She3-Myc (BDY8) and then transformed with the Ribotrap components [pDB201 with either pIIIA/untranslated region (UTR) (MS2 + UTR) or pIIIA/MS2-1 (Vector) as control]. Following IP reactions using IgG beads, Western analysis was completed using anti-MYC (9E-10; Santa Cruz Biotechnology) and anti-Tubulin (YOL1/34; Harlan Sera-Lab) primary antibodies. Results are shown for total cell lysate (as noted in section 3.2.3.1), supernatant (section 3.2.3.4), and IP eluates (section 3.3.1.9) for independent She2-Myc and She3-Myc strains. Note that the presence of She2-Myc or She3-Myc in the IP lanes is dependent on the presence of the ASH1 3′-UTR in the reporter transcript, and the ubiquitous protein, Tubulin, is not detected in the IP eluate. (**B**) To determine the specificity for mRNA precipitation, eluates were prepared by heating the beads to 95°C for 10 min in YBP-WB to release intact RNA. RNA was purified and concentrated using phenol : chloroform extractions and ethanol precipitation [*see (38)*] and then reverse transcribed using iScript (Bio-Rad) to make cDNAs. Standard PCR was used to amplify specific regions of cDNAs generated from the cell extract (input) and eluate (elution) using the following primer sets. Sequence elements from the actin (*ACT1*)-coding region (5′-GGTTATTGATAACGGTTCTGG-3′ and 5′-GGAACGACGTGAGTAACACC-3′) and the *ASH1* (5′-GTATTCGTTT CATTCGCTC-3′; 5′-TCGTTATTGCTGGATTTCC-3′)-coding region were not expected to precipitate in the IP reactions and were used as negative controls.

Therefore, we conclude that the Ribotrap method is capable of precipitating an entire RNP, including proteins directly bound to the RNA as well as more distal members of the complex.

To determine whether a single mRNA was being precipitated, RT–PCR analysis of IP eluates (*see* **Fig. 4B**) demonstrated that only the reporter construct is present in the purified RNP. The endogenously expressed ASH1 is discriminated from the reporter mRNA by targeting the PCR amplification to the coding region. Absence of the endogenous Ash1 mRNA from the IP eluate indicates that the She2p-dependant localization complex incorporates a single RNA into the localized RNP. Absence of the Act1 mRNA in the IP eluate confirmed that the Ribotrap method is specific, purifying a single RNP complex from the cell lysate.

Comparison of total proteins eluted from the ASH1 3′-UTR reporter construct against the vector control identified several proteins that are unique to each sample (*see* **Fig. 5**). Where the reporter mRNA contained the Ash1 3′-UTR, She2p (~28 kDa) was consistently apparent. Also, MS analysis of select bands confirmed the identity of She2p as well as identifying Poly(A)-binding protein (Pab1p, ~68 kDa; *see* **Fig. 5**). The identification of Pab1p in the eluate indicated that the reporter mRNA was being processed at least at the level of polyadenylation and confirms that novel proteins can be isolated using the Ribotrap method. These results confirm the function, specificity, and integrity of the Ribotrap method. As a general approach, the Ribotrap method using the MS2 CP as well as methods using RNA aptamers are amenable to in vivo investigations of a broad array of mRNA and associated factors not only in a yeast-based system but in mammalian systems and other model organisms as well.

2. Materials

1. Sterile, nuclease-free water: All solutions should be prepared with nuclease-free water either prepared by treatment with DEPC [*see (38)*] or purchased (Invitrogen; cat. no. 10977-015).

Fig. 4. (*Continued*) The reporter message was detected using primers specific to the RNAseP leader sequence (5′-TTACGTTTGAGGCCTCGTG-3′) and the ASH1 3′-UTR (5′-GAAAGAGATTCAGTTATCC-3′). Note that while all transcripts can be detected in the cell lysate (Input), only the reporter construct (MS2 + UTR) is detected in the elution (arrow). Molecular weight markers are in the center of the figure (bottom to top: 298, 344, 517, and 1000 bp).

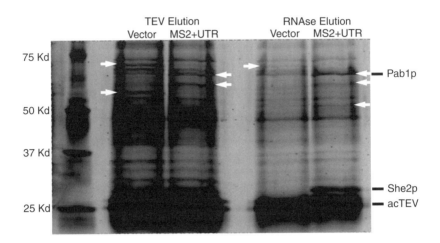

Fig. 5. Total protein from the Ribotrap immunoprecipitation (IT). Representative samples of eluates generated using the Ribotrap method and analyses for total protein as described. Cell extracts from yeast strains transformed with the Ribotrap components [pDB201 with either pIIIA/untranslated region (UTR) (MS2 + UTR) or pIIIA/MS2-1 (Vector) as control] were prepared as described in the protocol. The immunoprecipitation was scaled for the application of 250 mg of total protein. The lysate was not precleared over unconjugated beads. However, the IP beads were washed using 25 mM EDTA and 1 M urea. Elutions were performed serially with acTEV™ and then RNAse-A to demonstrate the amount of material that may remain after a single elution with acTEV™. The entire volume of each elution was precipitated as described and run on a 10% criterion PAGE gel (Bio-Rad) and then stained with SYPRO-Ruby as described. The broad band at approximately 25 K_d is residual acTEV™ protease which obscures the endogenous She2p band in the MS2 + UTR lane of the TEV elution. The acTEV™ signal is also present in the RNAse elutions due to incomplete washing of the beads before RNAse digestion; however, the She2p band is clearly visible. White arrows indicate bands that are specific to lanes containing vector or MS2 + UTR eluates. Select bands consistently strong in eluates from the MS2 + UTR samples were excised and identified by mass spectrometry as Poly(A)-binding protein (Pab1p) and She2p as shown. The sizes of molecular weight markers (leftmost lane) are indicated.

2. RNP-lysis buffer (RNP-LB): 20 mM Tris–HCl, pH 8.0, 140 mM KCl, 1.8 mM $MgCl_2$, and 0.1% Nonidet-40 (NP40).
3. Inhibiter mix: Add the following reagents when required. Concentrations shown are the final concentration for the buffer system. 1× Complete™ Protease Inhibitor Cocktail, 1 mM (phenylmethanesulphonyl fluoride) PMSF, 1 mM DTT, and 100 U/ml RNaseOut™ (*see* **Note 1**).

4. RNP-wash buffer(RNP-WB): 20 mM Tris–HCl, pH 8.0, 140 mM KCl, 1.8 mM MgCl$_2$, 0.01% NP40, 1× Compete™ Protease Inhibiter Cocktail, and 1 mM PMSF.
5. Complete™ Protease Inhibiter Cocktail: Complete™ EDTA-free (Roche Molecular Biochemicals; cat. no. 1-873-580). Follow manufacturer's recommendation for a 25× stock and store at −20°C. Use at 1×.
6. RNaseOut™ Ribonuclease Inhibitor: 40 U/μl stock. Store at −20°C (Invitrogen; cat. no. 17-0969-01).
7. Breaking beads: 0.5 mm Zirconia/Silica beads (Biospec Products Inc., cat. no. 11079105z) or 0.5 mm glass beads (Sigma, cat. no. G9268).
8. Bio-spin chromatography columns (Bio-Rad, cat. no. 732-6008).
9. Sepharose™ CL-6B: Store at 4°C (unconjugated beads; Sigma, cat. no. CL6B200-100).
10. IgG Sepharose™ 6 fast flow: Store at 4°C (Amersham Biosciences, cat. no. 17-0969-01).
11. AcTEV™ Protease: Store at −20°C (Invitrogen, cat. no. 12575-015).
12. RNAse A: 10 mg/ml. Store at −20°C [*see (38)*].
13. 100% Trichloroacetic acid (TCA) [*see (38)*].
14. 2% Deoxycholate solution (w/v) in water.
15. Bead stripping buffer: 1% formic acid (v/v) and 1% ethanol (v/v) in water.
16. SYPRO-ruby protein gel stain: 1 l (Bio-Rad, cat. no. 170-3125).
17. SYPRO fix/destain: 10% methanol (v/v) and 7% acetic acid (v/v) in water.

3. Methods
3.1. Cell Culture and Lysate Preparation
3.1.1. Cell Culture and Induction

1. Prepare a seed culture from a single colony of each desired strain previously transformed with either experimental or control reporter constructs and the zzCP fusion protein (*see* **Note 2**).
2. Inoculate 1−20 ml of selective media with each identified colony and grow at 30°C with shaking at 250 RPM overnight or to an optical density at 600 nm wavelength (OD$_{600}$) <4.0.
3. On the following day, use each seed culture to inoculate independent 1–1.5 l cultures of selective media in a large volume (Fernbach) flask and grow overnight at 30°C with shaking at 250 RPM to a final OD$_{600}$ of 1.0.
4. Pellet the cells by centrifugation in a precooled (4°C) Sorval GS-3 rotor at approximately 6000 × *g*. The culture must be split between several centrifuge bottles.
5. Resuspend the cell pellets into sterile water, using about 50 ml per individual pellet.
6. Pellet cells as in **step 4**.
7. If the cells require induction of one or more expression constructs, resuspend the pellets in inducing buffer (SD-MET for pDB201) to same final volume as the

original culture. Transfer to suitable culture flasks (Fernbach) and continue growth under the desired conditions and duration. Typically, 30°C for 1 h with shaking (*see* **Note 2**).

8. When the cells are ready to be harvested, pellet the induction cultures as in **step 4**. Wash the cells a total of three times with 100 ml ice-cold, sterile water. Pellet the cells as in **step 4** between each wash.

9. Resuspend the pellets in 100 ml ice-cold, sterile water, recombine the cells of each strain into a single bottle, and then transfer the cell slurry into 50-ml conical tubes.

10. Pellet the cells at 3000 RPM in a Beckman Tabletop GS-6R centrifuge with GH3.7 rotor (~2000 × *g*) at 4°C for 5 min.

11. Wash cells once in 25 ml of ice-cold RNP-LB and centrifuge as in **step 10**.

12. Resuspend cells in a volume of ice-cold RNP-LB + Inhibitor Mix equal to the pellet volume.

13. The cell slurry can be frozen in liquid nitrogen and stored at –80°C (*see* **Note 3**). At this point, the cells are still intact and can be stored frozen for several months.

3.1.2. Preparation of the Cell Lysate

1. Thaw cell pellets gently on ice. For the remainder of the protocol, cell lysates must remain on ice as much as possible.

2. Add one half of the volume of the cell suspension of breaking beads (*see* **Note 4**).

3. Vortex for 1 min.

4. Place on ice for 1 min.

5. Repeat **steps 3** and **4** for 10 cycles if using Zirconia beads, otherwise repeat for 20 cycles for glass beads.

6. Optionally, observe the degree of cell breakage in a microscope using a 40× or 100× objective. Cellular disruption should exceed 60–75%. More vortex cycles can by used to continue breakage, but this may degrade the quality of the RNPs produced.

7. Pellet the cells at 3000 RPM in a Beckman GS-6R centrifuge with GH3.7 rotor (~2000 × *g*) at 4°C for 5 min to pellet cell debris and breaking beads.

8. Transfer supernatant to a fresh high-speed centrifuge tube avoiding the debris in the pellet.

9. Centrifuge at 58,000 × *g* (30,000 RPM in Beckman TLA100.3 rotor) and 4°C for 30–60 min to pellet large particles and organelles.

10. Transfer supernatant to a clean high-speed centrifuge tube.

11. Optionally, pellet ribosomes by centrifuging the samples at >100,000 × *g* (50,000 RPM in a Beckman TLA100.3 rotor) and 4°C for 60 min (*see* **Note 5**).

12. Remove 100 µl of the supernatant to determine total protein or optionally test for efficient induction.

13. Freeze the cell lysate at −80°C until ready to use.

3.2. IP of Target mRNA

3.2.1. Preclear the Lysate (Optional)

1. Proteins within the cell lysate will nonspecifically adhere to the agarose matrix comprising the beads. It is advised to remove these proteins by adsorbing them to unconjugated beads (beads lacking the immunoaffinity antigen used for the IP) to minimize nonspecific background (*see* **Note 1** about removing free zzCP protein from the lysate).
2. Gently thaw cell lysate on ice while preparing the beads.
3. Measure a sufficient amount of the bead slurry to accommodate 0.5–1 ml packed beads into either a 2 ml Bio-Spin Column or Falcon 2063 culture tube.
4. Equilibrate the beads to RNP-LB by washing the packed beads with at least four washes using 1 ml ice-cold RNP-LB. If using the Bio-Spin columns, simply allow the buffer to flow through the column. Otherwise, pellet the beads at 3000 RPM in a Beckman GS-6R centrifuge with GH3.7 rotor ($\sim$2000 × g) at 4°C for 1 min between each wash.
5. Transfer an amount of the cell lysate containing 30 mg of total protein to a fresh tube and adjust the final volume to 1 ml using ice-cold RNP-LB including the Inhibiter Mix (resulting in a total protein concentration of 30 mg/ml).
6. Add the diluted lysate to the unconjugated beads equilibrated to RNP-LB.
7. Incubate the lysate with the beads for 1 h at 4°C while tumbling the sample head-over-tail on a rotisserie style lab shaker.
8. Collect the precleared lysate by allowing the supernatant to flow through the beads into a fresh microcentrifuge tube (column method) or pelleting the beads and removing the supernatant to a fresh microcentrifuge tube after centrifugation (culture tube) as in **step 4**.
9. When using the Bio-Spin columns, any residual lysate can be removed from the bead matrix by centrifugation. Insert the column into a clean Falcon 2063 tube and centrifuge in a tabletop centrifuge at 4°C for 5 min at 3000 RPM ($\sim$2000 × g).
10. Preferentially, the lysate is cleared immediately before being applied to the affinity column. However, the precleared lysate can be stored at –80°C until used.

3.2.2. Preparation of the Immunoaffinity Beads

1. If frozen, gently thaw cell lysate on ice while preparing the beads for IP.
2. Measure 100 µl IgG Sepharose™ 6 fast flow beads per IP reaction into 1.7-ml microcentrifuge tubes. The volume of bead slurry used to establish the packed bead volume may vary depending on the bead concentration in the stock solution. The volume of packed beads depends on the amount of total protein used in the reaction, with a target of approximately 3 µl of packed beads per milligram of total protein.

3. Wash the beads four times using 500 μl ice-cold RNP-LB per wash (~20 times the bead volume) to equilibrate the beads.
4. After the final wash, pellet the beads and remove any remaining buffer with a small bore pipette.

3.2.3. Immunoprecipitation

1. Add the cell lysate (from **Subheadings 3.1.2.** or **3.2.1.**) to the prepared IgG Sepharose™ 6 fast flow beads (prepared in **Subheading 3.2.2.**) equilibrated with RNP-LB.
2. Incubate the lysate with the beads for >3 h at 4°C while tumbling the sample head-over-tail to allow the affinity capture of the zzCP fusion protein and associated RNP.
3. Pellet the beads using a 10- to 30-s pulse in a microcentrifuge.
4. Transfer the supernatant to fresh tube and retain for analysis, removing as much of the supernatant as possible without disturbing the beads.
5. Add 1 ml ice-cold RNP-WB and vortex or invert sharply several times to resuspend the beads.
6. Repeat **steps 3–5** for a total of four times using 1 ml RNP-WB for each wash, retaining the fourth wash (W4) for analysis.
7. Wash the bead pellet another four times using 1 ml RNP-WB for each wash and allowing each wash to tumble head-over-tail for 10 min.
8. Collect the fourth tumbled wash (W8) for analysis.

3.2.4. RNP Elution

1. Remove as much of the WB from the beads as possible using a fine bore pipette without removing any beads.
2. Selective elution can be accomplished either using RNAse A (1 mg per ml) to digest the RNA component of the RNP, thus releasing the proteins of interest, or using TEV (200 units per ml) to cleave the zzCP. Both RNAse-A and acTEV™ Protease are active in the RNP-WB and will not require re-equilibrating the beads to a new buffer system.
3. Optionally, wash the beads to re-equilibrate to any buffer system required for the desired elution method. Note that high salt or chelating agents (EDTA) cause the release of some RNP components. Retain this WB for analysis to ensure that the RNP is not disrupted by the buffer.
4. Add 500 μl of the required elution buffer (RNP-WB) with diluted enzyme and incubate as necessary for the desired method. Tumble the beads head-over-tail to maintain the beads in suspension. Incubate RNAse-A reactions at 30° for 30–60 min or at 4°C for 60 min; acTEV™ digests at 30°C for 2 h.
5. Pellet the beads with a 30-s pulse centrifugation and remove and save the supernatant in a new tube. This supernatant contains the material released from the beads including any bound proteins or RNA.

6. Wash the pellet three times with 200 µl each ice-cold elution buffer (no enzyme is required). These washes remove additional material trapped in the bead bed. Combine all of the washes with the original eluate.
7. Optionally, strip the beads using three washes of 300 µl Bead Stripping Buffer. Collect all three washes into a single tube and retain for analysis. These samples can be used to determine the elution efficiency as well as the amount of zzCP adhering to the beads.

3.3. Analysis of Eluted Proteins

3.3.1. Precipitation of Proteins for Analysis

1. Proteins from the washes, elution, and the bead strip must be precipitated and re-suspended in a small volume before they can be loaded onto most PAGE gels. Approximately 90% of the material is used for a total protein analysis, and the remaining volume can be used for western blot or RT–PCR analysis (not discussed). Samples of the original input material (cleared lysate), IP supernatant, W4, W8, and any other washes retained throughout the procedure can be included on an analytical gel to determine the depletion of zzCP or RNP components from the cell lysate as well as the extent of washing as a potential source of background in the elution.
2. Transfer 900 µl of each sample into a new microcentrifuge tube.
3. Optionally, add 10 µl (1/100th final volume) 2% deoxycholate, mix well, and chill on ice for 10 min to help precipitate proteins present at low abundance.
4. Add 100 µl (1/10th final volume) 100% TCA and mix well.
5. Keep on ice for 15 min or store at −20°C overnight.
6. Centrifuge for 20 min in a microcentrifuge at maximum speed (~14,000 RPM) at 4°C. Discard the supernatant.
7. Wash the pellet twice with 200 µl ice-cold 25% acetone.
8. Remove as much of the acetone as possible without dislodging the pellet.
9. Dry the pellet in a Speed-Vac to remove residual acetone. The dried pellet can be used for PAGE or MS analysis.

3.3.2. SDS–PAGE and Western Blot

1. Resuspend the protein pellets in 30 µl PAGE gel-loading buffer supplemented with 0.2 M NaOH to neutralize acidification by TCA during precipitation.
2. For analysis of total proteins, the entire volume of each sample should be loaded onto the gel. For analysis by western blot, as little as 10% of the sample may provide suitable sensitivity to detect individual proteins.
3. Separate the samples by standard SDS–PAGE methodologies.
4. For total protein analysis, follow the staining protocol below (*see* **Subheading 3.3.3**). For western blotting, transfer proteins to membranes and probe using standard techniques [*see* **ref.** *(38)*].

3.3.3. SYPRO-Ruby Staining of PAGE Gels to Detect Total Protein

1. SYPRO-Ruby offers an alternative to standard silver staining for high-sensitivity applications *(39)*. Wear gloves whenever handling the gels as SYPRO-Ruby stain is very sensitive and will detect proteins from fingerprints on the gel. Note that unstained molecular weight markers should be used when staining with SYPRO-Ruby.
2. After SDS–PAGE electrophoresis, transfer the gel to a plastic or glass dish.
3. Add enough SYPRO Fix/Destain solution to cover the gel and incubate for >30 min at room temperature with gentle agitation. This fixation is optional for 1D PAGE gels.
4. Remove the fixing solution and replace with sufficient, undiluted SYPRO-Ruby stain to completely cover the gel in at least 1 gel thickness (typically 50 ml for a medium format gel).
5. Stain for 4 h to overnight in a closed container with gentle agitation at room temperature. Placing the container in a light tight box to inhibit photobleaching of the stain is advisable for long incubations.
6. Remove the stain and rinse the gel with water to remove residual stain (the used stain can be reused until depleted, but this may affect the sensitivity for later gels).
7. Destain the gel with three changes of >50 ml SYPRO Fix/Destain solution for 30 min each at room temperature with gentle agitation.
8. Wash the gel once in water to minimize drying of the gel as methanol in the destain solution evaporates.
9. Image the stained gel (*see* **Fig. 5**) as described by the SYPRO-Ruby product insert. SYPRO-Ruby-stained gels can be imaged in the same manner as an ethidium bromide-stained agarose gel on an UV transilluminator. Alternatively, some phosphoimaging devices are equipped with optics to image the RUBY stain. Follow the manufacturer's directions for optimal imaging conditions.

3.4. Analysis of Eluted RNAs

RNAs eluting from the column will be either additional mRNAs incorporated into an RNP particle or noncoding RNAs, such as miRNAs, associated with one or more of the precipitated mRNAs. Analysis and identification of mRNAs is amenable to a number of standardized methods including cloning or PCR amplification of cDNAs [*see* **Fig. 4B** and **ref.** *(38)*] or microarray analysis on a number of different platforms. The same approaches can be used to identify microRNAs in the eluate though more specialized protocols are required. Analysis of unknown microRNAs can be facilitated by cloning methods *(40,41)* as well as specialized microarray platforms *(42)*. In contrast, miRNAs predicted to bind the reporter transcript can be detected by RT–PCR *(43)* and Northern methods *(40)* as well as microarrays.

3.5. Data Interpretation and Troubleshooting

In practice, the researcher must remain wary of several artifacts that may be present in the final set of eluted proteins. As with any purification protocol, background or nonspecific bands may be present. One common source of nonspecific bands is insufficient washing. The porous nature of the agarose beads provides innumerable crevices for proteins to lodge. Vigorous washing early in the protocol is meant to release material trapped within the bead matrix as well as between the packed beads. Other proteins may nonspecifically adhere to the agarose itself or the conjugated immunoglobins. One common approach to decreasing the nonspecific binding is to block with bovine serum albumin (BSA). Pre-IP blocking of the beads is discouraged because the elution of BSA may overwhelm sensitive MS analysis of the eluate. To decrease the background, care should be taken to clear the lysate through high-speed centrifugation (*see* **Subheading 3.1.2.**) and preclearing the lysate (*see* **Subheading 3.2.1.**).

A second source of background results from nonspecific association with the mRNA and/or CP fusion protein. For this reason, cell lysates must be maintained on ice or stored at −80°C and maintained at 4°C to inhibit adventitious protein associations. The parallel use of a second reporter construct is recommended as a negative control. Essentially, a vector control or small RNA construct may be sufficient to identify these contaminants. When experimental and control eluates are run side by side on the same gel, common bands likely represent nonspecific proteins, whereas bands unique to each lane will represent proteins specifically binding those transcripts (*see* **Fig. 5**). In our experience, abundant proteins in the lysate are also the primary source of background. Furthermore, including a diluted sample of the lysate as well as samples from wash steps on the gel can also aid in the identification of common bands resulting from incomplete washing and nonspecific binding to the beads.

Optimization of the protocol to minimize background signal for specific RNPs may be necessary. Additional washes using salt solutions at different ionic strength can be included in the protocol to remove background material prior to the elution. Although sodium chloride (NaCl) is commonly used for stringency washes, urea or heparin may also be useful. For instance, lithium chloride (LiCl) releases specific ribosomal proteins as ionic strength is increased *(44)* and may serve as an alternative to EDTA treatment to remove ribosomes from the message. In all of these examples, one should expect loss of some RNP components. As an alternative to using salt washes to remove background material from the bead matrix, further treatments of the lysate may render a "cleaner" input material for the IP. Size exclusion columns *(21,45)* or density gradient centrifugation *(46)* can be utilized to enrich the lysate material for the RNP of interest, and removing nonspecific or unbound proteins. Incidentally, this may also remove free MS2 CP from the lysate enhancing the enrichment of bound RNP on the bead matrix (*see* **Note 1**).

Once the protocol has been optimized for the target RNP, the method must be scaled for proteomic analysis. As described, the protocol will generate sufficient material for PAGE and Western analysis. However, to prepare the amount of material required for MS analysis requires a substantial increase in the amount of starting material: approximately 250 mg of total protein or more. The method presented can serve as a template for increased scale either by repetition and pooling of the eluted material or by increasing the total volume of the IP reaction. A variety of MS methods are available for protein identification using different starting materials. Material separated by standard PAGE methods can be used for single-band extraction and identification as well as whole lane fragmentation *(47,48)*. Alternatively, the unmodified eluate can be submitted in its entirety for whole sample analysis *(47)*. In each of these examples, adequate controls must be used to ensure identification of proteins bound specifically to the mRNA of interest and remove background contaminants. We strongly recommend consultation with a MS technician before submitting samples to ensure that buffer conditions and control material satisfy the MS facility guidelines.

The interaction between newly identified proteins and the mRNA can be validated by inverting the IP (*see* **Fig. 1**). IP of the proteins identified in the Ribotrap method will copurify any bound mRNAs. The presence of the original RNA of interest in these target messages can then be verified using RT–PCR methods or the application of RIP-Chip methodologies can be used to identify all of the targeted messages. Thus, Ribotrap products can be re-introduced to the ribonomic discovery cycle (*see* **Fig. 1**) for continued identification and mapping of post-transcriptional regulatory networks.

4. Notes

1. Heparin has been used extensively to inhibit the nonspecific binding of RBPs to RNA following cell lysis and may be included in lysis and wash buffers. Some evidence suggests that the use of heparin in the lysis solution as well as the wash solutions may remove some components of the RNP *(21)*. Additionally, heparin-conjugated beads have been used for the purification of the MS2 CP for *Escherichia coli* cell lysates *(45)*. To minimize the amount of zzCP fusion protein not bound to an RNA from occupying IgG-binding sites on the affinity resin, agarose beads conjugated to heparin may be included while clearing the lysate (*see* **Subheading 3.2.1.**), though we have not yet attempted this approach.

2. The media used to culture the yeast will vary depending on the strain used. The chosen media should select for any markers desired in the strain, particularly the expression plasmids used for the study. For short inductions (<2 h), use the richest media possible as it is not strictly necessary to maintain selection during this time. The rich media will encourage yeast growth and recovery following the centrifugation to maximize induction. Any induction longer than 2 h should be done in media that maintains selection for the plasmids. A good description of different yeast media as well as a source for media components is the QBIOgene catalog (QBIOgene).

3. The cell slurry can be frozen as a single pellet or as small beads. Freezing the entire pellet as a single unit is faster than the drop method but takes much longer to thaw. However, frozen beads thaw more quickly than solid pellets, and samples of the frozen beads can be taken for small preparation during optimization. To prepare frozen yeast beads, slowly allow the cell suspension to drip from a pipette into liquid nitrogen in a 50-ml conical tube. Each drop will freeze into an individual bead. Allow excess liquid nitrogen to boil away, then store the tube at −80°C. WARNING: the tube will explode if any residual liquid nitrogen is sealed into the tube.

4. A variety of methods have been used to disrupt the cell wall and lyse yeast cells including grinding the frozen cell pellets with a mortar and pestle or blender, digesting the cell wall, passage through a French Press, and mechanically grinding the cells with small beads. The best preservation of the RNA is achieved with the most gentle treatment of the cell at the lowest temperature *(49)*. A fair compromise between cost, efficiency, and RNA quality is the use of a mechanical shaker ("Bead Beater") or a vortex mixer to provide a mechanical breakage of the cells using beads added to the cell slurry. Cell lysis is increased when high-density beads (Zirconia Silica) are used; however, simple glass beads of 0.5 mm will suffice (*see* **Materials item 7**).

5. In the model system presented, the reporter transcript is not expected to be translated as the CP sites at the 5′-end will block translation initiation. Therefore, pelleting the polyribosomes with a $100,000 \times g$ centrifugation does not cause significant loss of RNP material. As the assembly of some RNPs may be dependant on ribosome or polyribosome complexes, we can imagine the construction of a wide variety of reporter construct variants. Therefore, inclusion of this optional step is dependant on the experimental design and care should be taken not to discard the target RNP.

6. Standard laboratory practices should be followed for all protocols. Caution should be taken when working with material containing RNA to avoid contamination by RNAses [for more information *see* **ref. *(38)***].

Acknowledgments

We thank the members of the Keene laboratory for intellectual support and stimulation, especially Kyle Mansfield and Matt Friedersdorf for critical review of the manuscript. We thank Kerry Bloom, University of North Carolina at

Chapel Hill, for supplying yeast strains and plasmids. D.L.B. is supported by the Minority Opportunities in Research Division of the National Institute of General Medical Sciences (NIGMS) grant GM-000678.

References

1. Keene, J. D., and Lager, P. J. (2005) Post-transcriptional operons and regulons co-ordinating gene expression. *Chromosome Res* **13**, 327–37.
2. Moore, M. J. (2005) From birth to death: the complex lives of eukaryotic mRNAs. *Science* **309**, 1514–8.
3. Tenenbaum, S. A., Lager, P. J., Carson, C. C., and Keene, J. D. (2002) Ribonomics: identifying mRNA subsets in mRNP complexes using antibodies to RNA-binding proteins and genomic arrays. *Methods* **26**, 191–8.
4. Keene, J. D. (2001) Ribonucleoprotein infrastructure regulating the flow of genetic information between the genome and the proteome. *Proc Natl Acad Sci USA* **98**, 7018–24.
5. Keene, J. D., and Tenenbaum, S. A. (2002) Eukaryotic mRNPs may represent posttranscriptional operons. *Mol Cell* **9**, 1161–7.
6. Tenenbaum, S. A., Carson, C. C., Lager, P. J., and Keene, J. D. (2000) Identifying mRNA subsets in messenger ribonucleoprotein complexes by using cDNA arrays. *Proc Natl Acad Sci USA* **97**, 14085–90.
7. Keene, J. D., Komisarow, J. M., and Friedersdorf, M. B. (2006) RIP-Chip: the isolation and identification of mRNAs, microRNAs and protein components of ribonucleoprotein complexes from cell extracts. *Nat Protocols* **1**, 302–7.
8. Ainger, K., Avossa, D., Morgan, F., Hill, S. J., Barry, C., Barbarese, E., and Carson, J. H. (1993) Transport and localization of exogenous myelin basic protein mRNA microinjected into oligodendrocytes. *J Cell Biol* **123**, 431–41.
9. Rouault, T. A., Hentze, M. W., Haile, D. J., Harford, J. B., and Klausner, R. D. (1989) The iron-responsive element binding protein: a method for the affinity purification of a regulatory RNA-binding protein. *Proc Natl Acad Sci USA* **86**, 5768–72.
10. Waris, G., Sarker, S., and Siddiqui, A. (2004) Two-step affinity purification of the hepatitis C virus ribonucleoprotein complex. *RNA* **10**, 321–9.
11. Zielinski, J., Kilk, K., Peritz, T., Kannanayakal, T., Miyashiro, K. Y., Eiriksdottir, E., Jochems, J., Langel, U., and Eberwine, J. (2006) In vivo identification of ribonucleoprotein-RNA interactions. *Proc Natl Acad Sci USA* **103**, 1557–62.
12. Bachler, M., Schroeder, R., and von Ahsen, U. (1999) StreptoTag: a novel method for the isolation of RNA-binding proteins. *RNA* **5**, 1509–16.
13. Hartmuth, K., Urlaub, H., Vornlocher, H. P., Will, C. L., Gentzel, M., Wilm, M., and Luhrmann, R. (2002) Protein composition of human prespliceosomes isolated by a tobramycin affinity-selection method. *Proc Natl Acad Sci USA* **99**, 16719–24.

14. Srisawat, C., Goldstein, I. J., and Engelke, D. R. (2001) Sephadex-binding RNA ligands: rapid affinity purification of RNA from complex RNA mixtures. *Nucleic Acids Res* **29**, E4.

15. Srisawat, C., and Engelke, D. R. (2001) Streptavidin aptamers: affinity tags for the study of RNAs and ribonucleoproteins. *RNA* **7**, 632–41.

16. Tsai, D. E., Kenan, D. J., and Keene, J. D. (1992) In vitro selection of an RNA epitope immunologically cross-reactive with a peptide. *Proc Natl Acad Sci USA* **89**, 8864–8.

17. Takizawa, P. A., DeRisi, J. L., Wilhelm, J. E., and Vale, R. D. (2000) Plasma membrane compartmentalization in yeast by messenger RNA transport and a septin diffusion barrier. *Science* **290**, 341–4.

18. Brodsky, A. S., and Silver, P. A. (2000) Pre-mRNA processing factors are required for nuclear export. *RNA* **6**, 1737–49.

19. Jurica, M. S., Licklider, L. J., Gygi, S. R., Grigorieff, N., and Moore, M. J. (2002) Purification and characterization of native spliceosomes suitable for three-dimensional structural analysis. *RNA* **8**, 426–39.

20. Duncan, K., Grskovic, M., Strein, C., Beckmann, K., Niggeweg, R., Abaza, I., Gebauer, F., Wilm, M., and Hentze, M. W. (2006) Sex-lethal imparts a sex-specific function to UNR by recruiting it to the msl-2 mRNA 3′ UTR: translational repression for dosage compensation. *Genes Dev* **20**, 368–79.

21. Deckert, J., Hartmuth, K., Boehringer, D., Behzadnia, N., Will, C. L., Kastner, B., Stark, H., Urlaub, H., and Luhrmann, R. (2006) Protein composition and electron microscopy structure of affinity-purified human spliceosomal B complexes isolated under physiological conditions. *Mol Cell Biol* **26**, 5528–43.

22. Czaplinski, K., Kocher, T., Schelder, M., Segref, A., Wilm, M., and Mattaj, I. W. (2005) Identification of 40LoVe, a Xenopus hnRNP D family protein involved in localizing a TGF-beta-related mRNA during oogenesis. *Dev Cell* **8**, 505–15.

23. SenGupta, D. J., Zhang, B., Kraemer, B., Pochart, P., Fields, S., and Wickens, M. (1996) A three-hybrid system to detect RNA-protein interactions in vivo. *Proc Natl Acad Sci USA* **93**, 8496–501.

24. Hook, B., Bernstein, D., Zhang, B., and Wickens, M. (2005) RNA-protein inter-actions in the yeast three-hybrid system: affinity, sensitivity, and enhanced library screening. *RNA* **11**, 227–33.

25. Gonsalvez, G. B., Little, J. L., and Long, R. M. (2004) ASH1 mRNA anchoring requires reorganization of the Myo4p-She3p-She2p transport complex. *J Biol Chem* **279**, 46286–94.

26. Beach, D. L., Salmon, E. D., and Bloom, K. (1999) Localization and anchoring of mRNA in budding yeast. *Curr Biol* **9**, 569–78.

27. Bertrand, E., Chartrand, P., Schaefer, M., Shenoy, S. M., Singer, R. H., and Long, R. M. (1998) Localization of ASH1 mRNA particles in living yeast. *Mol Cell* **2**, 437–45.

28. Janicki, S. M., Tsukamoto, T., Salghetti, S. E., Tansey, W. P., Sachidanandam, R., Prasanth, K. V., Ried, T., Shav-Tal, Y., Bertrand, E., Singer, R. H., and Spector, D. L. (2004) From silencing to gene expression: real-time analysis in single cells. *Cell* **116**, 683–98.

29. Lim, F., and Peabody, D. S. (1994) Mutations that increase the affinity of a translational repressor for RNA. *Nucleic Acids Res* **22**, 3748–52.

30. Johansson, H. E., Dertinger, D., LeCuyer, K. A., Behlen, L. S., Greef, C. H., and Uhlenbeck, O. C. (1998) A thermodynamic analysis of the sequence-specific binding of RNA by bacteriophage MS2 coat protein. *Proc Natl Acad Sci USA* **95**, 9244–9.

31. Stripecke, R., Oliveira, C. C., McCarthy, J. E., and Hentze, M. W. (1994) Proteins binding to 5′ untranslated region sites: a general mechanism for translational regulation of mRNAs in human and yeast cells. *Mol Cell Biol* **14**, 5898–909.

32. Beach, D. L., and Bloom, K. (2001) ASH1 mRNA localization in three acts. *Mol Biol Cell* **12**, 2567–77.

33. Long, R. M., Singer, R. H., Meng, X., Gonzalez, I., Nasmyth, K., and Jansen, R. P. (1997) Mating type switching in yeast controlled by asymmetric localization of ASH1 mRNA. *Science* **277**, 383–7.

34. Shepard, K. A., Gerber, A. P., Jambhekar, A., Takizawa, P. A., Brown, P. O., Herschlag, D., DeRisi, J. L., and Vale, R. D. (2003) Widespread cytoplasmic mRNA transport in yeast: identification of 22 bud-localized transcripts using DNA microarray analysis. *Proc Natl Acad Sci USA* **100**, 11429–34.

35. Chartrand, P., Singer, R. H., and Long, R. M. (2001) RNP localization and transport in yeast. *Annu Rev Cell Dev Biol* **17**, 297–310.

36. Takizawa, P. A., Sil, A., Swedlow, J. R., Herskowitz, I., and Vale, R. D. (1997) Actin-dependent localization of an RNA encoding a cell-fate determinant in yeast. *Nature* **389**, 90–3.

37. Gu, W., Deng, Y., Zenklusen, D., and Singer, R. H. (2004) A new yeast PUF family protein, Puf6p, represses ASH1 mRNA translation and is required for its localization. *Genes Dev* **18**, 1452–65.

38. Sambrook, J., and Russell, D. W. (2001) *Molecular Cloning: A Laboratory Manual*, Third Edition. Cold Spring Harbor Laboratory Press, New York.

39. Lopez, M. F., Berggren, K., Chernokalskaya, E., Lazarev, A., Robinson, M., and Patton, W. F. (2000) A comparison of silver stain and SYPRO Ruby Protein Gel Stain with respect to protein detection in two-dimensional gels and identification by peptide mass profiling. *Electrophoresis* **21**, 3673–83.

40. Ambros, V., and Lee, R. C. (2004) Identification of microRNAs and other tiny noncoding RNAs by cDNA cloning. *Methods Mol Biol* **265**, 131–58.

41. Lau, N. C., Lim, L. P., Weinstein, E. G., and Bartel, D. P. (2001) An abundant class of tiny RNAs with probable regulatory roles in Caenorhabditis elegans. *Science* **294**, 858–62.

42. Thomson, J. M., Parker, J., Perou, C. M., and Hammond, S. M. (2004) A custom microarray platform for analysis of microRNA gene expression. *Nat Methods* **1**, 47–53.

43. Raymond, C. K., Roberts, B. S., Garrett-Engele, P., Lim, L. P., and Johnson, J. M. (2005) Simple, quantitative primer-extension PCR assay for direct monitoring of microRNAs and short-interfering RNAs. *RNA* **11**, 1737–44.

44. Malygin, A. A., Shaulo, D. D., and Karpova, G. G. (2000) Proteins S7, S10, S16 and S19 of the human 40S ribosomal subunit are most resistant to dissociation by salt. *Biochim Biophys Acta* **1494**, 213–6.

45. Jurica, M. S., and Moore, M. J. (2002) Capturing splicing complexes to study structure and mechanism. *Methods* **28**, 336–45.

46. Villace, P., Marion, R. M., and Ortin, J. (2004) The composition of Staufen-containing RNA granules from human cells indicates their role in the regulated transport and translation of messenger RNAs. *Nucleic Acids Res* **32**, 2411–20.

47. Peng, J., and Gygi, S. P. (2001) Proteomics: the move to mixtures. *J Mass Spectrom* **36**, 1083–91.

48. Shevchenko, A., Wilm, M., Vorm, O., and Mann, M. (1996) Mass spectrometric sequencing of proteins silver-stained polyacrylamide gels. *Anal Chem* **68**, 850–8.

49. Lopez de Heredia, M., and Jansen, R. P. (2004) RNA integrity as a quality indicator during the first steps of RNP purifications: a comparison of yeast lysis methods. *BMC Biochem* **5**, 14.

50. Peabody, D. S., and Ely, K. R. (1992) Control of translational repression by protein-protein interactions. *Nucleic Acids Res* **20**, 1649–55.

51. Mumberg, D., Muller, R., and Funk, M. (1994) Regulatable promoters of saccharomyces cerevisiae: comparison of transcriptional activity and their use for heterologous expression. *Nucleic Acids Res* **22**, 5767–68.

6

Advances in RIP-Chip Analysis

RNA-Binding Protein Immunoprecipitation-Microarray Profiling

Timothy E. Baroni, Sridar V. Chittur, Ajish D. George, and Scott A. Tenenbaum

Summary

In eukaryotic organisms, gene regulatory networks require an additional level of coordination that links transcriptional and post-transcriptional processes. Messenger RNAs have traditionally been viewed as passive molecules in the pathway from transcription to translation. However, it is now clear that RNA-binding proteins (RBPs) play a major role in regulating multiple mRNAs to facilitate gene expression patterns. On this basis, post-transcriptional and transcriptional gene expression networks appear to be very analogous. Our previous research focused on targeting RBPs to develop a better understanding of post-transcriptional gene-expression processing and the regulation of mRNA networks. We developed technologies for purifying endogenously formed RBP–mRNA complexes from cellular extracts and identifying the associated messages using genome-scale, microarray technology, a method called ribonomics or RNA-binding protein immunoprecipitation-microarray (Chip) profiling or *RIP-Chip*. The use of the RIP-Chip methods has provided great insight into the infrastructure of coordinated eukaryotic post-transcriptional gene expression, insights which could not have been obtained using traditional RNA expression profiling approaches *(1)*. This chapter describes the most current RIP-Chip techniques as we presently practice them. We also discuss some of the informatic aspects that are unique to analyzing RIP-Chip data.

Key Words: Ribonomics; RIP-Chip RNA-binding protein; immunoprecipitation; post-transcriptional gene regulation; microarray expression profiling; microarray; array.

From: *Methods in Molecular Biology, Vol. 419: Post-Transcriptional Gene Regulation*
Edited by: J. Wilusz © Humana Press, Totowa, NJ

1. Introduction

Most, if not all, genes are regulated by multiple mechanisms, the sum of which dictates the unique expression pattern of a gene under specific conditions. Many studies of gene expression have traditionally focused on transcription and are based on the idea that regulatory networks consist solely of transcription factors interacting with promoters and enhancer elements in a combinatorial manner. In eukaryotic organisms, transcription is uncoupled from translation and physically separated by the nuclear membrane. Thus, eukaryotic gene regulatory networks require an additional level of coordination that links transcriptional and post-transcriptional processes. Although there is no doubt that transcription contributes significantly to the regulation of eukaryotic gene expression, there is often a poor correlation between mRNA levels and protein production. This discordance occurs in part because eukaryotic mRNAs are subject to post-transcriptional processing and regulation. Not surprisingly, the number of RNA-binding proteins (RBPs) increases dramatically concurrent with the evolution of prokaryotes to eukaryotes and the development of the nuclear membrane.

Messenger RNAs have traditionally been viewed as passive molecules in the pathway from transcription to translation. However, it is now clear that RBPs play a major role in regulating multiple mRNAs to orchestrate complex and precise gene expression. On this basis, post-transcriptional and transcriptional gene expression networks appear to be quite similar. Although it is widely accepted that promoters residing on discrete genes participate in the coordinated production of gene products, there is little understanding of the post-transcriptional regulatory mechanisms that must also be maintained for coordinated protein production. In addition to eukaryotic genes being coordinately regulated by transcription factors that bind combinatorially to multiple promoter elements, there is also coordinated post-transcriptional control of multiple mRNAs by mRNA-binding proteins (mRBPs). Just as transcription factors control groups of promoters, distinct subsets of mRNAs are regulated as groups by RBPs at the post-transcriptional level *(1)*.

Our previous research focused on targeting RBPs to develop a better understanding of post-transcriptional gene-expression processing and the regulation of mRNA networks. We previously developed methods for purifying endogenously formed RBP–mRNA complexes from cellular extracts and identifying the associated messages using genome-scale, microarray technology (*see* **Fig. 1**). We subsequently termed this method *ribonomic profiling (2–5)*. The overall approach to ribonomic profiling is analogous to the more widely used

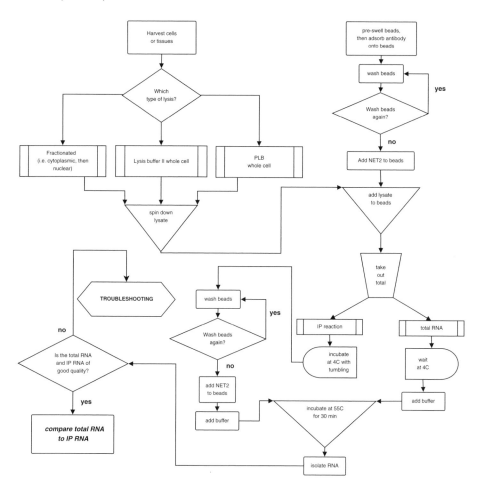

Fig. 1. Diagram of the RNA-binding protein immunoprecipitation-microarray (RIP-Chip) workflow.

methods for chromatin immunoprecipitation-microarray analysis or *ChIP-Chip*, as it has become known *(6–8)*. Accordingly, we and others have begun to refer to the methods involved in ribonomic profiling as RNA-binding protein immunoprecipitation-microarray (Chip) profiling or *RIP-Chip*. During the past several years, many groups have used RIP-Chip to identify the mRNA targets of selected RBPs *(9–25)*. Collectively, this research has greatly facilitated the analysis of structural and functional relationships of the proteins encoded by coregulated mRNAs. In fact, the use of RIP-Chip profiling has provided great insights into the infrastructure of coordinated eukaryotic post-transcriptional gene expression—insights, which could not have been obtained using traditional RNA expression profiling approaches *(1)*. Targeting RBPs lets us study RBPs integral involvement in compartmentalizing, processing and regulating their functionally related mRNA targets.

During the last several years, we have made significant modifications and improvements to the original methods that we developed. This chapter describes the most current RIP-Chip techniques as we presently practice them and provides a convenient protocol by which to concurrently examine both RNA targets of RBPs as well as other associated proteins in these complexes (*see* **Fig. 1**). We also discuss some of the informatic aspects that are unique to analyzing RIP-Chip data and note that Doyle et al. (Chapter 3) provides additional information on this subject in an accompanying chapter in this book.

2. Materials

The reagents described here have been used with mammalian cells and tissue samples, as well as *Caenorhabditis elegans* and *Drosophila melanogaster* cells. However, we imagine that RIP-Chip could be adapted or modified to samples from other organisms. Throughout these methods, we suggest that all cell-culture techniques use cell-culture grade media. Reagents for biochemical manipulations of nucleic acids and proteins should be purchased as DNase free, RNase free, pyrogen free, and 0.2 μm filtered. Generally, immunoprecipitations (IPs) are performed in a 1 mL final volume. This is based on experimental experience and enhances reproducibility, kinetics of reaction, and allows for ease of calculations (i.e., 100 μL = 10% of the IP).

1. All tips, tubes, and reagent bottles must be DNase and RNase free (*see* **Note 1**).
2. We recommend the use of nuclease-free water (Ambion, cat. no. 9932) to prepare buffers and solutions described below (*see* **Note 2**).
3. Tissue and/or cells of interest.
4. Phosphate-buffered saline (PBS).
5. Antibody to the RBP of interest.
6. Protein A sepharose beads: Sigma (P3391) or Amersham Biosciences (Fast Flow, cat. no. 17-0974-01). Store at 4°C.
7. Protein G agarose beads (Sigma, cat. no. P4691). Store at 4°C.
8. 1 M dithiothreitol. Store aliquots of 20 μL at –20°C and avoid freeze-thaw.
9. Glycogen, 5 mg/mL (Ambion, cat. no. 9510). Store at –20°C.
10. Sodium acetate, 3.0 M (Ambion, cat. no. 9740).
11. 10.0 M Lithium chloride: Dissolve 21.195 g lithium chloride (Sigma, cat. no. L-9650) in 40 mL nuclease-free water (the solution will get hot). Cool to room

temperature and dilute to 50 mL. Pass through 0.2-μm filter and store at room temperature.

12. Proteinase K (Ambion, cat. no. 2546, 20 mg/mL).

13. Buffer for proteinase K digestion (2×): 200 mM Tris–HCl, pH 7.5, 20 mM ethylenediaminetetraacetic acid (EDTA), 100 mM sodium chloride, and 2% sodium dodecyl sulfate. Store at room temperature.

14. TE-SDS buffer (2×): 20 mM Tris–HCl, pH 7.5, 2 mM EDTA, and 2% sodium dodecyl sulfate. Store at room temperature.

15. Acid-phenol : chloroform, pH 4.5 (with IAA, 125:24:1, Ambion, cat. no. 9720).

16. Complete tablets and proteinase inhibitor (Roche, cat. no. 1697498). Store at –20°C.

17. RNase OUT™ (Invitrogen, cat. no. 10777-019, 40 U/μL). Store at –20°C.

18. Superase•IN™ (Ambion, cat. no. 2696, 20 U/μL). Store at –20°C.

19. Bovine serum albumin (BSA)-fraction V proteinase free (Roche, cat. no. 8100350). Store at 4°C.

20. A microtip sonicator. We use a Fisher Scientific Model F60 sonicator with our experiments.

21. **Polysome lysis buffer** (PLB, 10×): 100 mM 4-(2-hydroxyethyl)-1-piperazineethanesulfonic acid (HEPES), pH 7.0, 1 M potassium chloride, 50 mM magnesium chloride, 0.25 M EDTA, and 5.0% Nonidet P-40. Store at room temperature.

22. **A ready-to-use PLB buffer** (1×) can be prepared by combining: 5 mL 10× PLB, 0.1 mL 1 M dithiothreitol, one tablet of complete proteinase inhibitor, 62.5 μL RNase OUT™, 125 μL Superase•IN™, and nuclease-free water to 50 mL. This solution is stable at –80°C for extended periods of time (>6 months).

23. **Fractionating lysis buffer** (FLB): 40 mM Tris base, 100 mM sodium chloride, and 2.5 mM magnesium chloride.

24. **Whole-cell lysis buffer** (WLB): 10 mM HEPES, pH 7.0, 100 mM potassium chloride, 5 mM magnesium chloride, 25 mM EDTA, 0.5% Nonidet P-40, 1% Triton X-100, 0.1% sodium dodecyl sulfate, and 10% glycerol.

25. **NT2 buffer:** 50 mM Tris–HCl, pH 7.4, 150 mM sodium chloride, 1 mM magnesium chloride, and 0.05% Nonidet P-40.

26. **NET2 Buffer:** Combine 850 μL NT2, 10 μL 0.1 M DTT, 30 μL 0.5 M EDTA, 5 μL RNase OUT™, and 5 μL Superase•IN™.

3. Methods

The present RIP-Chip method is meant to be highly adaptive and has many options available in several sections. For ease of navigation, a workflow chart for performing RIP-Chip is provided in **Fig 1**. Read through the method and consult the flowchart to familiarize yourself with the options available to you.

3.1. RBP Immunoprecipitation

3.1.1. Antibody Coating of Bead Matrix

1. Swell or resuspend the desired amount of protein G agarose beads (for most monoclonal antibodies) or protein A agarose beads (for most polyclonal antibodies) in 5–10 vol of NT2 containing 5% BSA or FLB (*see* **Notes 5** and **6**).
2. Add the immunoprecipitating antibody or serum and incubate for at least 1 h at room temperature on a rotating device or overnight (or longer) at 4°C (*see* **Notes 7–10**).
3. Beads coated with antibodies can be stored for up to 6 months with no deleterious effects at 4°C when buffers are supplemented to contain 0.05% sodium azide.

3.1.2. Preparation of Whole-Cell RNP Lysates from Cultured Cells

1. The whole-cell lysis protocol may use PLB or WLB supplemented with RNase and protease inhibitors (*see* **Note 11**).
2. Use 150 or 100 mm dishes to grow desired cell line.
3. Wash twice with ice-cold PBS, harvest using a scraper, and then transfer cells to a centrifuge tube using a pipette.
4. Pellet the cells by centrifugation at 2500–3000 × g for 5 min at 4°C.
5. Aspirate PBS from the cell pellet and add approximately 1 vol of PLB or WLB.
6. Use the pipette to agitate the cellular material (through aspiration and expulsion) several times until the extract looks uniform and spin in a microcentrifuge at 14,000 × g for 10 min at 4°C.
7. Freeze and thaw the cell lysate at least once in aliquots of 200–400 µL and store at –80°C (*see* **Note 3**). The extracts are ready to proceed to RBP immunoprecipitation.

3.1.3. Immunoprecipitation of mRNPs

1. Centrifuge the RNP lysate in a microcentrifuge at 14,000 × g for 10 min at 4°C.
2. Calculate the amount of antibody-coated beads necessary to perform the appropriate number of immunoprecipitations you are planning and wash the beads four to six times with room temperature NT2 buffer (*see* **Notes 8–10**).
3. For each IP, resuspend the antibody-coated beads in 850 µL NET2 buffer (*see* **Note 12**).
4. Mix the resuspended antibody-coated beads several times by inversion, then add 100 µL of your RNP lysate (prepared as described in method 3.1.2).
5. Give the tube a quick spin in a tabletop microfuge to settle out the beads and remove an aliquot (100 µL, i.e., 10%) to serve as the total RNA control (input), which can better assess RNase contamination (*see* **Notes 13–15**).
6. Tumble the immunoprecipitation reactions end-over-end at room temperature for 2–4 h at room temperature or at 4°C overnight (the latter is preferable because it gives better signal-to-noise ratio).

7. Following incubation, spin the beads down in a tabletop microfuge (a pulse is sufficient) and wash with 1 ml of ice-cold NT2 buffer, vigorously mixing (vortex) between each rinse. Repeat four to six times (*see* **Note 16**).
8. Proceed to RNP complex dissociation (*see* **Subheading 3.3.**).

3.2. Fractionated Cellular mRNP Isolation

1. This modification is specifically designed for the immunopurification of especially labile RNP complexes. The addition of protease and RNase inhibitors is to be avoided during the lysis step (*see* **Note 19**).
2. Typically, it is recommended to use 10 µg of antibody per immunopurification from 10-cm plate.
3. Wash the antibody–protein A complex five to six times with FLB.
4. Resuspend the antibody–protein A bead complex in FLB (*see* **Note 16**).
5. Mix the antibody–protein A complex gently by inverting the tube several times.
6. Add the RNP extracts from the cytoplasm or nucleus to the antibody–protein A agarose complex and incubate the complex on an orbital shaker at 4°C for no more than 10–20 min (*see* **Note 20**).
6. Following incubation, spin the beads by centrifuging at 5000 × g for 10–20 s.
7. Wash the beads five to six times with FLB, mixing the beads after each wash by inverting the tube several times.
8. Proceed to RNP complex dissociation (*see* **Subheading 3.3.**).

3.3. RNP Complex Dissociation

RNA may be harvested from the bound RNP complexes using one of the two following techniques: proteinase K digestion or heat denaturation.

3.3.1. Proteinase K Method

1. Resuspend the washed beads in 100 µL NET2 and 100 µL proteinase K buffer and then incubate for 30 min at 55°C, mixing occasionally. Transfer the buffer off the beads and into a new tube.

3.3.2. SDS-TE Denaturation Method

1. Resuspend the washed beads in 100 µL NET2 and 100 µL SDS-TE and then incubate for 30 min at 55°C, mixing occasionally. Transfer the buffer off the beads and into a new tube.

3.4. RNA Isolation

The specific RNA isolation you choose will depend on your downstream application. Generally either method is acceptable for microarray, RT–PCR, or northern blotting. The Qiagen spin-column cleanup offers the advantage of performing an optional DNase I digestion while purifying the RNA, so further processing is avoided. However, detection of RNA molecules of 200 bp or smaller will be limited while using the Qiagen cleanup procedure (see manufacturer's documentation).

3.4.1. Precipitation

1. Add 200 µL phenol-chloroform-isoamyl alcohol to the buffer (150–200 µL from the RNA isolation), vortex vigorously (10–30 s), and centrifuge at 14,000 × g for 3 min at 4°C.
2. Transfer the aqueous phase (150–200 µL) to a new microcentrifuge tube containing 750 µL ethanol (100%), 80 µL sodium acetate, 10 µL lithium chloride, and 5 µL glycogen. Mix well with pipette.
3. Store samples at –80°C until ready for gene expression analysis.
4. To harvest RNA, centrifuge samples at 14,000 × g for 30 min at 4°C. Decant supernatant. Add 1 mL 80% ethanol and then centrifuge at 14,000 × g for 15 min at 4°C (*see* **Note 18**). This is referred to below as IP RNA.

3.4.2. Qiagen RNEasy Micro-Cleanup

1. Add nuclease-free water to the 150–200 µL RNA from the RNA isolation to adjust the volume to 200 µL (*see* **Note 17**).
2. Perform the RNEasy micro-cleanup as per the manufacturer's protocol.

3.4.3. Assessment of RNA Quality

1. Using a NanoDrop® spectrophotometer, measure the optical absorbance characteristics of the sample (*see* **Note 22**). The A_{260}/A_{280} as well as the A_{260}/A_{230} ratio will ideally be close to 2.0, signifying the purification of nucleic acids away from protein and other organics, respectively. If either ratio is lower than 1.6, expect problems with downstream applications of the RNA (*see* **Note 23**).
2. Performance of a NanoChip assay using Agilent's BioAnalyzer allows for measurement of the molecular weight profile of the RNA isolated. In this way, you may evaluate the 28S/18S ratio measurements. A total RNA ratio between 1.6 and 2.0 is desireable. RNA samples from IPs may or may not have rRNA present depending on the targets of the RBP (*see* **Note 24**).
3. **Figure 2** depicts a BioAnalyzer output trace displayed as an electropherogram. Lanes contain examples of immunoprecipitations and their corresponding total RNA samples (*see* **Note 24**).

3.5. Expression Analysis of mRNA from mRNPs

1. Targets of mRBPs can be identified and quantified by several methods. RT–PCR is ideal due to the cost (*see* **Fig. 3**). The primers for mRNA targets depicted in **Fig. 3** are listed in **Table 1** *(26)*.

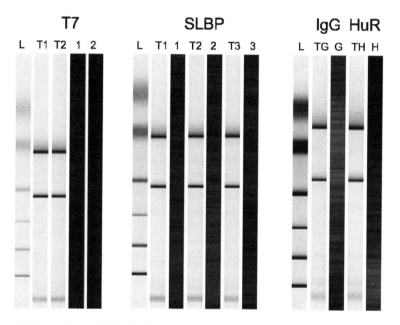

Fig. 2. Example of BioAnalyzer traces of successful RNA-binding protein immunoprecipitation-microarray (RIP) experiments. Lanes contain examples of immunoprecipitations from various experiments and their corresponding total RNA samples. The "T" containing lanes represent total (input) RNA. Immunoprecipitations were performed using anti-T7 (Novagen, cat. no. 69522) bacteriophage coat protein (negative control done in duplicate), anti-stem-loop binding protein (SLBP antibody, S-18, Santa Cruz, which targets most histone mRNAs and done in triplicate), immunoglobulin G_1 (mouse), and anti-HuR (Santa Cruz). The leftmost lane in each panel is the Ambion RNA 6000 control RNA ladder.

2. Multiprobe-based RNase protection assays (PharMingen) are an ideal alternative for the optimization and high-throughput analysis of mRNP immunoprecipitations.
3. We have identified many RBP-associated mRNAs using cDNA/genomic arrays and have found that the Affymetrix GeneChip® array platform and the BD Atlas Nylon cDNA Expression Array platform are excellent for conducting RIP-Chip analysis. However, other array platforms have worked with varying success (*see* **Note 21**).
4. If gene expression analysis is performed using glass arrays that utilize Cy3 and Cy5 labeling or on Affymetrix arrays, we typically increase the amount of extract by 2–3 times that required for atlas nylon arrays.

3.5.1. Synthesis of Labeled cRNA and Microarray Hybridization

1. We have successfully used 10–100 ng of immunoprecipitated RNA for microarray studies using the Affymetrix GeneChip® plaftform.

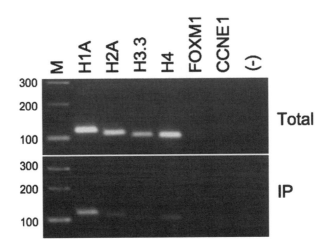

Fig. 3. RT–PCR detection of RNA-binding protein (RBP)-immunoprecipitated target mRNAs. *See* **Table 1** for list of primers used.

2. We recommend using 50 ng of immunoprecipitated material if available. The RNA is first converted to T7-oligo(dT) primed double-stranded cDNA using the Affymetrix two-cycle cDNA synthesis kit as per the manufacturer's protocol.
3. This is then converted to amplified RNA (aRNA) by in vitro transcription using MEGAscript® T7 polymerase (Ambion).
4. The aRNA is subsequently processed to double-stranded cDNA in a manner similar to the first strand cDNA, but uses random primers.
5. Then, biotinylated UTPs are incorporated in the second-IVT step (Affymetrix IVT Labeling kit) resulting in labeled, amplified cRNA.
6. The cRNA (15 µg) is fragmented using a metal-induced hydrolysis step to obtain 25–200 bp fragments that are then hybridized to the GeneChip® arrays as per manufacturer's protocol.

Table 1
Primer Sequences for RT–PCR Detection of Histones in Fig. 3

NCBI Symbol	NCBI Accession Number	GENE NAME	FRAGMENT LENGTH	Forward Primer	Reverse Primer	PolyA tail
HIST1H1A	NM_005325	histone H1A	104	5'-GGAGAAGAACAACAGCCGGCAT-3'	5'-TTGAGCTTGAAGGAACCCGAG-3'	no
HIST1H2AB	NM_003513	histone H2A	116	5'-ACAACAAGAAGACCCGCATCAT-3'	5'-TATTAGGCAAAACGCCACCCT-3'	no
HIST1H2BG	NM_003518	histone H2B	107	5'-ACAAGCGCTCGACCATTACCT-3'	5'-TGGTGACAGCCTTGGTACCTTC-3'	no
H3F3B	NM_005324	histone H3.3	102	5'-TGGTTTTTCGCTCGTCGACT-3'	5'-GGCCATTTTCTTTCACCCAAC-3'	yes
HIST1H4B	NM_003544	histone H4B	100	5'-GGATAACATCCAAGGCATCACC-3'	5'-CGCCACGAGTCTCCTCATAAAT-3'	no
FOXM1	NM_02′953	forkhead box M1	123	5'-GTGGCGATCTGCGAGATTTTG-3'	5'-TCTTTCCCTGGTCCTGCAGAAG-3'	yes
CCNE1	NM_00′238	cyclin E1	108	5'-GGAASAGGAAGGCAAACGTGA-3'	5'-TTATTGTCCCAAGGCTGGCTC-3'	yes

Primer sequences from Whitfield et al. (2004) *Nucleic Acids Res* 32(16): 4833–4842.

3.6. Analysis of RIP-Chip Data Obtained Using Microarrays

1. The difference in signal intensities obtained from hybridizing cRNA from immuno-precipitated samples and from total (a.k.a. input) RNA confounds traditional microarray analysis methods. The problem resides primarily in the normalization techniques used to distribute the signal intensities on the array.

2. We have developed a strategy to overcome this problem and have successfully identified targets of RBPs through RIP-Chip profiling of microarrays. To obtain a robustly confident list of genes associated with a given RBP, we use both MAS5 as well as GCRMA algorithms in data analysis to determine the subset of genes that stands out, regardless of the probe intensity normalization method.

3. While using MAS5 for analysis, we use a filter based on the requirement that the target must be called present in the input RNA.

4. Fold change-based filtering of these lists (we prefer a 4 fold or higher level of enrichment between input (total RNA) and IP RNA in all replicate sets) results in a final list of targets with a very high level of confidence.

5. We strongly recommend the use of replicates in the experiments using microarray technology for ribonomic profiling.

6. Additionally, we recommend the use of one or more control immunoprecipitation (i.e., IgG, T7, HA, FLAG, etc.) to estimate the degree of target "background" binding to proteins and/or agarose beads.

7. The output of many microarray protocols results in a high level of signal enrichment over traditional methods like northern blots. Consequently, the protocol creates the impression that a large number of targets are being immunoprecipitated by the "negative control."

8. In our analysis, we subtract the signal of those "negative" targets present to decrease the complexity of the resulting data and increase the confidence level.

9. In our analysis, we use the "negative" targets as an estimate of our background noise, subtracting their signals from the IP signals to adjust the confidence of high-background probes.

4. Notes

1. All instruments, glassware and plasticware, that touch cells or cell lysates should be certified Dnase free and Rnase free or should be pre-washed with RNase Zap (Ambion, cat. no. 9780; 9782) or RNase Away (Molecular BioProducts, cat. no. 7001) followed by DEPC water and allowed to air dry.

2. Generally, solutions that are certified Dnase free and RNase free from the manufacturer will make for easier solution preparation and allow for faster troubleshooting if they are handled properly. Ambion's buffer kit (cat. no. 9010) contains concentrated solutions of Tris (pH 7 and pH 8), EDTA, sodium chloride, magnesium chloride, potassium chloride, ammonium acetate, and DEPC-treated water.

3. Extracts typically range in concentration from 20–50 mg/mL of total protein, depending on the cytoplasmic volume of the cell type being used.

4. The number of cells required for each immunopurification can vary from cell type to cell type. Typically we utilize one to two, 10-cm dishes of confluent cells per immunopurification. This corresponds to about 2–20×10^6 cells.

5. The amount of lysis buffer used typically depends on the amount of protein. Typically, we recommend resuspending in a volume of lysis buffer that will have a protein concentration of 2–5 mg/ml of total protein. FLB is used to isolate cytoplasmic fraction while WLB is used to isolate nuclear material upon sonication. However, we recommend an intermediate lysis step with WLB, which releases the membrane and organelle fractions leaving behind a purely nuclear pellet.

6. IMPORTANT! The whole-cell variant uses NT2 buffer or WLB to resuspend and wash the antibody-coated beads. The cellular fractionation variant uses FLB for bead resuspension and washing.

7. As a general rule, we recommend Protein-A Sepharose 4 Fast Flow beads (Amersham Biosciences) or Protein-A Sepharose CL-4B (Sigma) if you plan to use rabbit or human polyclonal serum and Protein-G Agarose beads (Sigma) if you plan to use mouse monoclonal antibodies.

8. Check the binding capacity of the beads and the antibody concentration and ideally add the antibody in excess, therefore minimizing background-binding problems. Typically, 2–20 μL sera per immunoprecipitation reaction is used, depending on the concentration and specificity of the antibody.

9. Antibody-coated beads can be prepared in bulk and stored at 4°C with 0.02% sodium azide.

10. Depending on the antibody titer and mRNP concentration, use 50–100 μL packed antibody-coated beads and 100–400 μL RNP lysate (~2–5 mg total protein) for each immunoprecipitation reaction.

11. We recommend the use of PLB first to maintain the integrity of proteins and RNA in a sample. It is possible that the RNase inhibitors and protease inhibitors may interfere with the RIP protocol, in which case use of the WLB would be recommended.

12. IMPORTANT! To minimize reassortment potential, the final volume of resuspended beads in NET2 buffer should correspond to approximately 10 times the original volume of the RNP lysate being used (a 1:10 fold dilution of lysate). As we typically perform our reactions in 1 mL, we use 100 μL lysate. Performing the immunoprecipitation reactions in larger volumes can help decrease background problems.

13. We have noted that the temperature and duration of incubation can influence the efficacy and/or quality of the immunoprecipitation reaction. Longer incubation times may result in better RNP recovery.

14. Provided there are no RNA degradation or RNase problems, immunoprecipitations should be performed for a minimum of 2–3 h at room temperature or overnight at 4°C. A low background is occasionally observed, which is presumably the result of non-specific binding to the beads.

15. A concern when isolating mRNP complexes is the possibility of exchange of proteins and mRNAs. In principle, cross-linking agents, such as formaldehyde, could prevent this *(27)*. However, we have found mRNA exchange to occur at a minimal level using these methods described here and in *(3,5)* and cross-linking therefore, to be unnecessary. In some cases, formaldehyde actually can interfere with subsequent mRNA detection methods and increase the background *(5)*.

16. Several additional washes with NT2 buffer supplemented with 1–3 M urea can increase specificity and reduce non-specific binding. However, it is important to first determine whether urea disrupts binding of the antibody to the target protein and/or the RBP–mRNA interaction.

17. Typically, one can pipette all of the buffer containing the dissociated RNA from the bead slurry, but occasionally evaporation or poor bead compaction can limit the extraction of the complete 200 µL volume. It is frequently more favorable to extract lower volumes rather than contaminating isolated RNA with the agarose beads.

18. RNA pellets from precipitations can detach from the centrifuge tube very easily. Extra care should be taken when washing, decanting, and resuspending the RNA pellet.

19. The addition of RNase inhibitors may interfere with the binding of RBPs to their cognate targets, especially for transient or unstable RNP complexes.

20. Immunopurification of labile mRNP complexes is typically performed for a maximum of 30 min. Longer immunopurification times can result in the disruption of the mRNP complexes.

21. Depending on the quality of the antibody being used for ribonomic profiling, results and background can vary. Although non-specific binding can occur, background also may arise from specific mRNA–antibody interactions. Informative comparisons between total mRNA profiles and mRNP-associated mRNAs are frequently limited by the dramatic differences in signal intensity. There can be a large difference in the number of mRNA species detected in the total mRNA as compared with mRNP complexes, which makes most normalization approaches misleading. For this reason, we have typically not compared mRNP profiles with total RNA and suggest that totals be compared to other totals and mRNP immunoprecipitations compared with other mRNP immunoprecipitations. We have found that poly-A-binding protein profiles are frequently useful as a control mRNP profile to compare against.

22. As the total mass yield of RNA is small from immunoprecipitations (generally from 100 to 500 ng), conservation of material is paramount. If needed, one can analyze the RNA through NanoDrop® and then recover material to use for BioAnalyzer runs.

23. Ambion and Affymetrix protocols and technical literature (and our experience) suggest that samples failing to meet either (or both) of these criteria will very likely perform poorly in molecular techniques, which are based on reverse transcription

followed by amplification. This is likely due to the interference of protein, carbo-hydrate, or phenolic contaminants on the reverse transcription process.

24. The Agilent BioAnalyzer is a preferable substitute to MOPS-formaldehyde agarose gel analysis because of the reduced sample required, increased sensitivity, and reduced exposure to toxic reagents. Occasionally, we see ribosomal RNA peaks in an IP sample. The 28S/18S ratio of an IP is uninformative except to say to what degree rRNA is detectable. If both 28S and 18S subunits are absent and the total has a good rRNA ratio, we consider it a good sample for downstream analysis. This conclusion is based on the premise that rRNA subunits are not typically targets of the RBP being immunoprecipitated. Nevertheless, when small quantities of rRNA are occasionally present in an IP, they can be used as an internal standard by which to compare total RNA to IP RNA.

Acknowledgments

We thank Jack Keene, Luiz Penalva, Aparna Ranganathan, and the rest of the Tenenbaum laboratory members for helpful input. This research was supported by National Institute of Health/National Human Genome Research Institute grant R21HG003679.

References

1. Keene JD, Tenenbaum SA. Eukaryotic mRNPs may represent posttranscriptional operons. *Mol Cell* 2002, 9(6):1161–1167.
2. Tenenbaum SA, Carson CC, Lager PJ, Keene JD. Identifying mRNA subsets in messenger ribonucleoprotein complexes by using cDNA arrays. *Proc Natl Acad Sci USA* 2000, 97(26):14085–14090.
3. Tenenbaum SA, Lager PJ, Carson CC, Keene JD. Ribonomics: identifying mRNA subsets in mRNP complexes using antibodies to RNA-binding proteins and genomic arrays. *Methods* 2002, 26(2):191–198.
4. Tenenbaum SA, Carson CC, Atasoy U, Keene JD. Genome-wide regulatory analysis using en masse nuclear run-ons and ribonomic profiling with autoimmune sera. *Gene* 2003, 317(1–2):79–87.
5. Penalva LO, Tenenbaum SA, Keene JD. Gene expression analysis of messenger RNP complexes. *Methods Mol Biol* 2004, 257:125–134.
6. Katou Y, Kaneshiro K, Aburatani H, Shirahige K. Genomic approach for the understanding of dynamic aspect of chromosome behavior. *Methods Enzymol* 2006, 409:389–410.
7. Tsai HK, Huang GT, Chou MY, Lu HH, Li WH. Method for identifying transcription factor binding sites in yeast. *Bioinformatics* 2006, 22(14):1675–1681.
8. Wu J, Smith LT, Plass C, Huang TH. ChIP-chip comes of age for genome-wide functional analysis. *Cancer Res* 2006, 66(14):6899–6902.

9. Brown V, Jin P, Ceman S, Darnell JC, O'Donnell WT, Tenenbaum SA, Jin X, Feng Y, Wilkinson KD, Keene JD et al. Microarray identification of FMRP-associated brain mRNAs and altered mRNA translational profiles in fragile X syndrome. *Cell* 2001, 107(4):477–487.

10. Darnell JC, Jensen KB, Jin P, Brown V, Warren ST, Darnell RB. Fragile X mental retardation protein targets G quartet mRNAs important for neuronal function. *Cell* 2001, 107(4):489–499.

11. Eystathioy T, Chan EK, Tenenbaum SA, Keene JD, Griffith K, Fritzler MJ. A phosphorylated cytoplasmic autoantigen, GW182, associates with a unique population of human mRNAs within novel cytoplasmic speckles. *Mol Biol Cell* 2002, 13(4):1338–1351.

12. Lopez de Silanes I, Fan J, Yang X, Zonderman AB, Potapova O, Pizer ES, Gorospe M: Role of the RNA-binding protein HuR in colon carcinogenesis. *Oncogene* 2003, 22(46):7146–7154.

13. Lopez de Silanes I, Zhan M, Lal A, Yang X, Gorospe M. Identification of a target RNA motif for RNA-binding protein HuR. *Proc Natl Acad Sci USA* 2004, 101(9):2987–2992.

14. Gerber AP, Herschlag D, Brown PO: Extensive association of functionally and cytotopically related mRNAs with Puf family RNA-binding proteins in yeast. *PLoS Biol* 2004, 2(3):E79.

15. Gerber AP, Luschnig S, Krasnow MA, Brown PO, Herschlag D. Genome-wide identification of mRNAs associated with the translational regulator PUMILIO in Drosophila melanogaster. *Proc Natl Acad Sci USA* 2006, 103(12):4487–4492.

16. Hieronymus H, Silver PA. Genome-wide analysis of RNA-protein interactions illustrates specificity of the mRNA export machinery. *Nat Genet* 2003, 33(2): 155–161.

17. Inada M, Guthrie C. Identification of Lhp1p-associated RNAs by microarray analysis in Saccharomyces cerevisiae reveals association with coding and noncoding RNAs. *Proc Natl Acad Sci USA* 2004, 101(2):434–439.

18. Intine RV, Tenenbaum SA, Sakulich AL, Keene JD, Maraia RJ. Differential phosphorylation and subcellular localization of La RNPs associated with precursor tRNAs and translation-related mRNAs. *Mol Cell* 2003, 12(5):1301–1307.

19. Kim Guisbert K, Duncan K, Li H, Guthrie C. Functional specificity of shuttling hnRNPs revealed by genome-wide analysis of their RNA binding profiles. *RNA* 2005, 11(4):383–393.

20. Mazan-Mamczarz K, Galban S, Lopez de Silanes I, Martindale JL, Atasoy U, Keene JD, Gorospe M. RNA-binding protein HuR enhances p53 translation in response to ultraviolet light irradiation. *Proc Natl Acad Sci USA* 2003, 100(14):8354–8359.

21. Niranjanakumari S, Lasda E, Brazas R, Garcia-Blanco MA. Reversible cross-linking combined with immunoprecipitation to study RNA-protein interactions in vivo. *Methods* 2002, 26(2):182–190.

22. Quattrone A, Pascale A, Nogues X, Zhao W, Gusev P, Pacini A, Alkon DL. Posttranscriptional regulation of gene expression in learning by the neuronal ELAV-like mRNA-stabilizing proteins. *Proc Natl Acad Sci USA* 2001, 98(20):11668–11673.

23. Townley-Tilson WH, Pendergrass SA, Marzluff WF, Whitfield ML. Genome-wide analysis of mRNAs bound to the histone stem-loop binding protein. *RNA* 2006 12(10):1853–67.

24. Ule J, Jensen KB, Ruggiu M, Mele A, Ule A, Darnell RB. CLIP identifies Nova-regulated RNA networks in the brain. *Science* 2003, 302(5648):1212–1215.

25. Ule J, Ule A, Spencer J, Williams A, Hu JS, Cline M, Wang H, Clark T, Fraser C, Ruggiu M et al. Nova regulates brain-specific splicing to shape the synapse. *Nat Genet* 2005, 37(8):844–852.

26. Whitfield ML, Kaygun H, Erkmann JA, Townley-Tilson WH, Dominski Z, Marzluff WF. SLBP is associated with histone mRNA on polyribosomes as a component of the histone mRNP. *Nucleic Acids Res* 2004, 32(16):4833–4842.

27. Mili S, Steitz JA. Evidence for reassociation of RNA-binding proteins after cell lysis: implications for the interpretation of immunoprecipitation analyses. *RNA* 2004, 10(11):1692–1694.

7

Biosensors for RNA Aptamers—Protein Interaction

Sara Tombelli, Maria Minunni, and Marco Mascini

Summary

Aptamers are an alternative to antibodies in their role as biorecognition elements in analytical devices. RNA aptamers, specific for different proteins, have been exploited as biorecognition elements to develop specific biosensors (aptasensors). These recognition elements have been coupled to piezoelectric quartz crystals and surface plasmon resonance (SPR) devices as transducers. The procedure for fixing the aptamer onto these transducers and for monitoring the interaction with the target protein is shown here.

Key Words: RNA; aptamers; Tat protein; biosensor; quartz crystal microbalance; surface plasmon resonance.

1. Introduction

Aptamers are oligonucleotides (DNA or RNA) that can bind with high affinity and specificity to a wide range of target molecules *(1)*. They are generated by an in vitro selection process called systematic evolution of ligands by exponential enrichment (SELEX) which was first reported in 1990 *(2,3)*. Because of the extraordinary diversity of molecular shapes of all possible nucleotide sequences, aptamers have been selected for a wide array of targets, including many proteins, carbohydrates, lipids, small molecules, or complex structures such as viruses. The SELEX method has permitted the identification of unique RNA/DNA molecules, from very large populations of random sequence oligomers (DNA or RNA libraries), which bind to the target molecule with very high affinity and specificity. Selections are frequently carried out with

From: *Methods in Molecular Biology, Vol. 419: Post-Transcriptional Gene Regulation*
Edited by: J. Wilusz © Humana Press, Totowa, NJ

RNA pools due to the known ability of RNAs to fold into complex structures which can be a source of diversity of RNA function.

In comparison to other receptors such as antibodies, aptamers have a number of advantages that make them very promising in analytical and diagnostic applications. For this reason, aptamers have been used in numerous investigations as therapeutic or diagnostic tools *(4,5)* and for the development of new drugs. Moreover, aptamers have been used in analytical chemistry applications as immobilized ligands or in homogeneous assays *(6)*.

We show here some applications in which aptamers have been employed as recognition element in biosensors. An RNA aptamer specific for HIV Tat protein is coupled to optical [surface plasmon resonance (SPR)] and piezo-electric transduction for the development of a new generation of biosensors.

Tat is an HIV-1 RNA-binding protein that exhibits an inherent affinity for the TAR, *trans*-activating response element of the virus. Tat is 101 amino acids long, and it is encoded by two exons. It can be divided into five core domains among which the most important is the fourth one regulating its function and allowing its binding to its corresponding RNA element TAR. This region is positively charged and is rich in arginines *(7,8)*. A Tat-specific RNA aptamer has been reported in literature *(9,10)*. This aptamer featured a similar structure to TAR and exhibited a 133-fold increased affinity for Tat as compared to TAR (*see* **Fig. 1**).

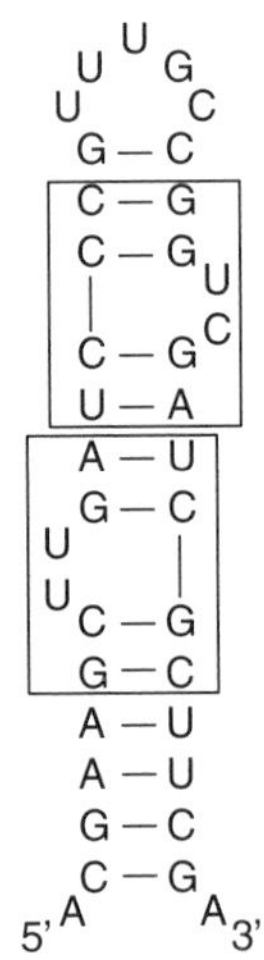

Fig. 1. Secondary structure of the aptamer-specific for Tat protein. The boxed sequence is the important region for Tat binding.

2. Materials

2.1. RNA Aptamer and Target Protein

1. Recombinant Tat HIV-1 IIIB (total length: 86 amino acids) (ImmunoDiagnostics Inc., Woburn, MA, USA). Because oxidation of Tat occurs rather easily and leads to inactivation and loss of biological activities *(11)*, Tat stock solution, 100 ppm in 0.2 M KCl and 5 mM glutathione, is stored at −80°C, and further dilutions are performed in the dark, on ice and with solutions containing 0.1% bovine serum albumin (BSA).

2. Biotinylated Tat-aptamer (IBA GmbH, Göttingen, Germany) with the following sequence: 5′ biotin ACGAAGCUUGAUCCCGUUUGCCGGUCGAUCGCUUCGA-3′ (*see* **Fig. 1**) *(9)*. The aptamer is received lyophilized and then diluted in diethylpyrocarbonate (DEPC)-treated water. The diluted aptamer can be stored at −40°C.

3. Immobilization buffer: 0.01 M *N*-2-hydroxyethylpiperazine-*N′*-2-ethanesulfonic acid (HEPES), 0.15 M NaCl, 3 mM EDTA, and 0.005% polyoxyethylene sorbitan monolaurate (Tween 20), pH 7.4.

4. Binding buffer: 0.01 M HEPES, 0.15 M NaCl, 3 mM EDTA, 0.005% Tween 20, and 0.1% BSA, pH 7.4.

5. HIV Rev (regulator of expression of viral proteins), a basic protein of 14 kDa (kindly donated by the UK Medical Research Council, Cambridge, UK). Due to its similarity to Tat, Rev can be used as negative control. The protein is diluted in binding buffer, divided into aliquots, and stored at −40°C.

2.2. Coupling and Analysis

1. Streptavidin solution I: 200 µg/ml streptavidin in 10 mM acetate buffer, pH 5.0.

2. Streptavidin solution II: 200 µg/ml streptavidin in acetate buffer 100 mM, pH 5.0.

3. Bromoacetic acid solution: 1 M bromoacetic acid (Sigma, Milan, Italy) in 2 M NaOH.

4. BSA.

5. Blocking solution: 1 M ethanolamine hydrochloride in water, pH 8.6.

6. 11-Mercaptoundecanol (Sigma): A 1 mM ethanolic solution of 11-mercaptoundecanol (2 mg of the thiol in 10 ml of ethanol) is freshly prepared before use.

7. DEPC (Sigma).

8. Boiling solution: H_2O_2 (33%), NH_3 (33%) and milliQ water in a 1:1:5 ratio (Merck, Milan, Italy).

9. Regeneration solution: 12 mM NaOH and 1.2 % EtOH.

10. Epichlorohydrin solution: 600 mM epichlorohydrin in a 1:1 mixture of 400 mM NaOH and bis-2-methoxyethyl ether (diglyme) (Fluka, Milan, Italy).

11. Activating solution: 50 mM *N*-hydroxysuccinimide (NHS) (Fluka) and 200 mM
    N-(3-dimethylaminopropyl)-*N*´-ethylcarbodiimide hydrochloride (EDAC) (Sigma)
    in water.
12. Basic dextran solution: Add 3 g of dextran T500 (Amersham Biosciences, Little
    Chalfont, UK) in 10 ml of 100 mM NaOH.

2.3. Instrumentation

1. SPR measurements are performed using the BIACORE X™ instrument (Biacore
   AB, Uppsala, Sweden) and carboxylated dextran-coated chips (CM5 chip, Biacore
   AB). The SPR signal is expressed in resonant units (RU).
2. 9.5-MHz AT cut quartz crystals (14 mm, 165 µm) with gold evaporated on both
   sides (42.6 mm^2 area, Ø 7.4 mm) (International Crystal Manufacturing, Oklahoma
   City, USA). The quartz crystal is housed inside the measurement cell such that
   only one side of the resonator is in contact with the solution in the cell well; in
   this way, two measurement series can be performed on the same resonator, one for
   each side. The cell is made of methacrylate, which is resistant and inert toward the
   chemicals used in the experiment, rigid, allowing to fix it to a support with a pincer
   and transparent, so that it is possible to observe any anomalies (air bubbles) that
   could be present into the well.
3. The frequency variations are continuously recorded using a quartz crystal analyzer,
   QCMagic from Elbatech (Marciana, LI, Italy). The analytical data are the differences
   between two stable frequency values (±0.5 Hz).

3. Methods
3.1. Treatment of Instruments and Solutions for RNA Preservation

1. To preserve the RNA from RNase degradation, solutions and materials (glassware,
   tips, etc.), in direct contact with the aptamer, are treated with specific RNase
   inhibitors. RnaseZAP™ (Sigma) is used to clean glass and plastic surfaces (work
   areas, pipettes, equipment, and the piezoelectric cell), followed by rinsing with H$_2$O
   treated with 0.1 % DEPC.
2. All the solutions used in the experiments are prepared in DEPC-treated water,
   filtered (0.22 µm pore size filter; Nalgene, Milan, Italy), and degassed daily prior
   to use.
3. DEPC-treated H$_2$O is prepared adding 1 ml of DEPC to 1 l of MilliQ water and
   stirred overnight. The solution is then autoclaved at 1 atm for 20 min.
4. Owing to the extreme sensitivity of RNA to nucleases and base hydrolysis, a
   "nuclease-free" environment is necessary also for the SPR measurements. Before
   the modification of the Biacore chip, the BIAcore Desorb method is run and all the
   external instrument components (the needle and the tubing placed in the running
   buffer) are treated with RnaseZap™. Moreover, before docking the chip, the fluidic

part of the instrument is cleaned with an injection of 20 μl of RnaseZap™ solution in each flow-cell at 20 μl/min followed by ten 20-μl injections of DEPC-treated water *(12)*.

3.2. Immobilization of the Aptamer onto Piezoelectric Crystals

1. To preserve the RNA from RNase degradation, solutions and materials (glassware, tips, etc.) are treated with RnaseZAP™ and rinsed with DEPC-treated H_2O.
2. The immobilization chemistry adopted follows the approach described in Tombelli et al. *(13)*. In particular, before the immobilization of the aptamer, the electrode surface of the quartz crystal needs to be cleaned with a boiling solution described in **Subheading 2.2.** The crystals are immersed in the solution for 10 min. They are then thoroughly washed with distilled water and used immediately afterwards.
3. The freshly cleaned crystal is immersed in an unstirred 1 mM ethanolic solution of 11-mercaptoundecanol at room temperature, in the dark, for 48 h. The crystal is then washed with ethanol and milliQ water and sonicated for 10 min in ethanol to remove the excess of thiol. The hydroxylic surface is treated with epichlorohydrin solution for 4 h. After washing with water and ethanol, the crystal is immersed for 20 h in basic dextran solution. The surface is further functionalized with a carboxymethyl group using bromoacetic acid solution for 16 h. All the reactions are performed at room temperature. The coated crystals can be stored at 4°C immersed in DEPC-treated water for 15 days. For their use, the crystals are washed with water and placed in the cell.
4. For further functionalization, the surface of the crystal is activated prior to covalent coupling with 200 μl of activating solution (*see* **Note 1**). After 5 min, the activating solution is replaced by streptavidin solution I for 20 min. The residual reacting sites are blocked with 200 μl of an aqueous solution of 1 M ethanolamine hydrochloride (pH 8.6). After washing with the immobilization buffer, the biotinylated aptamer is added (1.0 μM 200 μl of a solution of the probe in immobilization buffer). The immobilization is allowed to proceed for 20 min (*see* **Fig. 2**).
5. The biotinylated aptamer (1 μM) is thermally treated before its immobilization. The thermal treatment can unfold the aptamer RNA strand making the biotin label at the 5′-end available for the interaction with streptavidin on the chip surface. Before the immobilization, the biotinylated aptamer is heated at 90°C for 1 min to unfold the RNA strand and then cooled in ice for 10 min to block the RNA in its unfolded structure *(14)* (*see* **Note 2**).
6. The cell with the immobilized aptamer is stored at 4°C with 200 μl of binding buffer.

3.3. Immobilization of the Aptamer onto SPR Chips

1. Before docking the chip, the fluidic part of the Biacore is cleaned with an injection of 20 μl RnaseZap™ solution in each flow-cell at 20 μl/min followed by ten 20-μl injections of DEPC-water *(12)*.

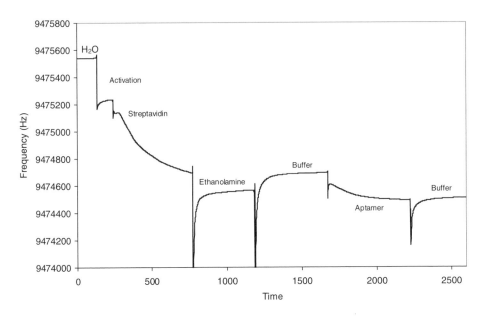

Fig. 2. Frequency variations during the immobilization of the biotinylated aptamer onto streptavidin linked to carboxylated dextran.

2. The immobilization procedure is performed at a constant flow rate of 5 µl/min and a temperature of 25°C. Immobilization buffer is used as running buffer. For the immobilization of the aptamer, the dextran surface of the CM5 chip is further modified with 35 µl of streptavidin solution II after treatment with 35 µl of activating solution. The remaining carboxylated sites on dextran are blocked with 35 µl of 1 M ethanolamine.
3. The biotinylated aptamer (1 µM) is injected (35 µl) after the thermal treatment (*see* **Fig. 3**). The aptamer is heated at 90°C for 1 min and then cooled in ice for 10 min to block the RNA in its unfolded structure *(14) (see* **Note 3**).

3.4. Binding Measurements (Quartz Crystal Microbalance)

1. After the immobilization of the aptamer, the frequency is stabilized by keeping the crystal in contact with 100 µl of binding buffer. 100 µl of Tat protein is then added to the cell and left in contact with the immobilized aptamer for 15 min. The surface is then washed with buffer to remove the unbound protein.
2. The analytical data, expressed as frequency shift, are the differences in the frequency of the crystal before the addition of Tat and after the washing with buffer subsequent to the affinity interaction. Different concentrations of the protein can be used to build a calibration plot. An example of calibration plot is shown in **Fig. 4**.

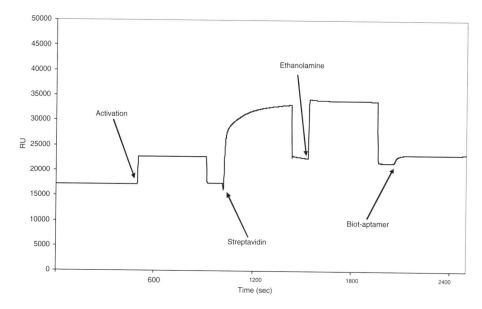

Fig. 3. Typical sensorgram recorded on the Biacore X™ instrument during the different steps (activation, streptavidin, blocking, and aptamer) of the biotinylated aptamer immobilization. Flow rate 5 μl/min, T 25°C.

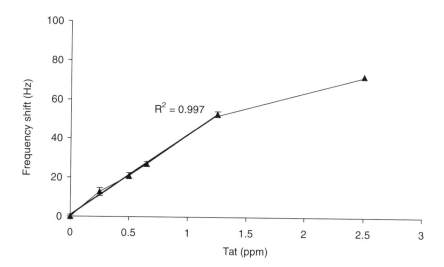

Fig. 4. Calibration plot obtained with the QCM-based biosensor. Interaction of Tat at different concentrations (0.25–2.5 ppm) with the immobilized Tat-specific aptamer.

3. Regeneration solution containing NaOH and ETOH allows a complete regeneration with two steps of 30 s, and in general, on the same crystal surface, 15 measurements can be performed without loss in sensitivity.

4. Negative controls can be tested to prove the specificity of the interaction. Rev represents a rather similar molecule to the analyte Tat, having the same molecular weight and being an RNA-binding protein as well due to its basic domain (*See* **Note 4**).

5. The response of the sensor to blank solutions has to be tested to make sure that the signal is generated by the interaction between the protein and the aptamer and not a matrix effect. The blank solution should contain the same concentration of BSA which is present in the binding buffer.

3.5. Binding Measurements (SPR)

1. After the immobilization of Tat-aptamer onto the SPR chip (*see* **Note 5**), the interaction between the immobilized receptor and the protein is monitored with

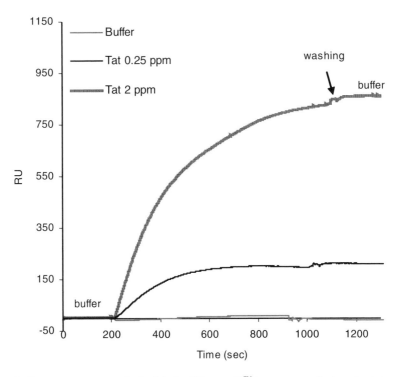

Fig. 5. Sensorgrams recorded with the Biacore X™ instrument during the interaction between the immobilized aptamer and Tat at different concentrations (0.25 and 2 ppm). Flow rate 5 μl/min, T 25°C.

an association time of 15 min followed by washing with running buffer. Binding interactions are monitored at a constant flow rate of 5 µl/min at a temperature of 25°C. An example of the signal obtained with different concentrations of Tat is reported in **Fig. 5**.
2. Also in the SPR experiments, the surface of the chip can be regenerated to remove the protein and obtain the aptamer free again for another binding measurement. This step can be conducted using the same regenerating solution as for **Subheading 3.4, step 3**, and more than 50 cycles can be performed on the same chip without loss in sensitivity.
3. A typical calibration plot which can be obtained from the interaction of the immobilized aptamer and different concentrations of Tat is shown in **Fig. 6**.

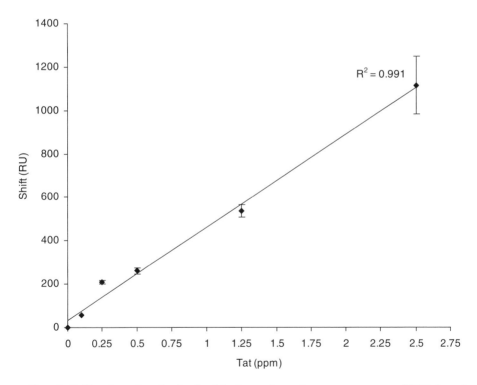

Fig. 6. Calibration plot obtained with the surface plasmon resonance (SPR)-based biosensor. Injections of Tat at different concentrations (0.12–2.5 ppm). Flow rate 5 µl/min, T 25°C.

4. Notes

1. The NHS and EDAC solution must be prepared immediately before the use to avoid loss of activity.
2. Ten crystals were modified following the described immobilization procedure but without treating the aptamer before its immobilization. On four of these crystals, the frequency shift because of the immobilization of the aptamer was not observed. On the other six crystals, an average value of 26 ± 20 Hz was obtained for the crystal modification, indicating a very low reproducibility for the aptamer immobilization step. To improve the reproducibility both for the immobilization of the receptor and for the interaction between the aptamer and Tat protein, the aptamer has been thermally treated before the immobilization. Six different crystals were modified with this treated aptamer and the frequency shift because of the immobilization step was recorded. An average value ($n = 6$) of 63 ± 10 Hz was obtained, confirming the efficacy of the thermal treatment on the aptamer in significantly improving the reproducibility of this step (CV% = 16). Considering the frequency shift resulting from the immobilization of the aptamer, the density of immobilized aptamer can be estimated to be around 0.5 ng/mm^2.
3. Three different chips have been modified with the RNA aptamer, and the reproducibility of the immobilization has been evaluated. An average value of 5410 ± 94 RU (CV = 2%) for the immobilization of streptavidin led to an aptamer immobilization value of 1639 ± 75 RU (CV = 5%), proving a high reproducibility of the immobilization method.
4. If two different concentrations of Rev protein, 0.65 and 2.5 ppm, are tested, the signals recorded are 5 ± 1 and 12 ± 1 Hz, respectively, much lower than the signal due to the interaction of Tat at the same concentrations, demonstrating the high specificity of the interaction.
5. The Biacore X$^{™}$ instrument has two cells. The two cells can be modified with different receptors to have a working and a reference cell. Otherwise, one of the cells can be kept with only carboxylated dextran to check the effect of a "blank" to eliminate the signal due to eventual non-specific adsorption of the target onto the medication layers.

References

1. Luzi, E., Minunni, M., Tombelli, S., and Mascini, M. (2003) New trends in affinity sensing: aptamers for ligand binding. *TrAC Trends in Analytical Chemistry* **22**, 810–818.
2. Tuerk, C. and Gold, L. (1990) Systematic evolution of ligands by exponential enrichment. *Science* **249**, 505–510.
3. Ellington, A.D. and Szostak, J.W. (1990) In vitro selection of RNA molecules that bind specific ligands. *Nature* **346**, 818–822.
4. White, R.R., Sullenger, B.A., and Rusconi, C.P. (2000) Developing aptamers into therapeutics. *Journal of Clinical Investigations* **106**, 929–934.

5. Brody, E. and Gold, L. (2000) Aptamers as therapeutic and diagnostic agents. *Reviews in Molecular Biotechnology* **74**, 5–13.

6. Tombelli, S., Minunni, M., and Mascini, M. (2005) Analytical applications of aptamers. *Biosensors & Bioelectronics* **20**, 2424–2434.

7. Chang, H.C., Samaniego, F., Nair, B.C., Buonaguro, L., and Ensoli, B. (1997) HIV-1 Tat protein exits from cells via a leaderless secretory pathway and binds to extracellular matrix-associated heparan sulfate proteoglycans through its basic region. *AIDS* **11**, 1421–1430.

8. Mucha, P., Szyk, A., Rekowski, P., and Barciszewski, J. (2002) Structural requirements for conserved Arg52 residue for interaction of the human immunodeficiency virus type 1 trans-activation responsive element with the trans-activator of transcription protein (49–57). Capillary electrophoresis mobility shift assay. *Journal of Chromatography A* **968**, 211–220.

9. Yamamoto, R., Katahira, M., Nishikawa, S., Baba, T., Taira, K., and Kumar, P.K.R. (2000) A novel RNA motif that binds efficiently and specifically to the Tat protein of HIV and inhibits the trans-activation by Tat of transcription in vitro and in vivo. *Genes to Cells* **5**, 371–388.

10. Yamamoto, R. and Kumar, P.K.R. (2000) Molecular beacon aptamer fluoresces in the presence of tat protain of HIV-1. *Genes to Cells* **5**, 389–396.

11. Ensoli, B., Buonagurio, L., Barillari, G., Fiorelli, V., Gendelman, R., Morgan, R.A., et al. (1993) Release, uptake and effects of extracellular human immunodeficiency virus type 1 Tat protein on cell growth and vial transactivation. *Journal of Virol* **67**, 277–287.

12. Davis, T.M. and Wilson, W.D. (2001) Surface plasmon resonance biosensor analysis of RNA-small molecule interactions. *Methods in Enzymology* **340**, 22–50.

13. Tombelli, S., Mascini, M., Braccini, L., Anichini, M., and Turner, A.P.F. (2000) Coupling of a DNA piezoelectric biosensor and polymerase chain reaction to detect apolipoprotein E polymorphisms. *Biosensors & Bioelectronics* **15**, 363–370.

14. Ducongè, F., Di Primo, C., and Toulmè, J.J. (2000) Is a closing GA pair a rule for stable loop-loop DNA complexes? *Journal of Biological Chemistry* **275**, 21287–21294.

8

A Tethering Approach to Study Proteins that Activate mRNA Turnover in Human Cells

Sandra L. Clement and Jens Lykke-Andersen

Summary

The regulation of mRNA turnover occurs in part through the action of mRNA-binding proteins that recognize specific nucleotide sequences and either activate or inhibit the decay of transcripts to which they are bound. In many cases, multiple mRNA-binding proteins, including those with opposing functions, bind to the same RNA sequence. This can make the study of the function of any one of these proteins difficult. Furthermore, monitoring endogenous mRNA decay rates using drugs that inhibit transcription (e.g., actinomycin D) can introduce pleiotropic effects. One way to circumvent these problems is to tether the protein of interest (POI) through a heterologous RNA-binding domain to an inducible reporter mRNA and measure the effect of the bound protein on mRNA decay. In this chapter, we illustrate the use of the tethering technique to study the role of a particular mRNA-binding protein, TTP, on the decay of an otherwise stable mRNA to which it is tethered through a fusion to the bacteriophage MS2 coat protein.

Key Words: Pulse-chase mRNA decay assay; tethering; MS2 coat protein; mRNA turnover.

1. Introduction

The control of mRNA stability is an important component of post-transcriptional gene regulation. Various proteins can modulate the stability of a given transcript by binding to *cis*-elements and either activate or impair decay. One of the most commonly studied *cis*-elements known to regulate mRNA decay is the AU-rich element (ARE), which is recognized by both activators and inhibitors of mRNA decay *(1–3)*. As many endogenous proteins can bind the ARE and exert synergistic or opposing effects, it can be difficult to discern

From: *Methods in Molecular Biology, Vol. 419: Post-Transcriptional Gene Regulation*
Edited by: J. Wilusz © Humana Press, Totowa, NJ

the precise role of a specific protein in the context of the ARE. This problem can be avoided by changing the sequence context by tethering the protein of interest (POI) to an otherwise stable reporter transcript through a heterologous RNA-binding domain. This tethering technique has been used with several RNA-binding proteins including the bacteriophage λN peptide and MS2 coat protein (MS2cp) as well as the mammalian splicing factor, U1A, to study mRNA processing, translation, localization, and decay *(4,5)*. In many studies of mRNA turnover, however, only steady-state mRNA levels are observed, which may not accurately represent effects on mRNA decay rates. Here, we describe the use of the MS2cp to tether an activator of ARE-mediated decay, TTP, to a stable reporter mRNA. This method is combined with a pulse-chase assay, which uses the "tet-off" system to activate and then subsequently repress reporter mRNA transcription *(6)*. The rate of mRNA decay is subsequently followed by measuring the decrease in mRNA levels over time by northern blotting. This technique permits the specific repression of the transcript of interest, thereby avoiding the potential side effects associated with global transcriptional shutdown.

2. Materials

2.1. Generation of mRNA Reporter and Tethered Protein Expression Constructs

1. pcDNA3 (Invitrogen).
2. SDS-PAGE and western blotting materials.
3. Mouse monoclonal anti-FLAG M2 antibody (Sigma).
4. Anti-FLAG M2 agarose (optional) (Sigma).

2.2. Pulse-Chase mRNA Decay Assay

2.2.1. Cell Culture and Transfections

1. HeLa tet-off cells (Clontech).
2. HeLa Monster Transfection reagent (Mirus).
3. Double-distilled water (ddH$_2$O).
4. Cell-culture medium: Dulbecco's modified Eagle's medium (DMEM) supplemented with 10% fetal bovine serum (FBS) and 10 ml/l of penicillin/streptomycin (pen/strep) (Invitrogen).
5. Trypsin (Invitrogen).
6. Sterile phosphate-buffered saline (PBS).
7. Tetracycline, 1 mg/ml in ethanol. Store in the dark at −20°C. Make fresh monthly.
8. Protein and mRNA expression plasmids (described in **Subheading 3.1.**).
9. Six-well tissue culture plates (Falcon).
10. Sterile 14 ml or 50 ml conical tubes (Falcon).

2.2.2. RNA Isolations

1. TRIzol reagent (Invitrogen).
2. Chloroform.
3. Isopropyl alcohol.
4. 70% ethanol.
5. Formamide.

2.2.3. Denaturing Formaldehyde/Agarose Gel Electrophoresis

1. Agarose.
2. Recirculating large horizontal electrophoresis system.
3. Formaldehyde (should be used in a fume hood).
4. 10× MOPS: 0.2 M MOPS, 40 mM sodium acetate, and 5 mM EDTA, pH 7.0.
5. Formamide-loading buffer: Deionized formamide, 5 mM EDTA, 0.1% bromophenol blue, and 0.1% xylene cyanol.
6. 2× MOPS/formaldehyde-loading buffer (prepare fresh every time): 2× MOPS, 30% formaldehyde, 5 mM EDTA, 0.2% bromophenol blue, and 0.2% xylene cyanol.

2.2.4. Northern Blotting and Hybridization

1. Whatman 3MM chromatography paper.
2. Paper towels.
3. Nitrocellulose (Whatman Optitran BA-S83) (*see* **Note 1**).
4. 10× SSC: 87.65 g NaCl and 44.1 g of sodium citrate in 800 ml H_2O; adjust pH to 7.0 with several drops of 10 N NaOH.
5. Stratalinker UV Crosslinker (Stratagene).

2.2.5. Hybridization

1. Strip-EZ probe kit (Ambion) to make a radiolabeled riboprobe following manufacturer's instructions (*see* **Note 2**).
2. 50× Denhardt's: 5 g Ficoll (type 400), 5 g polyvinylpyrrolidone, 5 g BSA (Fraction V), and H_2O to 500 ml. Filter sterilize and store aliquots at −20°C.
3. Hybridization buffer: 20 ml total per blot containing 10 ml deionized formamide, 4 ml 10× SSC, 1 ml 10% SDS, 0.5 ml 1 M Na-phosphate pH 6.5, 0.3 ml 10 mg/ml total yeast RNA (denatured at 95°C, 2 min), and 2 ml 50× Denhardt's 2.2 ml H_2O.
4. Low-stringency wash buffer: 0.5× SSC/0.1% SDS.
5. High-stringency wash buffer: 0.1× SSC/0.1% SDS.

2.2.6. Calculation of Reporter mRNA Half-Life

1. Phosphorimager (Molecular Dynamics) and ImageQuant or other machinery or program to measure band intensity.
2. Spreadsheet program with graphing ability, such as Microsoft Excel.

3. Methods

3.1. Plasmid Constructs

3.1.1. mRNA Decay-Reporter Plasmids

Two reporter transcripts are used in the pulse-chase mRNA decay assay (*see* **Fig. 1A**). Both reporters are based on the human β-globin gene, which is inserted between the HindIII and XbaI sites of pcDNA3 (Invitrogen) *(7)*. The use of the β-globin sequence allows a common probe to detect both transcripts in northern blots. In addition, as the HeLa tet-off cells used in this assay do not express β-globin, only the reporter transcripts are detected. The control reporter, β-GAP, is transcribed from the constitutively active CMV promoter and serves as an internal transfection and loading control. This plasmid has parts of the GAPDH open reading frame and all of the GAPDH 3-untranslated region (UTR) replacing the 3-UTR of the β-globin mRNA to make this control transcript larger than the test transcript. In the expression plasmid encoding the experimental reporter mRNA, pcTET2-β-6bs, the CMV promoter is replaced by a CMV minimal promoter with six upstream tetracycline-responsive elements

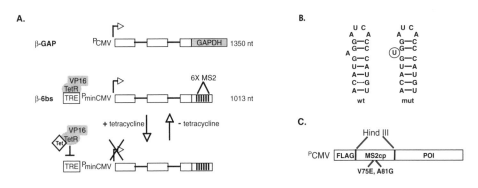

Fig. 1. (**A**) Diagram of the β-globin reporters used in the tethered pulse-chase assay. Transcript size in nucleotides (nt) (excluding the poly-A tail) is indicated on the right. β-GAP, internal control; β-6bs, test reporter containing six copies of the MS2cp-binding site (6× MS2) in the 3′-UTR; TRE, tetracycline responsive element; and TetR-VP16, tetracycline-repressible transcription activator expressed in HeLa tet-off cells (Invitrogen). (**B**) Nucleotide sequence and secondary structure of the wild-type (wt) and mutant (mut) MS2cp-binding site. The mutated nucleotide is circled. (**C**) Diagram of the FlagNMS2 expression vector. The MS2cp is expressed as an N-terminal fusion to the protein of interest (POI). An N-terminal FLAG epitope tag is included for detection and/or purification of the fusion protein. Residues V75 and A81 have been mutated to E and G, respectively, to prevent mulitmerization *(8)*.

(TREs). This results in the repression of reporter transcription in HeLa tet-off cells (Clontech) when tetracycline is present due to the expression of the tetracycline repressor protein fused to the VP16 *trans*-activation domain in these cells. Consequently, this transcriptional activator binds to the TRE to activate transcription only in the absence of tetracycline (*see* **Fig. 1A**). pcTET2-β-6bs contains six MS2cp-binding sites inserted into the β-globin reporter 3-UTR between XbaI and ApaI sites although more MS2cp sites can be introduced to increase the sensitivity of the assay. To generate a negative control reporter mRNA, a single mutation can be introduced into the MS2cp-binding sequence (*see* **Fig. 1B**), which greatly reduces MS2cp binding *(8,9)*.

3.1.2. MS2 Fusion Protein Expression Plasmids

The plasmid pcFlagNMS2 is based on pcDNA3, which has the Flag epitope tag (DYKDDDDK) inserted immediately upstream of the HindIII site (*see* **Fig. 1C**). A mutant MS2cp fragment, which contains two point mutations (V75E;A81G) *(8)* to avoid protein multimerization, is inserted into the HindIII site. The open reading frame of human TTP is inserted between the EcoRI and NotI sites to generate FlagNMS2TTP *(10)*.

3.1.3. Preliminary Experiments

Preliminary experiments should be performed to test for protein expression using western blot with an α-Flag antibody and to optimize the coexpression of the protein and reporter plasmids. The ability of the MS2-tethered protein to bind the β-6bs reporter mRNA can be confirmed by α-Flag immunopre-cipitation using anti-Flag M2 agarose (Sigma) followed by northern blotting. It is also important to carefully titrate the amount of MS2-POI expression plasmid used in the pulse-chase assay, as overexpression can lead to spurious effects. In our hands, the half-life of the β-6bs reporter transcript is >600 min. Coexpression of MS2-TTP decreases the half-life approximately 5- to 10-fold, depending on the expression level of MS2-TTP. Although TTP is known to be a potent activator of mRNA decay, overexpression can lead to inhibition of mRNA decay, perhaps by titrating decay enzymes away from β-6bs mRNA already saturated with MS2-TTP. An initial pulse-chase assay should be performed to approximate the half-life of the mRNA ± the MS2-POI so that appropriate time-points can be selected for the full-scale experiment. When designing the experiment, it is important to remember that the inclusion of more time points will lead to a more accurate calculation of half-life. We recommend to have the first two time points scheduled to occur at some point within the first half-life of the β-6bs transcript and to let the chase proceed through at least two half-lives. Finally, it is essential to ensure that transcription is completely repressed (*see* **Note 3**). This can be done by taking one time point after the mRNA has undergone multiple half-lives, at which point the levels of reporter mRNA should be greatly reduced or undetectable.

3.2. mRNA Decay Assays

The pulse-chase assay takes several days from start to finish (delineated below), with one day of intense activity in which the actual pulse and chase is performed. To be certain that the effect on mRNA decay observed is due to the tethered protein, several control experiments should be performed. These controls include a set of transfections with no exogenous protein, a set each with the expression vector encoding only the Flag-MS2cp, and one with the POI lacking the MS2cp fusion, as there could be cryptic POI-binding sites in the reporter mRNA. A representative set of northern blots is shown in **Fig. 2A**.

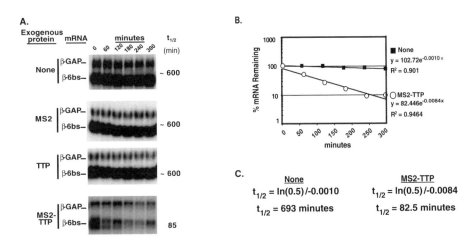

Fig. 2. TTP activates mRNA decay when tethered through the MS2 coat protein. (**A**) Northern blots showing decay rates of β-6bs in the absence of coexpressed protein (None) or in the presence of exogenous MS2, TTP, or MS2-TTP as indicated. Time points above the panels refer to minutes after transcriptional repression. Half-lives are indicated on the right. (**B**) Semi-logarithmic scatter-plot of percent β-6bs mRNA remaining over time in the absence of exogenous protein (None) and with MS2-TTP. The equation and R^2 value for each exponential trendline is shown on the right. (**C**) Calculation of mRNA half-lives from the slope of the trendlines shown in (**B**). As the calculated $t_{1/2}$ for "None" is longer than the duration of the time course, it should not be considered accurate.

3.2.1. Day 1: Plate Cells

Using standard procedures, trypsinize a 10-cm tissue-culture plate containing 80–90% confluent HeLa tet-off cells and resuspend in a final volume of 10 ml of cell-culture medium. Pipet up and down repeatedly to break up cell clusters. Plate 0.1 ml of cells in a total of 2 ml (1:20 dilution) of cell-culture medium per 3.5-cm well in a six-well plate. Plate one well per transfection per time point. One transfection can be harvested for protein isolation and analysis through western blot. Alternatively, both protein and RNA can be isolated following the TRIzol protocol described by the manufacturer (Invitrogen). However, we have found that protein prepared in this manner can be difficult to resuspend. For consistency, it is important to make a "mastermix" of cells with DMEM in one or more sterile Falcon tubes for all wells and then plate 2 ml in each well. Incubate overnight.

3.2.2. Day 2: Transfections

(This protocol is designed using Mirus TransIT HeLaMonster. Check manufacturer's protocols if using other reagents). Cells should be no more than 30% confluent.

For each set of six wells (six time points or five time points plus one protein sample) assemble the following:

1. Add a total of approximately 12 µg plasmid to an Eppendorf tube. For β-globin reporters use:

   a. 0.3 µg pcβGAP (internal control reporter).
   b. 3.0 µg pcTET2 β-6bs (mRNA decay reporter).
   c. 9.0 µg of other plasmid(s)—including FlagNMS2-POI or negative control plasmids (*see* **Note 4**).

2. Mix 675 µl O-MEM with 36 µl TransIT HeLa reagent and add 0.675 µl of 1 mg/ml Tetracycline (in ethanol) (*see* **Note 5**). Incubate for 5–20 min at room temperature.
3. Make 140 µl of HeLaMonster diluted 1:10 in ddH₂O. Store on ice (*see* **Note 6**).
4. Add 650 µl of the O-MEM/TransIT HeLa mix to the plasmid mix and incubate 5–20 min at room temperature.
5. Carefully add 100 µl of the O-MEM/TransIT/Plasmid mix dropwise to each 3.5-cm well of cells.
6. Add 20 µl of diluted HeLaMonster reagent to each 3.5-cm well of cells.
7. Gently rock plates back and forth and place plates in incubator for 30–40 h.

3.2.3. Day 4: Pulse-Chase Assay

1. Pulse transcription by removing tetracycline from the medium (early morning):

   a. Wash wells carefully with 2 ml PBS.

    b. Add 2 ml cell-culture medium, pre-warmed to 37°C.

    c. Incubate for approximately 6 h (*see* **Note 7**).

2. Stop transcription (the complete stop will take ∼20 min):

    a. To each well add 50 µl of O-MEM/40 µg/ml tetracycline.

    b. Place back in incubator.

    c. Take the first time point (*t* = 0) 20–30 min after addition of tetracycline.

3. For each time point:

    a. Carefully wash cells with 2 ml PBS.

    b. Add 1 ml TRIzol and pipet up and down until non-viscous and transfer to pre-labeled Eppendorf tube.

    c. Store at −20°C until ready to perform RNA isolations.

3.2.4. RNA Isolations

1. Perform following manufacturer's instructions for the TRIzol reagent.
2. Air dry briefly (*see* **Note 8**).
3. Dissolve in 20 µl formamide-loading buffer and store at −20°C.

3.2.5. Denaturing Formaldehyde/Agarose Gel Electrophoresis

1. Prepare 2 liters of 1× MOPS and place in cold room.
2. Prepare a 1.2% agarose/formaldehyde gel (*see* **Note 9**) by dissolving 3.31 g agarose in 200 ml ddH$_2$O by boiling for 3-5 min in a microwave (*see* **Note 10**). After cooling down approximately 15 min at room temperature, add 27.6 ml 10× MOPS and 48 ml formaldehyde in the hood. Swirl and pour gel, carefully removing bubbles with a pipette tip. Allow gel to set for at least 1 h.
3. Add 50 ml formaldehyde to the 2 liters of cold 1× MOPS buffer and add to chamber with hardened agarose gel.
4. Mix RNA samples needed for one run in Formamide LB 1:1 with 2× MOPS/formaldehyde LB. We usually load 10 µl total, or one-fourth of the original sample, per well.
5. Heat samples at 80°C for 2-5 min and load.
6. Run at 3 V/cm for 1 h and then increase to 3.75-5 V/cm for approximately 6 h or until the bromophenol blue is within 1 cm of the bottom to ensure separation between poly-adenylated and de-adenylated RNA (*see* **Note 11**).

3.2.6. Northern Blot by Downward Capillary Transfer (see Note 12)

1. After electrophoresis, trim gel to a size that includes the two reporter mRNAs, typically 20 cm wide x 13–15 cm long (*see* **Note 13**).
2. Wash gel for 30-60 min in 10× SSC, slowly agitating. Reserve buffer for transfer.
3. Cut eight pieces of Whatman 3MM paper and one piece of nitrocellulose membrane to the same size and shape as the trimmed gel. Label the nitrocellulose in the

upper left-hand corner (lane 1) with pencil. Cut two longer strips of Whatman 3MM paper (∼30–40 cm) to the width of the gel (20 cm).

4. Place an approximately 10 cm stack of paper towels on a bench. Place four pieces of Whatman paper on top of the paper towels, followed by one piece of 3MM paper wet in 10× SSC.

5. Wet the nitrocellulose membrane in 10× SSC and place on top of the damp filter paper.

6. Place the gel on top of the membrane, keeping the underside in contact with the membrane. Roll a pipette across to remove bubbles.

7. Cover exposed paper towel with saran wrap or parafilm to prevent buffer bypass of the gel during the transfer.

8. Wet three pieces of 3MM paper and place on top of gel, smoothing out bubbles.

9. Place a dish of 10× SSC on a support placed higher than the paper stack containing the gel (empty pipette tip boxes work well). Place the two larger pieces of 3MM paper together and soak in transfer buffer. Place one end of the wetted strip in the glass dish containing 10× SSC and smooth the other end over the gel to remove bubbles.

10. Place a glass plate and a weight (e.g., a 500-ml bottle of buffer) on top of gel stack. Transfer overnight.

11. The next day, wash membrane gently and briefly with water to remove salt.

12. Place membrane on top of 3MM paper, RNA side facing up, and crosslink RNA to membrane in a Stratalinker (Stratagene) using the "auto crosslink" setting. (Other crosslinking methods can be used as well.) The membrane can be stored damp at 4°C.

3.2.7. Hybridization

1. Pre-hybridize membrane in 20 ml pre-warmed hybridization buffer, reserving 0.8 ml hybridization buffer for probe, for ≥2 h at hybridization temperature. We routinely use 60°C for a >100 nucleotide riboprobe.

2. Add 10^6–10^7 cpm of probe, as measured in a scintillation counter, to 0.8 ml hybridization buffer. Denature at 95°C for 2 min.

3. Add denatured probe to hybridization bottle and incubate over night at hybridization temperature.

4. Wash membrane three times for 15 min each with 500 ml of low-stringency wash buffer at room temperature with gentle agitation and one time with high-stringency buffer pre-warmed to 60°C. Wash membrane briefly in water to remove salt and wrap in saran wrap.

5. Expose to film or phosphorimager screen (*see* **Note 14**).

6. After obtaining the desired exposure(s), keep membrane damp and at 4°C for future stripping and re-probing.

3.2.8. Calculation of mRNA Half-Lives

The half-life of the reporter mRNA is calculated by quantifying the radioactive signal obtained for the reporter mRNAs on the northern blot using a phosphorimager. The intensity of the signal obtained for the β-6bs transcript is then divided by that obtained for the internal control β-GAP mRNA, after subtracting from each a background signal obtained elsewhere on the same lane. This will adjust for variations in transfection efficiency and gel loading for each time point. The boxes for a given transcript should be drawn to the same size for each set of time points. Calculate the percent β-6bs mRNA remaining for each time point by dividing the value of that time point by that of $t = 0$ and multiply by 100. Plot the percent mRNA remaining on the Y-axis and the time in minutes on the X-axis using the XY scatter-plot function of Excel as shown in **Fig. 2B**. Adjust the scale of the Y-axis to logarithmic and add an exponential trendline for each set of time points, including both the equation of the line and the R^2 value on the graph. As demonstrated in **Fig. 2C**, the $t_{1/2}$ is calculated using the function $\ln(0.5)/b$, where b is the slope of the trendline obtained from the equation $y = n \times e^{-bx}$ (*see* **Note 15**). It is important to use only bands with signals at least threefold over the background signal. Late time points can often have very low, and therefore unreliable, signals and should be omitted from the analysis. In addition, if the calculated half-life is longer than the duration of the time course (such as for the "None" experiment in **Fig. 2C**), it should not be considered accurate and a longer time course should be performed to obtain an accurate $t_{1/2}$ (*see* **Note 16**).

3.3. Perspectives

In this chapter, we have illustrated the use of a tethered pulse-chase assay to study the ability of the protein TTP to activate the decay of an otherwise stable mRNA to which it is artificially tethered through the bacteriophage MS2 RNA-binding domain. Although tethering can be a powerful technique, it is important to be aware of its potential limitations. For example, it is possible that the POI needs to bind mRNA through its own RNA-binding domain to be active or that the MS2 protein can alter the conformation of the POI such that its active domains are no longer exposed. Furthermore, fusion to MS2 may block recruitment of accessory factors that may be required by the POI for its function. Conversely, the tethered protein may artificially stimulate mRNA decay, for example, by disturbing the mRNP structure. Consequently, tethered pulse-chase assays should also be complemented with a separate set of pulse-chase assays using reporters that encode the native *cis*-element recognized by the POI.

Even with the limitations described above, the tethering assay provides a unique context for studying the role that a given POI plays in mRNA decay in isolation of other proteins that may influence the turnover of its endogenous targets. Moreover, tethering studies can be extended to learn more about the mechanism by which the tethered protein affects mRNA turnover. For example, this assay has been used to identify mRNA activation domains in TTP *(10)* and to study the effect of proteins for which the specific RNA-binding sequence has yet to be identified *(7)*. In addition, coexpression of interaction partners or modifiers of the tethered protein can yield insight into how they might regulate the ability of the tethered protein to activate mRNA decay. Alternatively, the effect of depletion of various mRNA decay proteins on the activation of the decay of the β-6bs reporter transcript can be assayed by knocking down endogenous proteins using RNA interference. As described elsewhere, the tethering assay can also be used to study the effect of a tethered protein on translation by using luciferase or CAT reporters *(4)*. Furthermore, the localization of reporter mRNA and/or tethered protein can be monitored through in situ hybridization and indirect immunofluorescence, respectively.

4. Notes

1. Other nitrocellulose membranes or charged nylon membranes can be used although we have had the best luck with the Optitran membrane.
2. This kit uses modified CTP to generate a riboprobe that is easily removed from the blot following the manufacturer's stripping procedure, which allows for subsequent probings. Other types of RNA or DNA probes may be used instead, but the hybridization conditions may need to be adjusted accordingly.
3. Failure to block transcription is usually the result of a bad stock of tetracycline. When using β-globin reporter mRNAs, this can be diagnosed by the presence of a precursor mRNA, which runs above the internal control, that persists throughout the time course. It is critical that the tetracycline be made fresh at least once a month and that it is kept in the dark at $-20°C$.
4. To achieve consistent transfection efficiencies, it is important that each transfection contains the same total amount of DNA. This can be achieved by adding an empty vector, such as pcDNA3, to bring the total up to 12 µg.
5. A final concentration of 50 ng/ml Tet ensures no expression from the tetracycline-responsive promoter.
6. Thaw HeLaMonster reagent just before use, keep on ice, and return to freezer immediately.
7. The pulse can be as short as 4 h or as long as 15 h. Shorter pulses will yield a small pool of relatively uniform transcripts, whereas longer pulses will result in a large pool of heterogeneous transcripts in various stages of mRNA decay.

8. Overdrying the pellet can make it difficult to resuspend the RNA. To aid in resuspension, the dissolved pellet can be incubated at 55°C for 10–20 min with periodic vortexing.
9. This recipe is for a 20× 25 cm horizontal agarose gel. Scale volume up or down as needed. A 1.2% gel will give good separation of mRNAs from 400 to 3000 nucleotides; change percentage as needed according to size of the reporter transcripts.
10. Measure volume or weight before and after heating, add water as needed to bring up to original volume.
11. For higher voltage runs, it is important to manually re-circulate the buffer. Alternatively, samples can be run overnight at 2-2.5 V/cm for 15–20 h.
12. Alternatively, vacuum transfer or electrophoretic transfer methods can be performed.
13. Trimming the gel makes it easier to handle and reduces the amount of nitrocellulose membrane needed for the northern blot. In a 1.2% agarose gel, both of the reporter mRNAs described in **Fig. 1A** run between the xylene cyanol and bromophenol blue bands, which will be between 13–15 cm apart, depending on the duration of the run. If working with multiple gels at one time, it is useful to cut off a small corner(s) to tell them apart later. Make corresponding cuts in the nitrocellulose membrane.
14. We typically first expose the membrane to a phosphorimager screen for 2-20 h, depending on how "hot" the blot sounds through Geiger counter. Afterwards, one or more exposures on film should be obtained, as these tend to make better quality images for figures.
15. Ideally, the value for n should be 100, since it is the value of y at X = 0. A high R^2 value (close to 1) indicates that the trendline fits the data relatively well.
16. For mRNAs with long half-lives (>8 h), the half life can be hard to measure accurately because the cells divide and loose the internal control plasmid. Using an endogenous mRNA as an internal control can be helpful in this case.

Acknowledgments

The authors thank Dr. Christian Damgaard, Christy Fillman, Tobias Franks, and Guramrit Singh for discussions. This work was supported by a Postdoctoral Fellowship PF-06-156-01-GMC from the American Cancer Society to S.L.C. and a National Institutes of Health grant GM 066811 to J.L.

References

1. Chen, C. Y., Xu, N., and Shyu, A. B. (2002) Highly selective actions of HuR in antagonizing AU-rich element-mediated mRNA destabilization. *Mol Cell Biol* **22,** 7268–78.
2. Bakheet, T., Williams, B. R., and Khabar, K. S. (2006) ARED 3.0: the large and diverse AU-rich transcriptome. *Nucleic Acids Res* **34,** D111–4.

3. Brennan, C. M., and Steitz, J. A. (2001) HuR and mRNA stability. *Cell Mol Life Sci* **58**, 266–77.
4. Baron-Benhamou, J., Gehring, N. H., Kulozik, A. E., and Hentze, M. W. (2004) Using the lambdaN peptide to tether proteins to RNAs. *Methods Mol Biol* **257**, 135–54.
5. Coller, J., and Wickens, M. (2002) Tethered function assays using 3' untranslated regions. *Methods* **26**, 142–50.
6. Loflin, P. T., Chen, C. Y., Xu, N., and Shyu, A. B. (1999) Transcriptional pulsing approaches for analysis of mRNA turnover in mammalian cells. *Methods* **17**, 11–20.
7. Lykke-Andersen, J., Shu, M. D., and Steitz, J. A. (2000) Human Upf proteins target an mRNA for nonsense-mediated decay when bound downstream of a termination codon. *Cell* **103**, 1121–31.
8. LeCuyer, K. A., Behlen, L. S., and Uhlenbeck, O. C. (1995) Mutants of the bacteriophage MS2 coat protein that alter its cooperative binding to RNA. *Biochemistry* **34**, 10600–6.
9. Witherell, G. W., Gott, J. M., and Uhlenbeck, O. C. (1991) Specific interaction between RNA phage coat proteins and RNA. *Prog Nucleic Acid Res Mol Biol* **40**, 185–220.
10. Lykke-Andersen, J., and Wagner, E. (2005) Recruitment and activation of mRNA decay enzymes by two ARE-mediated decay activation domains in the proteins TTP and BRF-1. *Genes Dev* **19**, 351–61.

9

RNA Analysis by Biosynthetic Tagging Using 4-Thiouracil and Uracil Phosphoribosyltransferase

Gusti M. Zeiner, Michael D. Cleary, Ashley E. Fouts, Christopher D. Meiring, Edward S. Mocarski, and John C. Boothroyd

Summary

RNA analysis by biosynthetic tagging (RABT) enables sensitive and specific queries of *(a)* how gene expression is regulated on a genome-wide scale and *(b)* transcriptional profiling of a single cell or tissue type in vivo. RABT can be achieved by exploiting unique properties of *Toxoplasma gondii* uracil phosphoribosyltransferase (TgUPRT), a pyrimidine salvage enzyme that couples ribose-5-phosphate to the N1 nitrogen of uracil to yield uridine monophosphate (UMP). When 4-thiouracil is provided as a TgUPRT substrate, the resultant product is 4-thiouridine monophosphate which can, ultimately, be incorporated into RNA. Thio-substituted nucleotides are not a natural component of nucleic acids and are readily tagged, detected, and purified with commercially available reagents. Thus, one can do pulse/chase experiments to measure synthesis and decay rates and/or use cell-specific expression of the TgUPRT to tag only RNA synthesized in a given cell type. This chapter updates the original RABT protocol *(1)* and addresses methodological details associated with RABT that were beyond the scope or space allotment of the initial report.

Key Words: RABT; 4-thiouracil; 4-thiouridine monophosphate; microarray; gene regulation; cell-specific profiling; RNA; synthesis; decay; RNA purification; *Toxoplasma gondii*; cre; lox.

1. Introduction

Toxoplasma gondii is a ubiquitous protozoan able to infect most if not all warm-blooded animals. It is a member of the *Apicomplexa* family of obligate intracellular parasites, which include the etiologic agents of malaria

From: *Methods in Molecular Biology, Vol. 419: Post-Transcriptional Gene Regulation*
Edited by: J. Wilusz © Humana Press, Totowa, NJ

(*Plasmodium* spp.) and chicken coccidiosis (*Eimeria* spp.). *Toxoplasma* shows a complex developmental biology that has been the subject of much investigation, as has its intimate interaction with the host cell within which it resides and depends for survival. The parasite employs two mechanisms to satisfy its requirements for pyrimidine nucleotides, all of which involve synthesis of uridine monophosphate (UMP) *(2)*. The first mechanism is de novo UMP synthesis by a conventional pathway. The second is a pyrimidine salvage pathway, where free uracil is transported from the host to the parasite cytosol *(3)*. The enzyme uracil phosphoribosyltransferase (UPRT; EC 2.4.2.9) within the parasite then catalyzes the transfer of ribose 5-phosphate from α-D-5-phosphoribosyl-1-pyrophosphate to the N1 nitrogen of free uracil, which yields the nucleotide UMP.

The mammalian and avian hosts of *T. gondii* generally lack UPRT activity, a fact that has been exploited in several ways. First, *T. gondii* growth inside host cells can be monitored through the incorporation of ^{3}H-uracil analogs *(3)*. Second, owing to its central role in pyrimidine salvage, *Toxoplasma gondii* uracil phosphoribosyltransferase (TgUPRT) has been considered as a target for chemotherapeutic treatment of *T. gondii* infection through the use of subversive anti-toxoplasma substrates and uracil analogs that are substrates for the enzyme *(4,5)*. One such analog is a compound originally thought to be 2,4-dithiouracil *(4)*.

Thio-substituted nucleotides are not a natural component of nucleic acids, and there are a variety of commercially available reagents for the covalent tagging, detection and purification of sulfhydryl-containing macromolecules. Cleary et al. *(1)* built upon the observation of Iltzsch and Tankersley *(4)* and asked whether thio-substituted nucleotides, once generated, could be incorporated into *T. gondii* mRNAs and whether such RNAs could be subsequently coupled to biotin-HPDP [(*N*-(6-(biotinamido)hexyl)-3´-(2´-pyridyldithio)-propionamide], isolated by standard streptavidin bead chromatography and profiled by microarray analysis.

The results showed that the addition of 2 mM preparations of commercially purchased 2,4-dithiouracil to growing cultures of *T. gondii* does indeed yield thio-substituted *T. gondii* mRNAs. Unexpectedly, however, mass spectrometry showed that 2,4-dithiouracil itself either was not a substrate for TgUPRT or that its resultant nucleotide monophosphate could not be utilized by *T. gondii* nucleotide kinases and/or RNA polymerases *(1)*. Rather, 4-thiouracil, which was present as a contaminant in trace amounts ($\sim$0.1%) in the commercially purchased 2,4-dithiouracil, appeared to be the sole thio-substituted nucleotide species incorporated into *T. gondii* mRNA. Subsequent experiments were therefore performed with 2 μM 4-thiouracil (4-sU) in place of 2 mM 2,4-dithiouracil. This resulted in high-efficiency incorporation into *T. gondii* mRNA with approximately 4% of all uridines in mRNA being 4-thiouridine after an 8-h labeling period *(1)*.

Cleary et al. further demonstrated that purified 4-thiouridine-containing mRNAs can be quantitatively biotinylated with biotin-HPDP, which creates a reducible disulfide bond linking the biotin directly to the base. This biotin-tagged mRNA was readily purified by binding to immobilized streptavidin-coated magnetic beads, washing to remove non-biotinylated mRNAs, and eluting by reduction from the streptavidin column with either DTT or β-mercaptoethanol (BME). The elution products were 4-thiouridine-containing mRNAs that were free of any modifications resulting from the biotinylation. For labeling periods of 1–16 h with even 2 mM 2,4-dithiouracil, there was little if any toxicity on either *T. gondii* or on the host cells, and there were negligible effects on *T. gondii* steady-state mRNA levels as assayed by microarray profiling. In vitro translation of purified 4-sU-containing mRNAs showed that tagging has no effect on protein synthesis. Thus, there are no detected biological consequences to the presence of 4-sU in mRNA, which makes RNA analysis by biosynthetic tagging (RABT) ideally suited to assay gene expression.

Three main proof-of-concept experiments were performed to demonstrate the utility and specificity of RABT *(1)*. First, RABT can identify the mechanisms that regulate gene expression. A modified 4-sU pulse-chase was performed, which demonstrated that developmentally regulated *T. gondii* mRNAs had different synthesis and decay rates, as measured by microarray analysis. Second, RABT can be readily adapted to studies with other cell types: the engineered expression of TgUPRT in HeLa cells grown in culture followed by addition of 4-sU resulted in efficient 4-thiouridine labeling of HeLa cell RNA. Third, RABT is highly specific and can be applied in vivo. BALB/c mice infected with *T. gondii* were intraperitoneally injected with 4-sU. A lymph node was removed from the mouse, and total RNA was extracted. Only the cells expressing UPRT, in this case the parasites themselves, yielded RNA that was biosynthetically tagged with the thiouridine; the mouse RNA was unlabelled.

In combination with other commonly used techniques such as microarrays and methods for cell- or tissue-specific gene expression, RABT enables a variety of experiments that were previously extremely difficult or impossible due to technical limitations. For example, using RABT in combination with microarrays enables the efficient genome-wide profiling of RNA synthesis and decay rates using short 4-thiouracil pulse labeling. Additionally, mixed samples of RNA from different species or cells in a different state can be separated and profiled on microarrays thereby overcoming the problem of analyzing RNA that is derived from a minority of cells in the starting sample. For example, viruses engineered to carry a copy of *TgUPRT* will allow sensitive transcriptional profiling of the host response, even in cases where there are a relatively small number of infected cells within a larger background of uninfected host tissue. Alternatively, some infectious organisms can be studied without complications associated with an excess of contaminating host RNA, such as expression profiling of rare, intracellular *T. gondii* present in large tissue masses *(1)*.

It should also be possible to use RABT to purify tagged RNA from specific cell or tissue types using engineered plants or animals. This would allow developmental biologists to study model systems like *Caenorhabditis elegans* and *Drosophila melanogaster* that could easily be engineered to express TgUPRT under the control of tissue-specific promoters. Similarly, mice engineered with *TgUPRT* that is transcriptionally inhibited by an upstream cassette that is flanked by lox sites could be crossed to a variety of existing mouse strains that express Cre recombinase under the control of tissue-specific promoters; among the resulting F1 progeny should be mice that have de-repressed *UPRT* transcription in the cell types that express the Cre recombinase. These engineered metazoan systems will enable transcriptional profiling in a variety of biological contexts, such as development, infection, stress response, or oncogenesis.

2. Materials

Important: All reagents must be RNAse free!

2.1. RNA Analysis by Biosynthetic Tagging

1. Deionized water.
2. 4-thiouracil (Acros, cat. no. 591-28-6). Store at room temperature in the dark.
3. DMSO, RNAse free (Sigma, cat. no. D2650).
4. TRIzol® (Invitrogen). Store at 4°C.
5. Chloroform.
6. 2-Propanol.
7. 75% Ethanol.
8. EZ-Link® Biotin-HPDP (Pierce, product no. 21341). Store at 4°C.
9. Dimethylformamide (DMF) (Pierce, product no. 20673). Store at room temperature.
10. 1 M Tris–HCl, pH 7.4.
11. 0.5 M EDTA, pH 8.0.
12. 5 M NaCl.
13. Agarose.
14. 10× TAE buffer: 10 mM Tris–acetate, pH 8.5, and 1 mM EDTA.
15. 10 mg/ml Ethidium bromide—Caution: This chemical is a carcinogen.

16. Gel Loading Buffer II (Ambion, cat. no. 8546G).
17. 1 kb DNA ladder (Invitrogen).
18. 10× SSC buffer: 1.5 M sodium chloride and 150 mM sodium citrate.
19. Hybond-N+ membrane (Amersham).
20. Blocking solution: 125 mM NaCl, 17 mM Na_2HPO_4, 7.3 mM NaH_2PO_4, and 1% SDS (w/v). Blocking solution may need to be heated to 37°C to solubilize SDS.
21. Wash I solution: 1:10 (v/v) dilution of blocking solution in water.
22. 1× Wash II solution: 10x stock is 100 mM Tris-HCl, 100 mM NaCl, and 21 mM $MgCl_2$, pH 9.5
23. Streptavidin-HRP (Endogen).
24. SuperSignal West Pico Chemiluminescent Substrate (Pierce).
25. Dynabeads MyOne Streptavidin C1 (Dynal Biotech, cat. no. 650.01).
26. 1× MPG buffer: 1 M NaCl, 10 mM EDTA, and 100 mM Tris–HCl, pH 7.4, in RNAse-free H_2O.
27. 5% BME.
28. Kodak Biomax MR film.
29. Dulbecco's modified Eagle medium (Invitrogen, cat. no. 11960).
30. Fetal calf serum (Invitrogen, cat. no. 10437-028).
31. 200 mM L-glutamine (Invitrogen, cat. no. 25030).
32. Dynal MPC-S magnetic stand (Invitrogen, cat. no. 120-20D).

2.2. Optional Reagents (see Heading 4.)

1. RNAeasy® RNA purification kit (Qiagen).
2. Phenol : chloroform (1:1).
3. 4-Thiouridine nucleoside.
4. RNAseZAP® (Ambion).
5. 95% Ethanol.

3. Methods
3.1. Labeling Cellular Transcripts with 4-Thiouracil

1. Make a 10,000× working stock solution of 200 mM 4-thiouracil (mw = 128.15) in DMSO. This can be aliquotted and stored at –80°C.
2. Grow TgUPRT-expressing cells in an appropriate medium for the cell type, supplemented with 20 µM 4-thiouracil for 1–16 h. We routinely grow human foreskin fibroblasts in DMEM supplemented with 10% fetal calf serum, 2 mM glutamine, 50 µg/ml penicillin, 50 µg/ml streptomycin, and 20 µg/ml gentamicin. *See* **Note 1** for suggested controls.

3.2. RNA Extraction

1. To extract total RNA, we use TRIzol according to the manufacturer's protocol. *See* **Note 2** on avoidance of denatured protein at the organic/aqueous phase interface and for suggestions on RNA cleanup.

2. Ensure that the $A_{260/280}$ ratio is greater than 2.0. If not, re-extract the RNA. *See* **Note 3** on secondary RNA extraction. If recovery of small (20–100 nt) RNA is desired, *see* **Note 4** about alcohol and wash selection.

3.3. Biotinylation of RNAs

1. Prepare a 1 mg/ml solution of biotin-HPDP in DMF. *See* **Note 5**.
2. Assemble the RNA (total or polyA purified)-labeling reaction in a 1.7-ml microfuge tube. *See* **Note 6** for important parameters and considerations on the biotinylation reaction conditions. Example reaction:

   a. RNAse-free H_2O: 172 µl.
   b. 1 M Tris–HCl (pH 7.4): 2.5 µl.
   c. 0.5 M EDTA: 0.5 µl.
   d. 1 µg/µl 4-sU RNA: 25 µl.
   e. 1 mg/ml Biotin-HPDP: 50 µl (must be added last).
   f. Total volume: 250 µl.

3. Incubate the biotinylation reaction at room temperature for 3 h in the dark. *See* **Note 7** on biotinylation reaction times.
4. Precipitate the labeled RNA by addition of 1/10 reaction volume of 5 M NaCl and 1.1 vol of 2-propanol. *See* **Note 4** on the precipitation of small RNAs. For the above example add:

   a. 5 M NaCl: 25 µl.
   b. 2-propanol: 275 µl.

5. Incubate 2-propanol precipitates at room temperature for 5 min.
6. Spin in a microcentrifuge at 20,000 × *g* for 20 min at room temperature. Discard the supernatant.
7. Add 500 µl of 75% ethanol and re-spin in a microcentrifuge at 20,000 × *g* for 5 min at room temperature. Discard the supernatant. *See* **Note 4**.
8. Resuspend the RNA pellet in 20 µl RNAase-free H_2O.
9. Optional—remove a 1-µl sample of RNA for assessing the efficiency of 4-sU incorporation on a Nanodrop spectrophotometer. *See* **Note 8** for details of this optional procedure.
10. Total RNA can now be used to detect 4-sU RNA in a northern blot (*see* **Subheading 3.4.**). Northern blotting as a method of measuring overall incorporation into purified mRNAs is not advised as it requires a large amount of mRNA. 4-sU-labeled RNA can be purified from total RNA or mRNA (*see* **Subheading 3.5.**).
11. Store RNA at –80°C if you are not proceeding directly to **Methods 3.4. or 3.5.**

3.4. Northern Blotting

3.4.1. Electrophoresis and Transfer of RNA

1. Clean all gel boxes, combs, and trays with RNAseZap. Rinse with deionized water.
2. Pre-soak hybond N+ nylon membrane, transfer wick, and two Whatman slices in 10× SSC while gel is running.
3. Pour 0.8–1.0% gel in 1× TAE buffer with 0.5 µg/ml ethidium bromide (*see* **Note 9**).
4. Assemble apparatus and put gel box, running buffer, and gel in a 4°C cold room. Allow 1 h for the gel and buffer to equilibrate to 4°C (*see* **Note 10**).
5. Prepare 0.5–2.5 µg samples (*see* **Note 11** on amount of RNA to load in gel) of total RNA into well in Ambion Gel Loading Buffer II. The sample volume should be low enough to fit in the well (this will be determined by comb size, but typically the total volume is under 20 µl).
6. Before loading the samples on the gel, spin the tubes of RNA in loading buffer for 1 min at 20,000 × *g* in a microcentrifuge at room temperature to pellet any biotin left in the sample.
7. Pipette the samples into the wells. Load 1 µg of 1 kb DNA ladder to provide an approximate size marker in an outside lane.
8. Run the gel at 14 V/cm for 20 min in the cold room. Disassemble the apparatus and photograph the gel under UV light. Include a ruler as a mobility/size reference.
9. Set up a capillary transfer apparatus with materials from **step 2** of this method. Perform the capillary northern transfer according to your northern blotting protocol of choice (we routinely used the protocol outlined in *Current Protocols in Molecular Biology*). Allow the transfer to proceed overnight at room temperature.
10. Disassemble the transfer apparatus and cross-link the RNA to the membrane using a UV cross-linker at 254 nm wavelength according to the manufacturer's instructions.

3.4.2. Streptavidin-HRP Detection

1. Place the membrane face-up in a plastic container large enough to allow the membrane to lie flat and block membrane in excess 1× blocking solution for 20 min at room temperature on an orbital shaker at low speed. After incubation, pour off the blocking solution.
2. Incubate the membrane with streptavidin-HRP (Endogen) diluted 1:1500 in just enough 1× blocking solution to completely cover the membrane for 5 min at room temperature on an orbital shaker (10 ml is usually sufficient). Remove the streptavidin-HRP solution and discard.
3. Wash the membrane in 50 ml of Wash I (Wash I is a 1:10 dilution of 1× blocking solution in water) for 10 min at room temperature on an orbital shaker.
4. Repeat **step 3** once.

5. Wash the membrane in 50 ml of 1× Wash II for 5 min at room temperature on an orbital shaker.
6. Repeat **step 5** once.
7. Mix 2.5 ml of each ECL component together and add 5 ml of the complete ECL reagent to the membrane for 1 min at room temperature, rocking back and forth by hand to coat the membrane with ECL solution.
8. Place membrane between two plastic printer transparencies (Saran-wrap also works) and expose the membrane to a piece of film at room temperature for 30 s in a darkroom. Develop film and gauge the intensity of rRNA bands. Perform a longer exposure if necessary.

3.5. Purification of 4-sU Biotinylated RNA

3.5.1. RNA Capture on Beads

1. Determine the volume of Dynabeads MyOne Streptavidin C1 to use. The volume will depend on the length of the 4-sU-labeling time. *See* **Note 12**. Pre-warm a sample of MPG buffer to 65°C (to be used in **Subheading** 3.5.10.). The volume of this 65°C MPG buffer is equivalent to 5 vol of the total amount of beads used in **step 1** of **Subheading 3.5.2**.
2. Block the beads with yeast tRNA. Use 1 μl of 10 mg/ml tRNA for every 5 μl of beads for 20 min at room temperature. Rotate tube end-over-end slowly for this and all subsequent incubations.
3. Place the tube in a Dynal magnetic stand and let beads collect at side of tube for 1 min. Decant the supernatant, leaving 50 μl of excess liquid on bead pellet.
4. Resuspend the bead pellet in MPG buffer at room temperature with slow end-over-end rotation for 5 min and allow the beads to pellet by placing in a Dynal magnetic stand for 1 min at room temperature. Once beads are settled, remove all but 50 μl of liquid from pellet. Washes are performed in a volume equal to three times the original volume of beads. For example, if 100 μl bead slurry was used in **Method 3.5.1**, wash beads in 300 μl MPG buffer.
5. Repeat **step 4** two more times and resuspend the beads in a MPG buffer volume equal to original volume of beads from **step 1** of this method.
6. Spin the biotinylated RNA (from **step 8** of **Subheading 3.3.**) at 20,000 × *g* in a microcentrifuge for 1 min at room temperature to pellet out any excess biotin.
7. Avoiding the excess pelleted biotin, decant the biotinylated RNA supernatant and add 10 μl of this to 100 μl of beads. *See* **Note 12**.
8. Incubate the RNA-bead mixture with rotation at room temperature for 15 min.
9. Collect the beads in Dynal magnetic stand for 1 min at room temperature, and remove all but 50 μl of the supernatant. Keep the supernatant, which contains unbound (non-thiolated) RNA.

3.5.2. Washing of the Beads

1. Add three times the original bead volume (from **step 1** of **Subheading 3.5.1.**) of 65°C MPG buffer. Pipette this solution up and down to mix. Collect beads in

magnetic stand for 1 min and decant the wash supernatant into a separate disposable 15-ml screw-cap tube. Washes can be saved to assess the amount of unbound biotinylated RNA by northern blotting. *See* **Note 13**.

2. Add three times the original bead volume of room temperature MPG buffer to the purified beads. Pipette solution up and down to mix. Collect the beads in magnetic stand for 1 min, decant wash supernatant, and add to wash supernatant from **step 1** of this method.

3. Repeat **step 2** two more times, pooling the wash supernatants each time with the supernatant from **step 1**.

4. To the collected beads, add three the original bead volume of room temperature MPG buffer diluted 1:10 in water. Pipette the solution up and down to mix. Collect the beads in a magnetic stand for 1 min and pool the wash supernatant with the supernatants from **step 2** of this method.

3.5.3. Elution of RNA from Beads

1. Elute the bound RNA from the beads by addition of a volume of 5% BME equal to original bead volume (from **step 1** of **Subheading 3.5.1.**). Incubate at room temperature with end-over-end rotation for 5 min.

2. Collect the beads in Dynal magnetic stand for 1 min at room temperature. Decant the supernatant, which contains pure thiolated RNA, to a new 1.7-ml microfuge tube.

3. Repeat **step 1** of this method, but this time incubate the tube at 60° C with occasional mixing for 10 min. Collect the supernatant and pool with the first elution from **step 2** of this method.

4. Precipitate the RNA from the unbound fraction (**step 9** of **Subheading 3.5.1.**), the pooled wash fractions (**step 1** of **Subheading 3.5.2.**), and the pooled, eluted, thiolated RNA fractions (**step 3** of **Subheading 3.5.3.**). For unbound and wash fractions add an equal volume of isopropanol. There is no need to add more salt, as the MPG buffer contains a sufficient amount of salt for RNA precipitation. For the eluted thiolated RNA fraction (from **step 3** of this method), add a 1/10 vol of 5 M NaCl, 1 vol of 2-propanol, and 2 µg of glycogen to aid in precipitation of the RNA. *See* **Note 13**.

5. Incubate all samples at room temperature for 5 min and spin in a microcentrifuge at $20{,}000 \times g$ for 20 min at 4°C. Discard supernatant.

6. Wash the precipitated RNA pellets in 75% ethanol by vortexing for 10 s and spin in a microcentrifuge at $20{,}000 \times g$ at 4°C for 5 min (for small RNAs, *see* **Note 4**).

7. Resuspend the pelleted RNA in RNAse-free H_2O, typically in a volume of 20 µl for the purified thiolated RNA.

8. Place the resuspended RNA samples in magnetic stand for 1 min to collect any beads carried over from the earlier steps. Decant the supernatants to fresh 1.7-ml microcentrifuge tubes. Store these RNA samples at –80°C.

9. Determine the RNA yields by UV spectrophotometry on the eluted thiolated RNA. If you have access to a Nanodrop spectrophotometer, assess the extent

of 4-thiouracil incorporation. 4-thiouracil maximally absorbs at 327 nm, with an extinction coefficient of 16,400. *See* **Note 8** for determination of RNA yield and 4-thiouracil incorporation rates.

10. Check the RNA from each wash fraction (not the eluted thiolated RNA) by northern blotting with streptavidin-HRP to determine the amount of unbound, biotinylated RNA. This only works with total RNA, as the eluted RNA no longer contains a biotin tag. *See* **Note 13**.

4. Notes

1. As an optional positive control for 4-thiouracil ribosylation and downstream detection of thiol-substituted RNA by northern blotting, add 20 μM 4-thiouracil to a 10 ml mid-log phase culture of *Escherichia coli* at 37°C. Pellet the cells, remove the supernatant, and extract and process the *E. coli* RNA as described above. A second positive control for 4-sU incorporation into RNA is to use 50 μM 4-thiouridine (the nucleoside) in place of 20 μM 4-thiouracil. 4-thiouridine will be incorporated into RNA in a UPRT-independent manner, as most uridine kinases can convert 4-thio-UMP into 4-thio-UTP. This enzyme activity is abundantly present in most mammalian systems.

2. When decanting the aqueous phase to a new centrifuge tube for 2-propanol precipitation, exercise extreme caution in staying away from protein at the interface between the aqueous and organic phases. Cysteines label very well with the biotin-HPDP which, through non-specific sticking to RNA, can result in co-purification of non-thiolated RNA or cause insoluble aggregates to form in later steps.

3. Re-extracting the aqueous phase with 5:1 acid phenol : chloroform pH 4.5 may be useful to ensure all protein is removed from the aqueous phase. If the final OD 260/280 ratio is less than 2.0, another option is to clean the extracted RNA with the RNAeasy kit (Qiagen).

4. For precipitation of RNAs that are 20–100 nucleotides long, use 3 vol of 95% ethanol in place of isopropanol. For RNAs that are 20–50 nucleotides long, do not wash the pellet in 75% ethanol at any step, as smaller RNAs are soluble in 75% ethanol and will be lost in the wash step.

5. Make the biotin-HPDP solution fresh each day it is used. Using pre-made, frozen biotin-HPDP results in more insoluble material after RNA biotinylation and precipitation.

6. Three-hour incubations appear to yield better RNA biotinylation than the original 1.5-h incubation.

7. Increasing the temperature of the biotinylation reaction to 60°C may also increase the efficiency of biotinylation by denaturing the RNA and providing improved access to the 4-sU.

8. If there is access to a Nanodrop spectrophotometer, only 1 μl of eluted RNA is necessary to check the concentration. As the eluted thiolated RNA is both precious and scarce, we recommend using a Nanodrop for all spectrophotometry.

The approximate extent of 4-thiouridine incorporation can be crudely calculated with the following equation:

First determine the ratio of 4-sU nucleotides to non-thio-containing nucleotides, R.

$$R = \frac{A_{327}}{A_{260}} \times \frac{9010 \text{cm}^{-1} M^{-1}}{16,600 \text{cm}^{-1} M^{-1}} \tag{1}$$

The number of 4-sU nucleotides per kb of RNA is $[R/(R + 1)] \times 1000$.

The extinction coefficient for 4-thiouracil is 16,600 at 327 nm, whereas for typical RNA, it is about 9010 at 260 nm. There is very little absorbance by either compound at the other's optimum, and so this contribution can be ignored. Note that the purified 4-sU RNA often gives lower A_{260}/A_{280} ratios than unlabeled RNA samples. This appears to be due to some BME still present in the sample and the fact that 4-thiouridine shows moderate absorbance at A_{280}. A_{260}/A_{280} ratios as low as 1.5 are common, and RNA with values as low as 1.4 can still be used for microarrays or other reactions.

9. A non-denaturing 1.2% agarose gel in 1× TAE is sufficient to check for labeling of total RNA, as rRNA is approximately 95% of total RNA and stains readily with 0.5 µg/ml ethidium bromide. There is no need to run a denaturing formaldehyde-agarose gel unless the exact size of RNAs needs to be determined.

10. Running a non-denaturing 1.2% agarose 1× TAE gel allows a qualitative visualization of biotinylation of rRNA, balancing assay speed against vulnerability to RNA degradation. Thus, RNA should be run very rapidly through the gel. When running the gel at 14 V/cm, the gel will heat quickly, so pre-cooling the gel and running it at 4°C helps to minimize overheating. Run gel in a box with a full reservoir of buffer to help disperse heat.

11. The amount of RNA will depend on the duration of 4-thiouracil labeling. Load 2.5 µg for labeling pulses of up to 6 h, load 1 µg for labeling pulses of greater than 6 h.

12. The general guidelines for determining how much magnetic bead slurry to use are as follows:

    For a 1- to 2-h pulse, use 1 µl of beads per 160 ng of biotinylated RNA.
    For a 3- to 6-h pulse, use 1 µl of beads per 80 ng of biotinylated RNA.

    When a mixture of 4-sU-labeled and 4-sU-unlabelled RNA is present, be sure to base bead volume on only the biotinylated 4-sU RNA. For example, if 50% of the RNA is from UPRT(–) cells, then only half the RNA is potentially biotinylated. If the RNA is too dilute in **Method 3.5.1, step 7**, re-precipitate the RNA and resuspend the RNA in a smaller volume.

13. RNA can be run on a gel and probed with streptavidin-HRP to compare the efficiency of the purification. Load equal mass amounts (~0.5 µg RNA) from the input, unbound, and eluted samples as well as equal fraction volumes of the washes

(i.e., if 5 µl of the unbound sample gives 0.5 µg RNA, load 5 µl of each wash) on a 1.2% agarose gel in 1× TAE containing 0.5 µg/ml ethidium bromide. There should be decreasing amounts of RNA in the wash fractions, with no visible RNA in the last two washes. When this RNA is transferred to a membrane and probed with streptavidin-HRP, the unbound RNA should only have approximately 1.0% of the signal intensity of the input (the biotin capture of the magnetic streptavidin bead slurry is quite efficient) and the eluate should have no signal (since the biotin has been reduced off).

Acknowledgments

We thank Sean Curran and Tammy Doukas for helpful suggestions and comments during the course of this work. G.M.Z. was supported by the NIH (5F32AI066538). M.D.C. was supported by the NIH (CMB GM07276) and the University of California Universitywide AIDS Research Program (D02-ST-405). C.D.M. was supported by the NIH (5T32AI07328 and 1F32AI056959). A.E.F. was supported by a Stanford Graduate Fellowship and Cell and Molecular Training Grant fellowship (GM07276). E.S.M. was supported by the NIH (AI30363 and AI20211). J.C.B. was supported by the NIH (AI41014 and AI21423).

References

1. Cleary, M. D., Meiering, C. D., Jan, E., Guymon, R., and Boothroyd, J. C. (2005) Biosynthetic labeling of RNA with uracil phosphoribosyltransferase allows cell-specific microarray analysis of mRNA synthesis and decay. *Nat Biotechnol* **23**, 232–7.
2. Iltzsch, M. H. (1993) Pyrimidine salvage pathways in Toxoplasma gondii. *J Eukaryot Microbiol* **40**, 24–8.
3. Pfefferkorn, E. R., and Pfefferkorn, L. C. (1977) Specific labeling of intracellular Toxoplasma gondii with uracil. *J Protozool* **24**, 449–53.
4. Iltzsch, M. H., and Tankersley, K. O. (1994) Structure-activity relationship of ligands of uracil phosphoribosyltransferase from Toxoplasma gondii. *Biochem Pharmacol* **48**, 781–92.
5. Donald, R. G., and Roos, D. S. (1995) Insertional mutagenesis and marker rescue in a protozoan parasite: cloning of the uracil phosphoribosyltransferase locus from Toxoplasma gondii. *Proc Natl Acad Sci USA* **92**, 5749–53.

10

Efficient 5′ Cap-Dependent RNA Purification
Use in Identifying and Studying Subsets of RNA

Edyta Z. Bajak and Curt H. Hagedorn

Summary

Microarray-based screening technologies have revealed a larger than expected diversity of gene expression profiles for many cells, tissues, and organisms. The complexity of RNA species, defined by their molecular structure, represents a major new development in biology. RNA not only carries genetic information in the form of templates and components of the translational machinery for protein synthesis but also directly regulates gene expression as exemplified by micro-RNAs (miRNAs). Recent evidence has demonstrated that 5′ capped and 3′ polyadenylated ends are not restricted to mRNAs, but that they are also present in precursors of both miRNAs and some antisense RNA transcripts. In addition, as many as 40% of transcribed RNAs may lack 3′ poly(A) ends. In concert with the presence of a 5′ cap (m^7 GpppN), the length of the 3′ poly(A) end plays a critical role in determining the translational efficiency, stability, and the cellular distribution of a specific mRNA. RNAs with short or lacking 3′ poly(A) ends, that escape isolation and amplification with oligo(dT)-based methods, provide a challenge in RNA biology and gene expression studies. To circumvent the limitations of 3′ poly(A)-dependent RNA isolation methods, we developed an efficient RNA purification system that binds the 5′ cap of RNA with a high-affinity variant of the cap-binding protein eIF4E. This system can be used in differential selection approaches to isolate subsets of RNAs, including those with short 3′ poly(A) ends that are likely targets of post-transcriptional regulation of gene expression. The length of the 3′ poly(A) ends can be defined using a rapid polymerase chain reaction (PCR)-based approach.

Key Words: RNA purification; 5′ RNA cap; m^7G cap; eIF4E; microarray; mRNA; non-coding RNA; 3′ poly(A); deadenylation; cytoplasmic processing bodies.

From: *Methods in Molecular Biology, Vol. 419: Post-Transcriptional Gene Regulation*
Edited by: J. Wilusz © Humana Press, Totowa, NJ

1. Introduction

The role that a larger than expected variety of RNA species plays in regulating gene expression represents a relatively recent fascinating surprise in biology (1). The eukaryotic ribotype (total pool of mRNAs, rRNAs, microRNAs, siRNAs, snoRNAs, and ncRNAs) defines the phenotype of a cell or organism and provides a "soft-wired" genomic system (2). Rather than being static templates of the genome, RNA molecules are now recognized as targets of regulation and active agents regulating developmental, physiological, and stress responses (3–5). In addition, misregulation of these processes can lead to disease states such as cancer (6).

Advances in studying gene expression patterns under normal and pathological conditions have shown a larger than expected diversity of gene expression for a given ribotype. For example, in humans, the number of RNA transcripts which do not encode proteins (23,218) appear to outnumber the quantity of protein-encoding transcripts (20,929) (7). The presence of 5′ capped and 3′ polyadenylated RNA ends was generally considered to be signatures of mature mRNAs. However, we have more recently learned that precursors of microRNAs have m^7G caps and 3′ poly(A) ends (8–10).

New technologies can open new doors to discovery. The identification of high-affinity variants of the mRNA cap-binding protein eIF4E led to the development of an efficient system to isolate mRNAs by binding their 5′-ends (11,12). Using this 5′ cap-dependent and a standard 3′ poly(A)-dependent RNA purification to differentially select mRNA led to the observation that an unexpectedly large number of mRNAs in human liver have short 3′ poly(A) ends at steady state (12). A regulatory purpose for the short 3′ poly(A) phenotype is suggested by observations in several developmental and model systems (13,14). Shortening of the 3′ poly(A) end of mRNAs takes place during mRNA decay. Complete degradation of the transcript generally occurs rapidly if it is marked for decay (15–17). However, recent studies showed that the length of 3′ poly(A) ends of mRNA and other species of RNA molecules undergo dynamic re-modeling in response to a variety of physiologic or pharmacological stimuli (18). This may involve the storage of translationally repressed mRNAs in stress granules or P-bodies following nutrient starvation. This is where cytoplasmic polyadenylation may re-activate the mRNA for translation after nutrients are re-supplied (19). In another system requiring a rapid response, NO synthesis in endothelial cells by eNOS, the polyadenylation state of eNOS mRNA is regulated by shear stress or pharmacologically by statins (20). Messenger RNAs with short 3′ poly(A) ends at steady state most likely reflect regulation by post-transcriptional mechanisms of gene expression.

However, one challenge is to identify subsets of mRNAs with this phenotype under different biological conditions. The system we describe to efficiently isolate RNA in a 5´ cap-dependent manner provides a means to identify these subsets of RNAs when combined with 3´ poly(A)-dependent methods and array technologies.

The activities of complex DNA loci with overlapping transcriptional units on both DNA strands result in antisense RNA transcripts *(21,22)*. Antisense transcripts play a role in the degradation of the corresponding sense transcript (RNA interference) or gene silencing at the chromatin level by undefined mechanisms. In a recent study of 10 human chromosomes, it has been estimated that approximately 61% of the loci studied have antisense transcripts and that 44% of RNA transcripts are never polyadenylated *(23)*. These RNAs cannot be isolated nor amplified using standard 3´ poly(A)-dependent approaches. Results from tiling arrays using genomic sequences depict a transcriptional output that is more complex than expected *(24–26)*. These new developments in RNA biology emphasize the need for additional approaches to purify and study RNA species.

Microarray-based technologies have made it possible to determine gene expression patterns in cells, tissues, and organisms under diverse physiological conditions. Nevertheless, the transcript profiles may represent biased or incomplete data sets depending on the technology used. The 3´ poly(A)-dependent isolation of mRNA, as well as cDNA synthesis using oligo(dT)$_n$ primers, in microarray experiments can result in the depletion and loss of a number of RNA transcripts with short or absent 3´ poly(A) ends. An efficient 5´ cap-dependent system to purify RNA can overcome this limitation and when combined with differential RNA selection methods can identify subsets of RNA species.

We have developed an efficient RNA purification system based on a variant of the mRNA cap-binding protein eIF4E, which binds the m^{7}GTP moiety with a 5- to 10-fold higher affinity than wild-type eIF4E. This approach can also be applied to isolate subsets of mRNA and other 5´ capped RNA species regardless of their 3´ poly(A) ends status. To define the specific length of the 3´ poly(A) tail of mRNAs, a relatively rapid polymerase chain reaction (PCR)-based method is described. These methods have applications in studying diverse RNA transcripts and the increasingly complex topic of RNA regulation.

2. Materials

The basic supplies include RNase-free water, Gene Mate 1.5 mL non-stick hydrophobic RNase-free microfuge tubes (ISC BioExpress, cat. no. C-3302-1), and RNase-free pipette tips with aerosol-barrier made of polyethylene filter matrix, where the P1000 denotes maximal pipetting liquid volume (1000 μL) (*see* **Note 1**).

2.1. Preparation of eIF4E$_{K119A}$-Beads

1. S-linked glutathione-agarose beads (Sigma, cat. no. G 4510): re-hydrated by washing three times with a minimum of 10 vol of RNase-free deionized Milli-Q®-filtered water and stored at 4°C for up to several months.
2. Recombinant GST-eIF4E$_{K119A}$ (4E$_{K119A}$) (*see* **Note 2**).
3. Sterile phosphate-buffered saline (PBS) (calcium and magnesium free) stored at 4°C.

2.2. Purification of 5′ Capped RNAs with eIF4E$_{K119A}$-Beads

1. Buffer A (1×): 10 mM potassium phosphate buffer, pH 8.0, 100 mM KCl, 2 mM EDTA, 5% glycerol, 0.005% Triton X-100, and 1.3% poly(vinyl) alcohol 98–99% hydrolyzed (Avg. Mw 31,000–50,000) (Aldrich, cat. no. 36,313). Add the following just before use: 100 μg/mL of *Escherichia coli* tRNA (Roche, cat. no. 10109550001), 6 mM DTT, and 20 U/mL RNasin Superase-In® (Ambion, cat. no. 2694) (*see* **Note 3**).
2. Buffer B (1×): Same as buffer A *except* that *E. coli* tRNA is omitted (*see* **Note 3**).
3. Buffer B supplemented with 0.5 mM GDP (Sigma, cat. no. G7127) (*see* **Note 3**).
4. Elution Buffer: Buffer B supplemented with 1 mM m^{7}GDP (Sigma, cat. no. M5883) (*see* **Note 3**).
5. Phenol/chloroform extraction of RNA: Acid phenol/chloroform (Ambion, cat. no. 9720) and chloroform (Sigma, cat. no. C2432) (*see* **Note 4**).
6. Precipitation of RNA: Linear acrylamide (Ambion, cat. no. 9520), sodium acetate 3 M (pH 5.2–5.5) and absolute isopropanol, absolute ethanol and 70% ethanol.
7. Analysis of isolated 5′ capped RNAs: Quantity and purity with NanoDrop or standard spectrophotometer (A$_{260}$/A$_{280}$ methods), quantification with the fluorescent RiboGreen® assay (Molecular Probes, cat. no. R-11491), non-denaturing polyacrylamide or agarose gel electrophoresis, and/or with the Agilent 2100 Bioanalyzer (Nano- or Pico-LabChip®). The Bioanalyzer technology makes it possible to analyze the quality and quantity of isolated RNA samples simultaneously.

2.3. Rapid Amplification of cDNA Ends Poly(A) Test

1. Reverse transcription for cDNA synthesis: Template RNA (*see* **Note 5**), DNase- and RNase-free water, oligo(dT)$_{12}$ primer with G/C-rich anchor (5′-GCGAGCTCCGCGGCCGCG(T)$_{12}$-3′), 200 U/μL SuperScript™ II reverse transcriptase (RT) (Invitrogen, cat. no. 18064-022), 5× SuperScript™ II RT first-strand buffer (250 mM Tris–HCl, pH 8.3, at room temperature, 375 mM KCl, and 15 mM MgCl$_2$), dNTP mix (10 mM each), 0.1 M DTT, and 2 U/μL RNase H (Invitrogen, cat. no. 18021-014) (*see* **Note 6**).

2. PCR: cDNA template generated during RT reaction, oligo(dT)$_{12}$ primer (as above), a target-specific primer (complementary to the sequence of selected RNA), PCR buffer 10× (200 mM Tris–HCl, pH 8.4, and 500 mM KCl), 50 mM MgCl$_2$, and *Taq* DNA polymerase.

3. Methods

The isolation of RNA from cells or tissues is a critical step in studying gene expression, preparing full length cDNA libraries, answering structure-function questions in RNA biology, and exploring the new area of non-coding regulatory RNAs. While studying the 5´ mRNA m^{7}G(5´)ppp(5´)N cap-binding site of eukaryotic initiation factor 4E (eIF4E), we identified a variant protein (eIF4E$_{K119A}$) that had a 5- to 10-fold increase in affinity for the 5´ mRNA cap as compared to the wild-type eIF4E *(11)*. In addition, eIF4E$_{K119A}$ is capable of binding trimethylated 5´ cap structures (3.5-fold increase for m^{2,2,7}GTP and 11-fold increase for m^{7}GpppG) and eIF4G peptides with a higher affinity than wild-type eIF4E *(27)*.The identification of this high-affinity variant of eIF4E enabled an efficient microtube batch method to purify RNA with 5´ caps to be developed using recombinant GST-4E$_{K119A}$ bound to agarose beads *(12)*. This is an efficient system where 70% of the starting 5´ capped RNA is recovered independently of its 3´ poly(A) status (*see* **Fig. 1**). This technology is applicable to the preparation of cDNA libraries that are enriched with full length clones and complete 5´-ends, the purification of RNA that is not efficiently recovered by standard oligo(dT) methods, and differential selection processes to identify subsets of RNAs.

To determine the length of the 3´ poly(A) ends of RNA efficiently purified with eIF4E$_{K119A}$ but underestimated following oligo(dT) methods, a rapid amplification of cDNA ends poly(A) test (RACE-PAT) can be employed *(12,28)*. This method represents a relatively fast and sensitive technique that enables the length of the 3´ poly(A) end of a specific target RNA to be measured. The RT–PCR based method uses an oligo(dT)$_{12}$ primer that contains a 3´ G/C-rich motif to more accurately measure the 3´ poly(A) end. The PCR products are analyzed by standard gel electrophoresis and direct nucleic acid sequencing *(12)*. Several variations of RACE have been developed and can be used depending on the requirements of a study *(29–31)*.

3.1. Preparation of eIF4E$_{K119A}$-Beads

1. Let the recombinant GST-eIF4E$_{K119A}$ bind to re-hydrated glutathione-agarose beads in sterile PBS for 1–2 h at room temperature (*see* **Note 7**) by continuous end-over-end mixing using an immunoprecipitation rotor. Use 1 mg of protein per 1 mL of beads (packed volume).

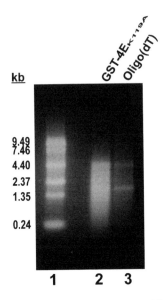

Fig. 1. Comparison of RNA purified by 5′ cap and 3′ poly(A)-dependent methods. Total RNA (315 μg) was mixed with GST-4E$_{K119A}$ beads in a microfuge tube or applied to an oligo(dC$_{10}$T$_{30}$) column *(12)*. The GST-4E$_{K119A}$ matrix was washed and mRNA was recovered by eluting with m^{7}GDP. The oligo(dT) column was used following the protocol of the manufacturer (Qiagen). One-tenth of the mRNA recovered from each purification product was separated by formaldehyde agarose gel electrophoresis. The photograph shows an ethidium bromide-stained gel: lane 1, RNA ladder; lane 2, RNA that was batch purified with GST-4E$_{K119A}$ agarose beads; and lane 3, RNA that was purified with an oligo(dT) column. Image has been reprinted with the permission of Copyright 2003 National Academy of Sciences, U.S.A.

2. Allow the agarose-GST-4E$_{K119A}$ beads (4E-beads) to settle by gravity (recommended) or briefly spin them down in the mini centrifuge at 2500 rpm (Costar, 10 MVSS-09024). Avoid packing the beads too tightly.
3. Wash the beads gently with 10 vol of PBS twice by inverting the tube and allowing the beads to develop into a loosely packed pellet.
4. Resuspend the 4E-beads in PBS to make a 10% (v/v) slurry and store at 4°C until use. Do not store them for more than 2 months.

3.2. Purification of 5′ Capped RNAs with eIF4E$_{K119A}$-Beads

1. Transfer 200 μL of 4E-beads (packed bead volume) to each microfuge tube.
2. To pellet the beads, spin the slurry at maximal speed (~5000 × *g*) for 10 s. Carefully remove as much storage buffer as possible without losing the beads.

3. Rinse the 4E-beads three times with 1 mL of buffer A as follows: Resuspend the beads in the buffer A by tapping with index finger. Mix by inverting the tube. Centrifuge beads at 2500 rpm ($\sim$3800 $\times$ g) in a microcentrifuge for 10 s. Remove supernatant with a P1000 pipette.

4. Heat denature the starting material (total RNA) for 10 min in a 70°C heating block and chill on ice for additional 2 min prior to mixing with 4E-beads.

5. Add 100–150 μg of RNA dissolved in RNase-free H_2O to each tube (0–50 μL). Bring the volume of RNA sample to 500 μL by adding enough buffer A to the RNA-containing tube (*see* **Note 8**). Do not include the volume of beads when measuring. The final volume at this stage should be 700 μL.

6. Mix gently end-over-end for 1 h at room temperature using a rocking or rotating platform mixer.

7. Centrifuge at 2500 rpm ($\sim$3800 $\times$ g) in a microcentrifuge for 10 s to pellet the 4E-beads with bound RNA (through the 5′capped end of RNA).

8. Remove supernatant with a P1000 pipette and save in a separate microfuge tube for analysis if desired.

9. Wash beads three times with 1 mL of buffer B as follows: Resuspend beads in buffer B with a P1000 pipette. Mix end-over-end on an immunoprecipitation rotor for 5 min. Spin for 10 s at 2500 rpm ($\sim$3800 $\times$ g) to pellet beads. Remove supernatant with a P1000 pipette. Save each wash in a separate tube for analysis if desired.

10. Wash beads two times with 1 mL of buffer B containing 0.5 mM GDP. Perform wash as in step 9.

Now, depending on study requirements continue with Option I or Option II for the recovery of 5′capped RNAs.

3.3. Option I: Direct Extraction of RNA from eIF4E$_{K119A}$-Beads (see Note 9)

1. Wash beads two times with buffer B, as in step 9, to remove GDP, which interferes with quantification of the purified RNA (*see* **Note 10**).

2. To the washed beads, add 400 μL 1× buffer B and 500 μL acid phenol/chloroform. Close the tube tightly. Mix gently by inverting the tube 10 times (recommended). Alternatively, mix by gentle vortexing for 10 s (*see* **Note 11**). Afterwards, chill for 15 min on ice (*see* **Note 12**).

3. Centrifuge the mixture in a pre-cooled centrifuge (4°C) at 14,000 rpm ($\sim$20000 $\times$ g) for 10 min. At this stage, three layers should appear: top clear layer, middle white layer, and bottom clear layer. Transfer top aqueous (clear) layer (contains 5′ capped RNA) to a new microfuge tube. Add an equal volume of chloroform to the aqueous layer and mix by inverting three times. Spin for 3 min and transfer only the aqueous layer to a new tube and save it. Discard the remaining layers.

4. To the saved aqueous layer add an equal volume of absolute isopropranol, salt (50 μL 3 M sodium acetate) and 4 μL of 1 μg/mL linear acrylamide. Precipitate the RNA over night at –80°C.

5. The following day, spin down the precipitation mixture at 14,000 rpm ($\sim$20000 × g) in a microcentrifuge for 30 min at 4°C. Carefully remove and discard the supernatant. Precipitated RNA (white pellet) should remain in the bottom of tube.
6. Wash the RNA pellet two times with ice-cold 70% ethanol (1 mL) by inverting the tube only and spin down at 14,000 rpm ($\sim$20000 × g) for 10 min at 4°C.
7. Remove supernatant and air dry pellets for approximately 5 min at room temperature.
8. Resuspend RNA in 50–100 µL of RNase-free water and store at –80°C. Alternatively, for long-term storage, overlay the RNA pellet with 70% ethanol and keep in –80°C freezer (*see* **Note 13**).

3.4. Option 2: Ligand Elution of RNA from eIF4E$_{K119A}$-Beads (see Note 14)

1. Elute the bound RNA (obtained from step 10) by mixing beads in 1 mL of buffer B containing 1 mM m^{7}GDP for 5–10 min as in **Method 3.2**, step 9.
2. Collect and combine the supernatant with the eluted RNA (weakly bound to the beads) in a new microfuge tube and ethanol precipitate the RNA as described in **Subheading 3.3** (Option I), steps 14–18.
3. Save the remaining beads and continue with elution and recovery of the 5´ capped RNA that is still present on the saved beads. Do so by using the phenol/chloroform extraction as described in steps 12 and 13 of **Subheading 3.3** (Option I).
4. Precipitate the extracted RNA by dividing the m^{7}GDP elutant into two tubes ($\sim$0.5 mL per tube) followed by adding 95% ethanol (1 mL) to each tube (no salt necessary). Then add 4 µL of 1 µg/mL linear acrylamide to each sample. This will help to pellet the RNA. Precipitate overnight at –80°C.
5. Continue with steps 15–18 of **Method 3.3** (Option I) (*see* **Note 15**).

3.5. Determination of the Size of the 3´ Poly(A) Ends with the RACE-PAT

1. For cDNA synthesis, use 1–5 µg of total RNA (*see* **Note 5**). Mix and briefly centrifuge each reagent before use.
2. Set up RNA/primer mixtures in sterile, nuclease-free 0.2- or 0.5-mL tubes as follows:

Component	Sample	No RT control	Control RNA*
RNA (up to 5 µg)	X µl	X µl	—
Control RNA (50 ng)*	—	—	1 µl
10 mM dNTP mix	1 µl	1 µl	1 µl
Oligo(dT)$_{12}$ (200 ng)	Y µl	Y µl	Y µl
RNase-free water	Up to 13 µl	Up to 13 µl	Up to 13 µl

*Optional.

3. Heat mixtures at 70°C for 10 min, then immediately chill on ice for at least 1 min. Centrifuge briefly to collect the contents of each tube.
4. Prepare reaction master mix by adding each component in the indicated order:

Component	One sample	Four samples
5× RT buffer	4 μl	16 μl
0.1 M DTT	2 μl	8 μl

5. Add 6 μL of reaction master mixture to each RNA/primer mixture, mix gently, and collect in the bottom of the tube by brief centrifugation. Incubate at 42°C for 2 min.
6. Add 1 μLof SuperScript™ II RT (final 20 U/μL) to each tube, but do not add to the RT control, mix by pipetting up and down. The final volume of RT reaction will be 20 μL. Incubate at 42°C for 60 min.
7. Terminate the reaction by heating at 70°C for 15 min, chill on ice, and collect the contents of each reaction tube by brief centrifugation.
8. Add 1 μL (2 units) of RNase H to each sample and incubate for 20 min at 37°C before proceeding to PCR (see **Note 16**).
9. Add the following to a PCR tube (see **Note 17**):

Component	One sample
10× PCR buffer (-Mg)	5 μL
50 mM MgCl$_2$	1.5 μL
10 mM dNTP mix	1 μL
25 μM oligo(dT)/2	1 μL
Taq DNA Polymerase	X μL
cDNA template from RT reaction	0.5 μL
DNase-free water	Up to 49 μL

10. Add 1 μL of RNA target-specific primer (final 0.5 μM) to each indicated tube with PCR mixture. Total volume of PCR will be 50 μL. Mix gently and heat to 94°C for 5 min to denature.
11. Continue with 15–40 cycles of PCR amplification with the recommended annealing and extension conditions specified for your template, primer sets, and type of *Taq* DNA polymerase used.
12. Analyze the RACE-PAT products by standard non-denaturing polyacrylamide or agarose gel electrophoresis or with an Agilent 2100 Bioanalyzer.

4. Notes

General considerations while creating RNase-free environment: reserve all reagents, micropipettes, forceps, and so on. for RNA-related procedures only. Apply proper aseptic techniques. Wear a long-armed lab coat and gloves, which should be frequently changed. Use certified RNase-free disposable, presterilized plasticware and pipette tips at all times. Keep the working area used for RNA isolation and analysis, marked as "FOR RNA WORK ONLY," ordered and clean. Intermittently clean surfaces (including external parts of micropipettes, rotors etc.) with RNase*Zap®* (Ambion, cat. no. 9780).

1. Be aware of "self-sealing" filters that may contain cellulose gum additives, which may be deposited into samples while pipetting (on contact with liquid and inhibit the RT–PCR) *(32)*.
2. Human recombinant GST-4E$_{K119A}$ (the nomenclature indicates the substituted amino acid, its position, and the replacing residue) was purified from *E. coli* containing a T7 polymerase-driven vector essentially as previously described *(33,12)*.
3. When making buffers A and B, you may first prepare a core solution containing KHPO$_4$, KCl, EDTA, glycerol, Triton X-100, and poly(vinyl alcohol). This solution can be stored at 4°C for months. Prior to usage, divide it into separate bottles and add buffer-specific components: tRNA, DTT, and RNasin for buffer A while only DTT and RNasin for buffer B and use within 1 week. During that time, store complete buffers A and B at 4°C. The same applies for buffer B containing GDP or m^{7}GDP.
4. Both phenol and chloroform are harmful. Before starting to work with any reagent, carefully read its material safety data sheets (MSDSs). Work in the fume hood and use protective clothing (long-armed lab coat, gloves, eye protection, and shoes covering toes). Phenol and/or chloroform-contaminated waste, as well as phenol and/or chloroform containing plasticware must be disposed of through the hazardous chemical waste route.
5. Specific RNA pools can be used for cDNA synthesis. For example, you can use the 5′ capped RNA isolated with 4E-beads as a template for RACE-PAT.
6. Store reagents for cDNA synthesis at –20°C. Thaw the 5× first-strand RT buffer and DTT at room temperature just prior to use and re-freeze immediately after use.
7. The recombinant GST-4E$_{K119A}$ or GST-4Ewt *(35,36)* may be used for 5′ cap-dependent RNA isolation. We generally use a packed volume of 200 μL, of high-affinity GST-4E$_{K119A}$ beads per 100–150 μg of starting material.
8. A change in buffer concentration because of partial dilution with total RNA in H$_2$O (as much as 50 μL) will not impair the purification process.
9. Direct phenol/chloroform extraction of RNA bound to beads provides a higher yield as compared to m^{7}GDP elution because the later does not remove all RNA.

10. During the first wash, transfer the beads and buffer to a new tube. This minimizes carry over of tRNA and ribosomal RNA bound to plastic tube during the earlier steps.

11. Gentle vortexing should not be a problem for most RNAs; however, it induces shear stress, which may have deleterious effect(s) on the susceptible subpopulation of RNA molecules. Thus, we strongly suggest optimization of the mode of mixing.

12. Cooling down the temperature of the mixture increases the efficiency of nucleic acid–protein complex dissociation.

13. In some instances, specific requirements of down-stream analysis or difficulties in dissolving RNA pellets, the denaturation of RNA by heating may be required (e.g., 5 min at 70°C on a heating block). However, the exposure of RNA to heat should be as limited as possible.

14. Use Option II for specific applications only. This approach gives lower yields of RNA, approximately 60–70% of Option I is recovered, but the product is less contaminated with ribosomal RNA.

15. The 5′ capped RNA recovered with the m^7GDP elution procedure will have increased background making it difficult to quantify using standard spectrophotometry methods. You may consider using the RiboGreen® and/or the Agilent 2100 Bioanalyzer.

16. The cDNA is ready for downstream applications. However, the sensitivity of PCR from cDNA can be increased, especially for long templates (>1 kb), by removal of RNA complementary to the synthesized cDNA. This is done by RNA digestion with RNase H.

17. The amount of $MgCl_2$ (final 1.5 mM) listed above is approximate. The optimal $MgCl_2$ concentration for your PCR depends on the type of DNA polymerase used. Moreover, the amount of enzyme we have used (2.5 U/μL as the final concentration), the number of PCR cycles, and the annealing and extension conditions are primer and template dependent.

Acknowledgments

This work was supported by NIH RO1 CA063640 (CHH).

References

1. Mattick, J. S. and Makunin, I. V. (2006) Non-coding RNA. *Hum Mol Genet* **15**, R17–R29.
2. Herbert, A. and Rich, A. (1999) RNA processing and the evolution of eukaryotes. *Nat Genet* **21**, 265–269.
3. Carninci, P. (2006) Tagging mammalian transcription complexity. *Trends Genet* **22**, 501–510.
4. David, L., Huber, W., Granovskaia, M., Toedling, J., Palm, C. J., Bofkin, L., Jones, T., Davis, R. W., and Steinmetz, L. M. (2006) A high-resolution map of transcription in the yeast genome. *Proc Natl Acad Sci USA* **103**, 5320–5325.

5. Bajak, E. Z. (2005) *Genotoxic Stress: Novel Biomarkers and Detection Methods. Uncovering RNAs Role in Epigenetics of Carcinogenesis.* Karolinska University Press, Stockholm, Sweden.

6. Calin, G. A., Dumitru, C. D., Shimizu, M., Bichi, R., Zupo, S., Noch, E., Aldler, H., Rattan, S., Keating, M., Rai, K., Rassenti, L., Kipps, T., Negrini, M., Bullrich, F., and Croce, C. M. (2002) Frequent deletions and down-regulation of micro-RNA genes miR15 and miR16 at 13q14 in chronic lymphocytic leukemia. *Proc Natl Acad Sci USA* **99**, 15524–15529.

7. Carninci, P., Kasukawa, T., Katayama, S., Gough, J., Frith, M. C., Maeda, N., et al. (2005) The transcriptional landscape of the mammalian genome. *Science* **309**, 1559–1563.

8. Cai, X., Hagedorn, C. H., and Cullen, B. R. (2004) Human microRNAs are processed from capped, polyadenylated transcripts that can also function as mRNAs. *RNA* **10**, 1957–1966.

9. Lee, Y., Kim, M., Han., J., Yeom, K. H., Lee, S., Baek, S. H., and Kim, V. N. (2004) MicroRNA genes are transcribed by RNA polymerase II. *EMBO J* **23**, 4051–4060.

10. Houbaviy, H. B., Dennis, L., Jaenisch, R., and Sharp, P. A. (2005) Characterization of a highly variable eutherian microRNA gene. *RNA* **11**, 1245–1257.

11. Spivak-Kroizman, T., Friedland, D. E., De Staercke, C., Gernert, K. M., Goss, D. J., and Hagedorn, C. H. (2002) Mutations in the S4-H2 loop of eIF4E which increase the affinity for m^7GTP. *FEBS Lett* **516**, 9–14.

12. Choi, Y. H. and Hagedorn, C. H. (2003) Purifying mRNAs with a high-affinity eIF4E mutant identifies the short 3′ poly(A) end phenotype. *Proc Natl Acad Sci USA* **100**, 7033–7038.

13. Wang, L., Eckmann, C. R., Kadyk, L. C., Wickens, M., and Kimble, J. (2002) A regulatory cytoplasmic poly(A) polymerase in *Caenorhabditis elegans*. *Nature* **419**, 312–316.

14. Barnard, D. C., Cao, Q., and Richter, J. D. (2005) Differential phosphorylation controls Maskin association with eukaryotic translation initiation factor 4E and localization on the mitotic apparatus. *Mol Cell Biol* **25**, 7605–7615.

15. Wilusz, C. J. and Wilusz, J. (2004)Bringing the role of mRNA decay in the control of gene expression into focus. *Trends Genet* **20**, 491–497.

16. Coller, J. and Parker, R. (2005) General translational repression by activators of mRNA decapping. *Cell* **122**, 875–886.

17. Mukherjee, D., Gao, M., O'Connor, J. P., Raijmakers, R., Pruijn, G., Lutz, C. S., and Wilusz, J. (2002). The mammalian exosome mediates the efficient degradation of mRNAs that contain AU-rich elements. *EMBO* **21**, 165–174.

18. Newbury, S. F., Mühlemann O., and Stoecklin, G. (2006) Turnover in the Alps: an mRNA perspective. Workshop on mechanisms and regulation of mRNA turnover. *EMBO Rep* **7**, 143–148.

19. Brengues, M., Teixeira, D., and Parker, R. (2005) Movement of eukaryotic mRNAs between polysomes and cytoplasmic processing bodies. *Science* **310**, 486–489.

20. Weber, M., Hagedorn, C. H., Harrison, D. G., and Searles, C. D. (2005) Laminar shear stress and 3´ polyadenylation of eNOS mRNA. *Circ Res* **96**, 1161–1168.

21. Engström, P. G., Suzuki, H., Ninomiya, N., Akalin, A., Sessa, L., Lavorgna, G., et al. (2006) Complex loci in human and mouse genomes. *PLoS Genet* **2**, e47.

22. Katayama, S., Tomaru, Y., Kasukawa, T., Waki, K., Nakanishi, M., Nakamura, M., et al. (2005) Antisense transcription in the mammalian transcriptome. *Science* **309**,1564–1566.

23. Cheng, J., Kapranov, P., Drenkow, J., Dike, S., Brubaker, S., Patel, S., et al. (2005) Transcriptional maps of 10 human chromosomes at 5-nucleotide resolution. *Science* **308**, 1149–1154.

24. Siddiqui, A. S., Khattra, J., Delaney, A. D., Zhao, Y., Astell, C., Asano, J., et al. (2005) A mouse atlas of gene expression: large-scale digital gene-expression profiles from precisely defined developing C57BL/6J mouse tissues and cells. *Proc Natl Acad Sci USA* **102**, 18485–18490.

25. Hayashizaki, Y. and Carninci, P. (2006) Genome Network and FANTOM3: assessing the complexity of the transcriptome. *PLoS Genetics* **2**, e63.

26. Kiyosawa, H., Mise, N., Iwase, S., Hayashizaki, Y., and Abe, K. (2005) Disclosing hidden transcripts: mouse natural sense-antisense transcripts tend to be poly(A) negative and nuclear localized. *Genome Res* **15**, 463–474.

27. Friedland, D. E., Wooten, W. N., LaVoy, J. E., Hagedorn, C. H., and Goss, D. J. (2005) A mutant of eukaryotic protein synthesis initiation factor eIF4EK119A has an increased binding affinity for both m^7G cap analogues and eIF4G peptides. *Biochemistry* **44**, 4546–4550.

28. Salles, F. J., Darrow, A. L., O'Connell, M. L., and Strickland, S. (1992) Isolation of novel murine maternal mRNAs regulated by cytoplasmic polyadenylation. *Genes Dev.* **6**, 1202–1212.

29. Liu X. and Gorovsky, M. A. (1993) Mapping the 5' and 3' ends of *Tetrahymena thermophila* mRNAs using RNA ligase mediated amplification of cDNA ends (RLM-RACE). *Nucleic Acids Res.* **21**, 4954–4960.

30. Schaefer, B. C. (1995) Revolutions in rapid amplification of cDNA ends: new strategies for polymerase chain reaction cloning of full-length cDNA ends. *Anal Biochem* **227**, 255–273.

31. Friedrich, M., Grahnert, A., and Hauschildt, S. (2005) Analysis of the 3' UTR of the ART3 and ART4 gene by 3' inverse RACE-PCR. *DNA Seq* **16**, 53–57.

32. Bustin, S. A. and Nolan, T. (2004) Template handling, preparation, and quantification, in *A-Z of Quantitative PCR* (Bustin, S. A., ed.), International University Line, La Jolla, CA, pp. 141–213.

33. Hagedorn, C. H., Spivak-Kroizman, T., Friedland, D. E., Goss, D. J., and Xie, Y. (1997) Expression of functional eIF-4E$_{human}$: purification, detailed characterization, and its use in isolating eIF-4E binding proteins. *Protein Expr Purif* **9**, 53–60.

34. Yang Z., Edenberg, H. J., and Davis, R. L. (2005) Isolation of mRNA from specific tissues of Drosophila by mRNA tagging. *Nucleic Acids Res* **33**, e148.

35. Sonenberg, N., Rupprecht, K. M., Hecht, S. M., and Shatkin, A.J. (1979) Eukaryotic mRNA cap binding protein: purification by affinity chromatography on sepharose-coupled m^{7}GDP. *Proc Natl Acad Sci USA* **76**, 4345–4349.

36. Edery, I., Chu, L. L., Sonenberg, N., and Pelletier, J. (1995) An efficient strategy to isolate full-length cDNAs based on an mRNA cap retention procedure (CAPture). *Mol Cell Biol* **15**, 3363–3371.

III

TECHNIQUES FOR SPECIFIC ASPECTS OF RNA BIOLOGY

11

Enrichment of Alternatively Spliced Isoforms

Julian P. Venables

Summary

Most metazoan genes are alternatively spliced, and a large number of alternatively spliced isoforms are likely to be functionally significant and expressed at specific stages of pathogenesis or differentiation. Splicing changes usually only affect a small portion of a gene, and these changes may cause significant mRNA degradation. After RT–PCR, minor variants can form heteroduplexes with the major variants. Affinity purification of these heteroduplexes using immobilized *Thermus aquaticus* single-stranded DNA-binding protein allows purification of alternative splice forms in a 1:1 ratio, which makes it easy to sequence the rare form. This chapter provides a detailed protocol of the technique I have developed to identify spliced isoforms called enrichment of alternatively spliced isoforms or EASI.

Key Words: Alternative splicing; affinity selection; *Thermus aquaticus* single-stranded DNA-binding protein.

1. Introduction

Alternatively spliced isoforms come from identical pre-mRNAs because of alternative splicing. Therefore, the isoforms differ in some internal region while being identical at their ends. These differences can be exploited for isoform identification in the following manner: two alternative spliced isoforms were separately amplified using RT–PCR and mixed in a 1:1 ratio, boiled for 1 min, and allowed to reanneal on the bench for 10 min in water or in low salt buffer. In all, six different molecular species were visualized on an agarose gel (*see* **Fig. 1A**). Directed RT–PCR performed on tissue RNA with gene-specific

From: *Methods in Molecular Biology, Vol. 419: Post-Transcriptional Gene Regulation*
Edited by: J. Wilusz © Humana Press, Totowa, NJ

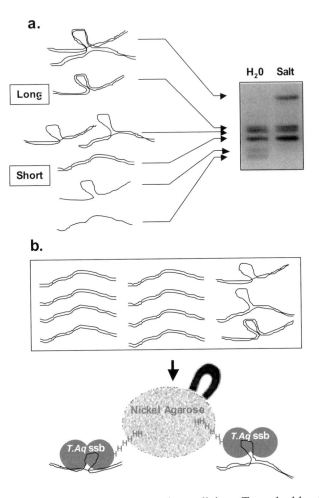

Fig. 1. **(A)** Multiple species of alternative splicing. Two double-stranded PCR products of different splice forms of a gene contain four different single strands, which can reanneal to one another in the combinations shown diagrammatically. The left lane shows the species that reanneal from a 1:1 mixture of splice forms in pure water. The right lane shows the presence of salt causes heteroduplexes to anneal to one another in four-stranded "octopus" structures *(7)*. **(B)** The principle of enrichment of alternatively spliced isoforms (EASI). When one splice form is abundant and another rare, their RT–PCR products are essentially the same as in **(A)**, either homoduplex or heteroduplex. *T.aq* ssb, tethered by magnetic nickel agarose particles, purifies the heteroduplexes that contain the common and rare splice forms in a 1:1 ratio.

primers also allowed heteroduplexes to form between alternative splice forms. As well as containing two alternatively spliced isoforms in a 1:1 ratio, their single-stranded regions are biochemically different from the rest of the mixture. Thus, these heteroduplexes can be purified on a *Thermus aquaticus (T.aq)* single-stranded DNA-binding protein (ssb) column (*see* **Fig. 1B**) *(1)*.

Three groups have attempted to use the single-stranded regions of heteroduplexes to enrich alternatively spliced isoforms from complex libraries *(2–4)*. These complicated procedures have had varying degrees of success because of inherent difficulties such as heteroduplex formation between paralogous genes and also between nuclear and cytoplasmic sequences from the same gene. In the EASI protocol described here, the use of gene-specific primer pairs removes the problem of paralogs whereas amplification across several introns in a section avoids the enrichment of pre-mRNA. Even so, a simple PCR leads to a significant amount of non-specific priming, and this is overcome here with nested PCR. The combination of the highly efficient *T.aq* ssb column and a gene-specific focus allows identification of all the alternative splices of a gene in a specific tissue and appears sufficiently powerful to identify significant alternative splicing in mixed tissue or pooled patient samples.

This article presents a detailed protocol for the preparation for and performance and downstream analysis of EASI. Briefly, recombinant His-tagged ssb is bound directly from recombinant *Escherichia coli* lysate to ultra-small magnetic particles of nickel-agarose and stored indefinitely in the fridge. These beads can be applied directly to nested PCRs that have had loose single-stranded cDNA ends removed with exonuclease VII. After a 5-min incubation in the PCR the beads are washed and the heteroduplexes are quickly eluted from the beads with imidazole. The eluate contains rare splice forms, approaching the common form in abundance. Sequence can be obtained after further PCR amplification either directly from gel-purified bands or by shotgun cloning.

2. Materials

2.1. Recombinant ssb Production

1. Plasmid pEASI. The pEASI plasmid encodes His-tagged *T.aq* ssb *(1)*.
2. *E. coli* strain BL21 or equivalent.
3. LB media and agar.
4. 100 µM ampicillin.
5. Isopropyl-beta-D-thiogalactopyranoside (IPTG): 0.5 M.

2.2. Determination of Expression Levels

1. Coomassie Protein Assay Reagent (Pierce) or any equivalent Bradford Reagent.

2.3. Preparation of EASI Beads

1. Magnetic Particle Concentrator for Microtubes of Eppendorf type (1.5 ml) (MPC-E) (Dynal, Oslo, Norway).
2. Dynabeads TALON slurry (Dynal) (*see* **Note 1**).
3. Phosphate-buffered saline (PBS).
4. EASI buffer (1×): 100 mM NaCl, 20 mM Tris–HCl pH 7.5.
5. Bead-prep salt buffer: 1×: 1 M NaCl, 20 mM Tris–HCl pH 7.5.

2.4. Reverse Transcription

1. "Transcriptor" Reverse Transcriptase (Roche, Basel, Switzerland) (*see* **Note 2**).
2. Random hexamers (20 mM).
3. dNTPs (10 mM).
4. Superasin—RNAse inhibitor (Ambion, Austin, Texas, USA).

2.5. Nested PCR

1. "Expand High-fidelity" PCR system (Roche) (*see* **Note 3**).
2. Nested primer pairs: If a control is required, it is recommended to enrich the HipK3-T splice form from human testis mRNA *(5)* using the following primers: CCTGTTCGGCAGCCTTACAGGGT and GGATG-GTTCAGGGTCTCAGCTGGAGT for the first PCR, then GGGTCGGCCAGT-CATGTATC and GCGCTACATCATCCAGACTGTTGA for the second PCR; both PCRs use an annealing temperature of 59°C. The final products are 391 bp for the constitutive form and 500 bp for the scarce variant. An alternative control that uses mRNA from whole blood rather than testis is the BRCA2 alternative splice lacking exon 3. This uses the following primers GTGGGTCGCCGCCGGGAGAA and GAACTAAGGGTGGGTGGTGTAGCTA for the first PCR, then CAGATTTGT-GACCGGCGCGGTT and GGTGTCTGACGACCCTTCACAAACT for the second PCR with the same conditions as above. The final products are 648 bp for the constitutive form and 399 bp for the scarce variant.

2.6. Single-Stranded End Cleanup

1. Exonuclease VII 10U/μl (USB) (*see* **Note 4**).
2. L buffer (10× from Roche) or home-made 1×: 10 mM Tris–HCl pH 7.5, 10 mM MgCl$_2$, 1 mM dithioerythritol (DTE).

2.7. Enrichment

Elution buffer: EASI buffer containing 200 mM imidazole.

2.8. Post-Enrichment Amplification

"HotStarTaq Plus" DNA polymerase (Qiagen, Venlo, The Netherlands) (*see* **Note 3**).

2.9. Sequencing

1. QIAquick PCR purification kit (Qiagen).
2. QIAquick Gel extraction kit (Qiagen).
3. "DYEnamic ET terminator kit (MegaBACE)" sequencing kit (Amersham, Little Chalfont, Buckinghamshire, UK).

2.10. Colony Sequencing—Standard M13 Primers

M13uni(-43) AGGGTTTTCCCAGTCACGACGTT; M13rev(-49) GAGCG-GATAACAATTTCACACAGG

3. Methods
3.1. Recombinant Single-Stranded DNA-Binding Protein Production

1. Transform. *E. coli* BL21 with pEASI and plate on LB ampicillin.
2. The next morning—scrape up half a plateful of colonies (~100) into 400 ml of LB media containing ampicillin.
3. Incubate with shaking for approximately 3 h at 37°C until OD 600 is 0.4–0.6.
4. Add IPTG to final concentration 0.5 mM and leave shaking for an additional 3 h at 37°C.
5. Centrifuge culture at 3000 × *g* for 5 min.
6. Resuspend pellet in 5 ml of PBS.
7. Sonicate three times for 30 s (without cooling).
8. Divide lysate into four microfuge tubes and heat to 70°C in a water bath (to precipitate *E. coli* proteins) for 5 min.
9. Micro-centrifuge at top speed for 5 min.
10. Re-pool the supernatants.

3.2. Determination of the Expression Level or Saturating Concentration

This procedure should be performed once for each batch. The data obtained from this method are a confidence booster and do not take long.

1. Pipette 40 µl Dynabeads TALON slurry into a 1.5-ml tube and place in magnetic separator until separated (approximately 30 s).
2. Remove liquid and resuspend beads in 500 µl PBS.
3. Put 50 µl washed beads into five microfuge tubes and separate.

4. Add 80, 40, 20, 10, and 5μl of cleared lysate to the numbered tubes. Leave for 5 min.
5. Wash beads three times in 500 μl PBS and remove all liquid.
6. Add 10 μl Bradford reagent to beads. Leave for 1 min and separate.
7. Remove eluate to microtitre plate wells and compare colors by eye. The amount of lysate needed to saturate the beads is the minimum amount to produce full blue intensity. For example, if the 40-μl tubes are roughly as blue as the 20-μl tube (as I have generally found), then the saturating amount is 20 μl—so you will need to load 20 μl of lysate per 4 μl of original slurry. I recommend storing the lysate in 400 μl aliquots at –80°C (enough for 20 enrichments). If the 10 or 40-μl tubes are instead found to contain the saturating amount, make 200 or 800 μl aliquots respectively for the correct loading in **Subheading 3.3.**

3.3. Preparation of EASI Beads

1. When required, thaw an aliquot of lysate (prepared as described in **Subheading 3.2.**) and further clear by micro-centrifugation for 3 min.
2. Magnetically separate beads from 80 μl Dynabeads-TALON slurry.
3. Resuspend beads in the cleared supernatant and leave on the bench for 5 min.
4. Separate beads and then incubate them in "bead-prep salt buffer" for 3 min (this dissociates sheared *E. coli* DNA).
5. Separate again and then resuspend in 1 ml of PBS. Store in a refrigerator for later use (the charged ssb-beads can be stored for at least 3 months at 4°C without loss of activity).

3.4. Preparing the Template for Enrichment by Reverse Transcription

1. Mix together 5 pmoles of the distal reverse strand primer for each section of the target gene and ≤1 μg RNA and water to 13 μl final volume.
2. Heat to 95°C for 1 min, 75°C for 1 min, 65°C for 1 min, and then to 55°C.
3. Then add the following mix:
   2 μl dNTPs (10 mM)
   4 μl 5× buffer (provided with the kit)
   0.5 μl RNAsin (Ambion)
4. Then add 0.5 μl (10 U) "Transcriptor" reverse transcriptase (*see* **Note 2**).
5. Mix well and leave at 55°C for 30 min. The sample can then be stored at –20°C if desired.

3.5. Nested PCR

Use 5 μl of the RT reaction (described in **Subheading 3.4.**) as template in a 50-μl manual hot-start PCR reaction as follows:

1. Heat PCR-containing template, buffer, magnesium, and primers to 95°C for 1 min.
2. Add 1 μl "Expand" polymerase (*see* **Note 3**).

3. Perform a PCR for 25 cycles as follows: 94°C 30 s, annealing temperature 30 sec, 72°C 60 s.
4. For the second (nested) reaction, use 0.1 μl of the first reaction as template with the same conditions as before but with 40 cycles and a final extension time of 3 min (PCRs can be frozen overnight if necessary at this point).

3.6. Removal of Stray Single-Stranded cDNA Ends with Exonuclease VII

1. Treat PCR (described in **Subheading 3.5.**) by adding 100 μl of 1× L buffer containing 1 μl Exonuclease VII and incubate for 30 min at 37°C.
2. Completed exonuclease reactions can be stored at –20°C if desired.

3.7. Enrichment of Alternatively Spliced Isoforms

All steps of the following enrichment procedure are performed at room temperature.

1. Pipette 50 μl of charged ssb beads (equivalent to 4 μl of the original uncharged slurry) into a microfuge tube for each separation.
2. Magnetically remove liquid and pipette 100 μl (*see* **Note 4**) of exonuclease-treated PCR into the tube until beads are resuspended. Leave on the bench for 5 min.
3. Wash beads three times in EASI buffer (*see* **Note 5**).
4. Resuspend in 10 μl of elution buffer and leave on the bench for 5 min.
5. Separate and remove liquid to another tube and place in magnetic separator (to ensure that every last bead particle is separated).

3.8. Amplifying the Enriched Material (see Note 6)

1. Dilute 1 μl of the exonucleased PCR (from **Subheading 3.6.**) 100-fold for "before" samples (see *Note 7*) and use 1 μl of the enriched eluate (from **Subheading 3.7.**) for an "after" sample.
2. Use 1 μl of "before" and "after" samples in a further 25-cycle PCR with a 3-min final extension.
3. Allow samples to cool and run 12 μl on a 1.25% agarose TBE gel.
4. Visualize with ethidium bromide and UV.

3.9. Sequencing of the New Splice Forms

There are two different ways of getting the sequences of PCR products that represent spliced isoforms. Which one to use depends on the electrophoresis results. If there is a single strong extra band use **Protocol 3.9.1**. If there are some weak novel bands, to discover all the alternative isoforms of the gene, use **Protocol 3.9.2**.

3.9.1. Gel Extraction Sequencing

1. Excise your band of interest and gel purify.
2. Amplify 1 μl of the purified product with the same primers used to generate it with a 25 cycle PCR.
3. Cleanup PCR (to remove buffers and primers).
4. Sequencing: Use 75 ng of cleaned PCR product and 4 pmoles of either of the primers in a standard 20-μl sequencing reaction.

3.9.2. Colony Sequencing

1. Create TA overhangs in the remaining 80 μl of amplified enriched material by adding 0.25 mM dATP and 1 μl of a basic non-proofreading *T.Aq* DNA polymerase (e.g. Promega, Madison, Wisconsin, USA). Incubate at 72°C for 15 min.
2. Clean-up the PCR (to remove buffers and primers).
3. To 4 μl of the cleaned PCR, add 1 μl 5× dilute salt solution (provided with the kit) and 1 μl of PCR 4 TOPO vector mix (Invitrogen, Carlsbad, California, USA). Leave on the bench for 15 min.
4. Transform 1 μl into electro-competent *E. coli* and plate on LB-amp plates.
5. The next day—use clean tips to resuspend a fraction of colonies in a number (*see* **Note 8**) of 100 μl PCRs with an inexpensive source of *T.Aq* DNA polymerase (e.g. Promega) and M13 primers. PCR conditions are as follows: 95°C 3 min (to break open the *E. coli*) followed by, (94°C 30 s, 59°C 30 s, 72°C 60 s) 35 cycles, 72°C 3 min.
6. Centrifuge PCR products briefly to remove cell debris.
7. Run PCR products on an agarose gel.
8. Decide which PCR products to pursue and clean-up the desired (*see* **Note 8**) PCR reactions.
9. Sequence as in **Subheading 3.9.1** but with the individual M13 primers.

4. Notes

1. Dynal-TALON is a suspension of brown magnetic nickel-agarose beads that are uniformly 1 μm in diameter. This small size compared with most other affinity resins allows the beads to stay in solution for hours without settling to the bottom of the tube. The beads are effectively in solution and do not need mechanical mixing. Magnetic separation makes them much more convenient than centrifugal separation beads.
2. It is very important to use a processive reverse transcriptase to avoid false alternative transcripts because of uncompleted reverse-transcripts reannealing at short complementary sequences *(6)*. For this reason "Superscript II" is definitely not suitable (data not shown). However, "Thermoscript" (Invitrogen) is presumably an alternative to "Transcriptor" *(6)*.
3. Processive, proofreading, and accurate thermostable DNA polymerases are preferred before enrichment as truncated products can cause problems. Hot-start is important

for specificity; a manual hot-start with "Expand High-fidelity" polymerase has been used effectively (even on very difficult templates) for the nested PCR steps and the self-activating "HotStarTaq" was used for the post-enrichment step for convenience.

4. Exonuclease VII is expensive. The protocol presented here, which was found to be consistently effective, uses a whole microlitre of enzyme for every enrichment. If high-throughput enrichment is being attempted then it would be desirable to reduce the amount of enzyme by treating and enriching a sub-fraction of the PCR.

5. Use 500, 600, and 700 μl for the respective washes to aid the memory in how many washes remain to be done. To save time—if six enrichments, for example, are being performed together in a magnetic rack then each sample can have liquid removed and then be resuspended before moving on to the next sample in a rotation system. Removing washings should be performed with a fresh 1-ml tip each time and subsequent resuspension should be performed with a fresh 1-ml filter tip.

6. The eluate typically contains <100 ng enriched material. PCR amplifying this serves both to increase its quantity and to separate the fully heteroduplexed splice forms.

7. After enrichment, the best form of comparison to see whether the protocol has worked is to amplify the enriched material and to compare it with an amplified dilution of the material that was applied to the column. This is better than using the applied material itself, as a reference, because some alternative splices, if they are smaller than the main isoform, can be enriched to a certain extent because of a minor advantage in PCR replication efficiency of the small form, magnified over-repeated PCRs. This new reference also avoids any problems of the respective "ages" of the samples including the build up of four-stranded double heteroduplex structures (*see* **Fig. 1 A**).

8. Screening 6–24 colonies is recommended depending on the strength of the bands being targeted and the exhaustiveness required. Just over half of the reactions will form a line of full-sized products across the gel. All the other bands are the desired alternative spliced isoforms. Sequence two bands of each size.

References

1. Venables, J. P., and Burn, J. (2006) EASI - enrichment of alternatively spliced isoforms. *Nucleic Acids Res* **34**, e103.
2. Schweighoffer, F., Ait-Ikhlef, A., Resink, A. L., Brinkman, B., Melle-Milovanovic, D., Laurent-Puig, P., Kearsey, J., and Bracco, L. (2000) Qualitative gene profiling: a novel tool in genomics and pharmacogenomics that deciphers messenger RNA isoform diversity. *Pharmacogenomics* **1**, 187–97.
3. Watahiki, A., Waki, K., Hayatsu, N., Shiraki, T., Kondo, S., Nakamura, M., Sasaki, D., Arakawa, T., Kawai, J., Harbers, M., Hayashizaki, Y., and Carninci, P. (2004) Libraries enriched for alternatively spliced exons reveal splicing patterns in melanocytes and melanomas. *Nat Methods* **1**, 233–9.

4. Thill, G., Castelli, V., Pallud, S., Salanoubat, M., Wincker, P., de la Grange, P., Auboeuf, D., Schachter, V., and Weissenbach, J. (2006) ASEtrap: a biological method for speeding up the exploration of spliceosomes. *Genome Res* **16**, 776–86.
5. Venables, J. P., Bourgeois, C. F., Dalgliesh, C., Kister, L., Stevenin, J., and Elliott, D. J. (2005) Up-regulation of the ubiquitous alternative splicing factor Tra2beta causes inclusion of a germ cell-specific exon. *Hum Mol Genet* **14**, 2289–303.
6. Cocquet, J., Chong, A., Zhang, G., and Veitia, R. A. (2006) Reverse transcriptase template switching and false alternative transcripts. *Genomics* **88**, 127–31.
7. Eckhart, L., Ban, J., Ballaun, C., Weninger, W., and Tschachler, E. (1999) Reverse transcription-polymerase chain reaction products of alternatively spliced mRNAs form DNA heteroduplexes and heteroduplex complexes. *J Biol Chem* **274**, 2613–5.

12

In Vivo Methods to Assess Polyadenylation Efficiency

Lisa K. Hague, Tyra Hall-Pogar, and Carol S. Lutz

Summary

Mammalian gene expression can be regulated through various post-transcriptional events, including altered mRNA stability, translational control, and RNA-processing events such as 3′-end formation or polyadenylation (pA). It has become clear in recent years that pA is governed by several core sequence elements and often regulated by additional auxiliary sequence elements. These regulatory events are frequently not reproducible in in vitro assays. Therefore, in vivo methods to measure mRNA pA were developed to meet this need and are described here.

Key Words: Polyadenylation; 3′ UTR; *cis*-acting core and auxiliary elements; RNase protection assays; transfection; luciferase assays.

1. Introduction

Almost all eukaryotic messenger RNAs (mRNAs) have an unencoded poly(A) tail at their 3′ ends. This tail influences translation, stability, and nucleo-cytoplasmic transport of the mRNA *(1–3)*. The process of acquisition of this poly(A) tail is called polyadenylation (pA) and is a two-step process that first involves endonucleolytic cleavage of the RNA at a site determined by recognition and binding of the core pA factors (*trans*-acting factors; reviewed in *4* and *5* and references therein) to specific *cis*-acting sequence elements. The second step involves polymerization of the poly(A) tail to a length of approximately 200–300 adenosine (A) nucleotides (nts) in mammalian cells. These two steps are tightly coupled; reaction intermediates are not normally detectable.

From: *Methods in Molecular Biology, Vol. 419: Post-Transcriptional Gene Regulation*
Edited by: J. Wilusz © Humana Press, Totowa, NJ

The entire pA signal is comprised of a growing number of defined sequence elements. A schematic of these elements is shown in **Fig. 1**. These sequence elements can be divided into two groups: the core elements and the auxiliary elements. The core upstream element is the hexamer AAUAAA or a close variant which is found 10–35 nts upstream of the cleavage site (*see* **Fig. 1**). The core downstream element is found 10–40 nts downstream of the cleavage site; although it does not have a clear consensus, it is usually characterized as U- or GU-rich (*see* **Fig. 1**). In addition to the core elements, auxiliary elements both upstream and downstream have been identified and characterized in viral and cellular systems (*see* **Fig. 1**, **ref. 6** and references therein). The existence and activity of such auxiliary elements suggests that pA can be regulated by elements in addition to the two core elements. Indeed, it has recently been noted that a reasonably large number of mammalian pA signals do not have one or both core elements, also reinforcing the importance of auxiliary elements *(6,7)*.

Many aspects of mammalian pA have been revealed using in vitro assays that rely upon nuclear salt wash extracts. These experiments have been key to our understanding of the mechanistic aspects of pA. However, the regulatory aspects of pA have not always been clear in these in vitro systems. We have developed the assays presented here to examine the requirements and regulatory functions for such *cis*-acting core and auxiliary pA elements in an in vivo setting.

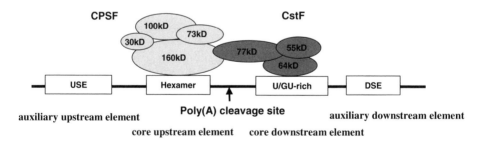

Fig. 1. Schematic of *cis* elements in a "canonical" mammalian polyadenylation signal. The cleavage site is marked with an arrow. Auxiliary upstream elements (USEs) and auxiliary downstream elements (DSEs) denote potential *cis*-auxiliary elements. Also shown are cleavage and polyadenylation specificity factor (CPSF) and cleavage stimulation factor (CstF), part of the basal polyadenylation machinery. Not shown here for simplicity: CFIm and CFIIm, Fip1, Poly (A) polymerase (PAP).

2. Materials

2.1. Plasmids

1. pCβS vector with BGH alone: *see* **Subheading 3.1** for detailed description of this vector.
2. pCβS with pA signal site of interest: **Subheading 3.1** explains how to obtain the pA signal of interest and how to clone it into the pCβS vector.
3. pUbc-luc: Modified pGL33 vector (Promega, Madison, WI), *see* **Subheading 3.7**.
4. phRL vector (Promega): *Renilla* luciferase vector.

2.2. Tissue Culture

1. DMEM serum free: Dulbecco's Modified Eagle's Medium (DMEM) with 4500 mg/ml glucose/L, L-glutamine, $NaHCO_3$, pyridoxine HCl (Sigma Aldrich, St. Louis, MO).
2. DMEM complete: DMEM with 4500 mg/ml glucose/L, L-glutamine, $NaHCO_3$, pyridoxine HC (Sigma Aldrich) with 10% fetal bovine serum (FBS, Sigma Aldrich) and 4 mM additional L-glutamine (Sigma Aldrich).
3. 10× phosphate-buffered saline (PBS), pH 7.2 (GIBCO™; Invitrogen Corp., Grand Island, NY): Dilute to 1× by adding 50–450 ml of sterile autoclaved dH_2O.
4. 1× PBS: Addition of 100 ml of 10× PBS described above to 900 ml of sterile dH_2O.
5. Trypsin-ethylenediaminetetraacetic acid (EDTA) (1×) (GIBCO™; Invitrogen Corp.).
6. TransIT-LT-1 transfection reagent (Mirus Bio Corp., Madison, WI).

2.3. Total RNA Isolation

1. 1× PBS
2. TRIZOL reagent (Invitrogen, Carlsbad, CA)
3. Chloroform
4. Isopropanol
5. 80% ethanol: To make 500 ml, dilute 400 ml of 200 proof absolute, anhydrous ethyl alcohol in 100 ml of sterile RNase-free double-distilled H_2O.
6. Sterile RNase-free double-distilled H_2O: Place double-distilled water in RNase-free containers that have been used for nothing but RNase-free double-distilled water. Autoclave the bottles filled with double-distilled water to sterilize them.

2.4. In vitro Transcription of a-³²P-UTP-Labeled RNA

1. pCβS vector with pAs of interest (*see* **Subheading 3.1.**).
2. T7 RNA polymerase (Promega).
3. RNasin: RNase Inhibitor (Promega).
4. Cap Analog: $(m^7G(5')ppp(5')G)$ is resuspended in sterile RNase-free double-distilled water to make a 5-mM stock solution. Store at $-20°C$ (GE Healthcare, Piscataway, NJ).

5. 10× rntps: 5 mM cytosine triphosphate (CTP), 5 mM adenosine triphosphate (ATP), 0.5 mM guanosine triphosphate (GTP), 0.5 mM uridine triphosphate (UTP) (final concentration). The concentration of GTP is lower than that of CTP and ATP because the cap analog is added and will be incorporated into the transcript with higher efficiency if the concentration of GTP is reduced. The concentration of UTP is reduced because of the addition of α-^{32}P-UTP to the transcription reaction.
6. α-^{32}P-UTP (Perkin Elmer, Boston, MA).
7. Phenol : chloroform : isoamyl alcohol (25:24:1): Combine 24 parts chloroform to one part isoamyl alcohol. This 24:1 mixture is then added to an equal amount of saturated phenol. This solution must be stored in a dark bottle at 4°C.
8. High-salt (pH ~7.4) column buffer (HSCB): 0.4 M NaCl, 0.1% SDS, and 50 mM Tris pH 8.0; bring to volume with sterile RNase-free double-distilled water.
9. 100% EtOH : 200 proof absolute, anhydrous ethyl alcohol.
10. RNA-loading buffer: 8 M Urea, 1× Tris–borate–EDTA (TBE), final concentration, 0.01% xylene cyanol and bromophenol blue.

2.5. Urea/Polyacrylamide Gel Electrophoresis

1. Urea: UltraPure™ Urea (Invitrogen).
2. 10× TBE: 890 mM Tris base, 890 mM granular boric acid, and 20 mM EDTA resuspended in dH$_2$O.
3. 1× TBE-running buffer: Add 100 ml of 10× TBE to 900 ml of dH$_2$O.
4. Acrylamide, 99.9% electrophoresis purity reagent (Bio-Rad Laboratories, Hercules, CA). (*see* **Note 1**)
5. Bis *N,N*′-methylene-bis-acrylamide electrophoresis purity reagent (Bio-Rad Laboratories).
6. TEMED: *N,N,N*′,*N*′-tetra-methyl-ethylenediamine (Bio-Rad Laboratories).
7. Ammonium persulfate (AP): A 10% solution is prepared by addition of 1 g of AP to 10 ml of dH$_2$O (Bio-Rad Laboratories) (*see* **Note 2**).

2.6. RNase Protection

There are other components in the RPA III™ Kit from Ambion including gel-loading buffer and controls. We routinely only use the components listed below. However, use of the Ambion RPA III™ Kit can be done exactly as the manufacturer's protocol as described if desired.

1. RPA III™ Kit (Ambion, Austin, TX) includes

  a. Hybridization III buffer.
  b. RNase digestion III buffer.
  c. RNase A/T1 mix.
  d. RNase inactivation/precipitation III solution.

2. Radiolabeled antisense probe

3. RNase-free water
4. Ethanol
5. 80% EtOH made with RNase-free water
6. RNA-loading buffer (see above)
7. DNA Size Marker (to assess approximate size of RNA fragments): pBR322 cut with MspI (New England BioLabs, Ipswitch, MA) that has been ^{32}P-end labeled by T4 polynucleotide kinase and [γ-^{32}P]ATP.
8. Total RNA prepared from (1) HeLa cells alone; (2) HeLa cells transfected with pCβS vector containing bovine growth hormone (BGH) pA alone; and (3) HeLa cells transfected with pCβS vector with pA signal of interest.

2.7. Dual Luciferase Assay

1. Promega Dual-Luciferase® Reporter Assay System: This kit includes the following reagents: Luciferase assay buffer II, luciferase assay substrate, Stop & Glow® Buffer, Stop & Glow® Substrate, and passive lysis buffer.
2. Extracts prepared from (1) HeLa cells alone; (2) HeLa cells transfected from Firefly and *Renilla* luciferase constructs.
3. Deionized H$_2$O.
4. Black 96-well luminescence plate (USA Scientific, Ocala, FL).

2.8. Detection and Analysis Equipment and Software

1. PhosphorImager Screen (Molecular Dynamics, GE Healthcare)
2. Typhoon™ PhosphorImager (GE Healthcare)
3. ImageQuant™ Software (GE Healthcare)
4. Luminescence plate reader: Synergy™ HT Multi-detection Microplate Reader with Bio-Tek® Synergy™ HT KC4 version 3.4 Software package (Bio-Tek® Instruments Inc., Winooski, VT)

3. Methods

Owing to the various complexities of the cellular environment, it is difficult to study pA site use in vivo. The in vivo tandem pA assay overcomes some of the difficulties of studying RNA in a cellular context. This method involves placing two separate pA signals in tandem in the pCβS vector (a gift of David Fritz, *see* **Fig. 2**). The vector is then transfected into a mammalian cell line of choice. The cellular machinery transcribes this vector, and the resulting mRNAs are subsequently processed. The processed mRNA products can be then analyzed and quantified using an RNase protection assay. The high sensitivity of the RNase protection assay allows for excellent analysis of the products produced by the cellular RNA processing mechanisms.

The pCβS vector has sites to insert one or more pA signals. However, the second site often contains a pA signal that serves as an internal control, such

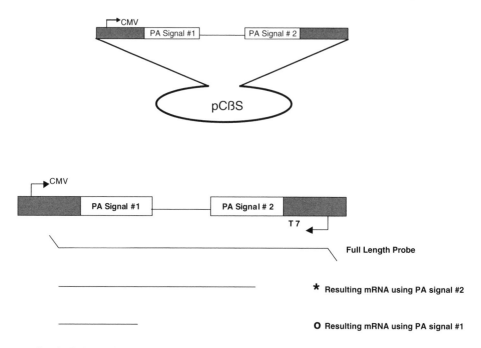

Fig. 2. Schematic of tandem in vivo polyadenylation vector and resulting mRNAs produced. Schematic of our in vivo polyadenylation reporter vector (top) and resulting mRNAs from RNase protection experiments (bottom) (*see* also **refs.** *8* and *9*).

as BGH (*see* **Fig. 2**). The BGH pA signal has been used and characterized extensively. The utilization of the inserted pA signal can therefore be analyzed and compared with the utilization of the BGH pA signal.

An α-^{32}P-radiolabeled antisense probe transcribed from the pCβS vector with an inserted pA signal of interest will hybridize to both the pA signal of interest and the BGH pA signal (*see* **Fig. 2**). In addition, the amount of probe that hybridizes to each of the processed RNAs is proportional to the efficiency with which the cell utilizes the pA signal. Therefore, this assay can aid in determining the strength of a pA signal and begin to elucidate mechanisms behind pA signal choice.

Many of the complexities of performing in vivo assays still apply to this method, so it is vital that this assay be accompanied by many controls. An RNase protection using RNA prepared from cells transfected with the pCβS vector containing only the BGH pA signal must be included. This control will demonstrate the level of BGH that will be produced if the cellular machinery only has one pA signal to use and process, and can be used as a baseline. In addition, RNase protection using RNA from non-transfected cells is also necessary to account for background endogenous expression levels of the pA signal of interest.

3.1. Isolation and Cloning of the pA Signal of Interest Into the pCβS Vector

The pCβS vector contains intron 1 of the rabbit β-globin gene in conjunction with the splice donor and acceptor sites *(8, 9)*. These intronic sequences and splicing sequences are necessary to ensure proper processing of the transcript. The pCβS vector has a cytomegalovirus (CMV) promoter and the entire multiple cloning site (MCS) from the pBluescript SK plasmid. pA signals are typically inserted into BamHI and PstI sites contained in the MCS. However, the other restriction sites can also be used for insertion of the pA signal of interest, if needed. Approximately 200 bp downstream of the MCS is the BGH pA signal. Located on the antisense strand downstream of the BGH pA signal is a T7 promotor that can drive synthesis of an antisense probe (*see* **Subheading 3.3.**).

1. To obtain a pA signal of interest, forward and reverse primers (~20–25 nts) are generated and should flank sequences ±150-bp upstream and downstream of the pA signal core (hexamer) element, respectively. Amplification of approximately 300 bp surrounding the core upstream element ensures that both core and auxiliary pA sequences are included. Be sure that both the core hexamer and the core DSE are included the predicted PCR fragment.
2. So that this sequence can be inserted into the BamHI and PstI sites of the pCβS vector, place a BamHI site at the 5′ end of the forward primer and a PstI site at the 5′ end of the reverse primer. If necessary or desired, other restriction sites present in the MCS can be used.
3. A standard PCR using HeLa genomic DNA and primers generated in steps 1 and 2 will yield an approximately 300-bp fragment with the pA signal of interest and will contain BamHI and PstI restriction sites.
4. This PCR product can be cloned into the pCβS vector using the BamHI and PstI sites located in the vector upstream of the BGH pA signal using standard methods.

3.2. Transfection of Cells

For simplicity, we describe this protocol by using HeLa cells and a Mirus transfection reagent. However, our lab has successfully transfected both the pCβS and the pUbs-luc constructs in various other cell lines and with other transfection reagents. Ability to assess the same pA signal in different cell lines can show cell type-specific changes in pA signal choice (*see* **Note 3**).

1. Plate cells in 60-mm dishes to 60–80% confluency after a 16-h incubation at 37°C in a 5% CO_2 atmosphere. For example, HeLa cells are plated at 8×10^5 cells/plate.
2. The pCβS vector is introduced into cells by transient transfection using the Mirus TransIT-LT1 transfection reagent and is performed according to the manufacturer's instructions. The transfection reagent is diluted by addition of 180 μl of serum-free DMEM to autoclaved sterile 1.5-ml microcentrifuge tubes followed by addition of 4–6 μl of LT-1 transfection reagent. The mixture is then mixed by pipetting up and down and gentle tapping of the tube.
3. The serum-free DMEM : LT-1 mixture is then incubated at room temperature for 15–20 min.
4. Plasmid DNA is added to the serum-free DMEM : LT-1 mixture and mixed by gentle pipetting. Optimizing the amount of DNA added to the transfection must be done for each construct. We typically use 2–3 μg of the pCβS construct per 60-mm dish of cells for this assay.
5. Incubate DMEM : LT-1 : plasmid mixture at room temperature for 15–20 min.
6. Remove media from plates, and add 1 ml of fresh complete media to cells.
7. Add transfection complexes to cells dropwise and place at 37°C in a 5% CO_2 atmosphere.
8. Incubate cells for 24 h in a 37°C and 5% CO_2 incubator and then proceed to harvesting total RNA.

3.3. Harvesting of Total RNA

Other methods of harvesting RNA are also able to be used in this assay. In addition, the RNA can be prepared and then selected for poly(A)+ RNA or can be prepared to assay cytoplasmic and/or nuclear pools of RNA.

1. After removal from the incubator, cells are examined to check for confluency and possible cytotoxicity from the LT-1.
2. After it has been established that the cells are healthy and have grown to the correct confluency (60–80%), the media is removed and the cells are then washed with 1× PBS.
3. TRIZOL reagent (2 ml) is added to each 60-mm dish and left on the cells for 5 min at room temperature (*see* **Note 4**). This step is used to remove adherent cells from the dish, to lyse the cells, and also to protect the RNA from degradation.
4. The TRIZOL : cell mixture is evenly resuspended and split into two 1.5-ml micro-centrifuge tubes (1 ml of Trizol : RNA suspension in each).
5. The RNA is separated from DNA and proteins by a phenol : chloroform extraction. First, chloroform (200 μl) is added to the tubes, which are then tightly capped and then vigorously shaken. The thoroughly mixed solution is then incubated at room temperature for 10 min followed by centrifugation ($\sim$15,700 × g) in a microcen-trifuge for 15 min.
6. After centrifugation, the resulting mixture is separated into three phases: the pink organic phase that contains the DNA, the white interphase that contains the proteins,

and the clear aqueous phase that contains the RNA (*see* **Note 5**). This clear upper phase is removed and placed into a new 1.5-ml microcentrifuge tube, and 500 µl of isopropanol is used to precipitate the RNA from this aqueous solution.

7. Incubate tubes on dry ice for 15 min and then centrifuge at ∼15,700 × g in a microcentrifuge for 10 min.
8. Aspirate the isopropanol, and the RNA should be in a cloudy white pellet at the bottom of the tube. Wash this pellet in 80% sterile RNase-free ethanol. Remove 80% ethanol and allow pellet to dry for about 5 min. Be sure to not overdry pellet. The RNA-precipitated pellet will be very hard to dissolve if it is overdried. Each sample is resuspended in 15 µl of sterile RNase-free double distilled H_2O.

3.4. Preparation of Radiolabeled Antisense Probe

There are other methods, besides the one described here, to prepare a radio-labeled probe (*see* **Note 6**). Any method used is fine as long as the resulting probe is uniformly labeled and has been thoroughly purified.

1. Digest the pCβS vector with the restriction enzyme BamHI to linearize the vector. The BamHI restriction site is the 5′ site where the pA signal of interest was cloned. Digesting the vector at this site will ensure that the transcription of the antisense probe will terminate directly after the insert containing the pA signal of interest has been transcribed.
2. Generate an α-^{32}P-radiolabeled transcript by mixing 500 ng of linearized DNA plasmid, 2 µl of 5× transcription buffer, 1 µl of 5-mM cap analog, and 1 µl of 10× rNTPs in a 1.5-ml sterile microcentrifuge tube.
3. Add the RNA polymerase (1 µl), and an RNase inhibitor is subsequently added to the reaction mixture [0.5 µl (5–10 units) of RNasin].
4. To this transcription reaction, 4.5 µl of α-^{32}P UTP (∼45 µCi) is added, and the reaction is incubated at 37°C for 1 h. Be sure to cover the tubes with Plexiglas or block off area with a shield.
5. To remove unincorporated rNTPs and remove added proteins and enzymes, perform phenol : chloroform : isoamyl alcohol extraction and ethanol precipitation.
6. The dry RNA pellet is resuspended in 6 µl of RNA-loading buffer and is loaded on a 5% polyacrylamide-8 M urea gel (*see* **Notes 7** and **8**), which is subjected to electrophoresis at 500 V, 60 mA, and 15 W for about 1–2 h.
7. The gel is then subjected to autoradiography in the darkroom for 30 s to 1 min to visualize where the radioactive bands are located on the gel.
8. The gel and the X-ray film are aligned and the bands corresponding to the full-length product of the transcription reaction are excised with a razor blade. Each excised gel slice is placed in a 1.5-ml microcentrifuge tube containing 400 µl of HSCB. To elute the probe into solution, the gel slice is incubated in the HSCB overnight while it remains behind a shield and in a Plexiglas container.
9. To purify the probe after overnight elution, it is subjected to phenol: chloroform: isoamyl alcohol extraction and ethanol precipitation.

10. The probe pellet is resuspended in enough RNase-free sterile double-distilled water to make sure the probe is in excess in the RNase protection assay. In many cases, 20,000–50,000 cpm per reaction is sufficient. However, using up to 100,000 cpm or more may be necessary in some cases.

3.5. RNase Protection (see Note 9)

1. Hybridization of the RNA sample and the probe is accomplished by adding 5–10 µg of total RNA and approximately 50,000 cpm of probe to a 1.5-ml sterile microcentrifuge tube. Then, 10 µl of the hybridization buffer is added. This mixture is placed in a 90°C heat block to denature the RNA and the probe. The tubes are placed in a 50°C water bath overnight to allow for hybridization of the probe to the RNA, yielding a probe : RNA duplex.
2. Any single-stranded RNA that has not hybridized to the antisense probe can be digested using RNase A/T1 mixture diluted to 1:300 in RNase Digestion III Buffer. One hundred and fifty microlitre of aliquots of diluted RNase T1/A solution is added to each sample and then placed at 30°C for 15 min (*see* **Notes 10 and 11**).
3. To stop the degradation activity of the RNases, 225 µl of inactivation solution is added to the samples. As an optional step, addition of extra ethanol at this step can be added to increase recovery of small protected fragments.
4. Protected fragments are recovered by centrifugation. The resulting pellet is rinsed with 80% ethanol, dried, and resuspended in 6 µl of RNA-loading buffer.
5. The protected fragments are separated by electrophoresis on a 8-M urea/5% polyacrylamide gel.

See **Fig. 3A** for representative RNase protection data.

3.6. Analysis and Quantification of Data

1. The RNase protection gel is placed on a used X-ray film or Whatman 3-mm paper and covered in plastic wrap.
2. To visualize bands, place the gel on a PhosphorImager Screen overnight. Then, scan the screen using an imager capable of detection of radioactivity from a storage phosphor screen. In this case the Typhoon Imager was used.
3. The PhosphorImager data can be analyzed using computer software capable of densitometry measurements. The bands of interest are analyzed by using the software to calculate the volume of each radioactive band. For example, in **Fig. 3**, the bands indicated by the stars and the circles are selected for analysis. These bands are selected because of two factors. The first factor is the size of the radiolabeled probe is known and the size of the fragments that would be produced when the probe is hybridized to either BGH or the pA signal of interest is known as well. Second, the radiolabeled RNA marker aids in determination of the size of the protected fragments.

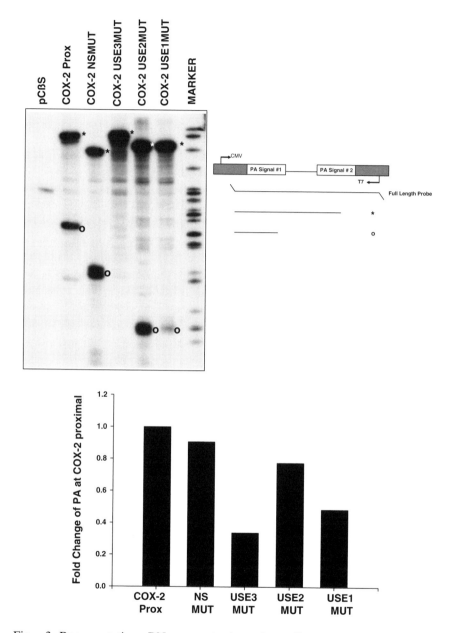

Fig. 3. Representative RNase protection data. Sequences surrounding the cyclooxygenase-2 (COX-2) proximal polyadenylation signal was cloned into our pCßS vector in polyadenylation signal site 1 in tandem with bovine growth hormone (BGH) in polyadenylation signal site 2 (**ref.** *9*; *see* also **Fig. 2**). Various mutant versions of the

4. Once densitometric volume analysis is performed on all the selected fragments, each sample is displayed in a bar graph as follows. The BGH signal is used as an internal control. Therefore, the value designated for the BGH in a given sample is used to normalize the experimental pA signal. The *y*-axis is expressed as the ratio of value of the pA signal of interest/value of BGH signal. *See* **Fig. 3B** for quantification of the representative data.

3.7. In vivo Luciferase Assay to Assess Effect of pA on Luciferase Protein Production

The strength and composition of a pA signal is able to affect 3´-end processing as demonstrated in a luciferase assay. Regulating formation of mature mRNA ultimatley influences the amount of protein that is expressed from the transcript. Thus, post-transcriptional regulatory events can be observed as an effect on protein expression. A specific pA signal and any additional sequence elements within the 3´ UTR can be cloned into the pUbc-luc luciferase reporter plasmid and used in a standard dual luciferase assay to assess expression. Thus, the wild-type pA signal can, as in the RNase protection assays, be an internal control for normalization between experiments and constructs and can eliminate many of the issues that arise from in vivo experiments. Co-transfection of a *Renilla* luciferase vector in addition to the test pUbs-luc vectors provides a baseline with which to normalize the experimental data.

3.7.1. Generation of pUbc-Luc and the pUbc-Luc Vectors

1. The pGL3-Basic vector (Promega) was modified for the 3´ UTR luciferase assays (gift of Catherine Newnham). The ubiquitin promoter was obtained through PCR from human genomic DNA and inserted upstream of the luciferase open-reading frame (ORF) at the SacI and NheI sites, and this vector was called pUbc-luc. The luciferase expression is driven by a ubiquitin promoter to ensure that expression levels remained within detectable limits.

Fig. 3. *(Continued)* COX-2 polyadenylation signal were also independently cloned in this vector. The vector was then transfected into HeLa cells and after 24 h, total RNA was harvested and RNase protection assays were performed, and the resulting RNA fragments were separated on a 5% polyacrylamide/8M urea gel (Panel A). The radiolabeled protected fragments were visualized using autoradiography. Marker, ^{32}P-labeled pBR322 digested with MspI. Protected fragments marked with an asterisk (*) represent usage of the BGH polyadenylation signal; fragments marked with an open circle (o) represent usage of the COX-2 polyadenylation signal (see also diagram to the right). Panel B depicts the quantification of this particular experiment.

2. The SV40 pA signal was removed from the pUbc-luc construct by digesting the vector with the restriction enzymes XbaI and HpaI. The remaining parental vector was re-ligated to create pUbc-luc-no pA.
3. The pCβS vector with the pA signal of interest (a COX-2 pA signal is used in **Fig. 4**) is completely digested with BamHI and SalI and the resulting fragments are isolated.
4. The pUbc-luc-no pA vector was digested with BamHI and SalI and followed with ligation of the fragment in this vector to generate the pUbc-luc vectors that contain the pA signal of interest.

3.7.2. Transfection of the Luciferase Constructs

1. Cells were seeded 12 h before transfection in 6-well culture dishes at 0.3×10^6 cells/well.
2. The modified firefly luciferase pUbc-luc expression constructs and the *Renilla* vector (phRL, Promega) were transiently transfected by combining the appropriate

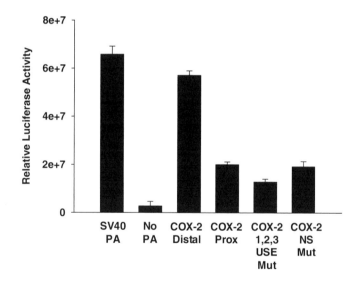

Fig. 4. Quantification of representative luciferase assays. The luciferase reporter was used as purchased with the SV40 polyadenylation signal (SV40 pA) or without (No pA) for transfection of HeLa cells and subsequent luciferase assays. The COX-2 distal or proximal polyadenylation signals, or various mutants of the COX-2 proximal polyadenylation signal, were cloned independently into the luciferase reporter vector, were transfected into HeLa cells, and subsequent luciferase assays were performed. Quantification of multiple independent experiments is shown here. Error bars represent ±SD.

amount of each DNA with 4 μl of the TransIT-LT1 (Mirus, Madison, WI) transfection reagent and incubating for 20 min at room temperature. Typically, 300 ng of pUbc-luc vectors and 60 ng of phRL vector are added yielding a 5:1 ratio, respectively. The DNA complexes were then diluted in 800 μl of DMEM (complete) and then added to the cells.

3. The reagents used in this assay were described above in **Subheading 2.2**. Twenty-four hours after transfection, the dual transfected cells wereharvested and the cell lysate was analyzed for luciferase activity by using the dual luciferase assay reagents (LARs) (Promega).

3.7.3. Analyzing Gene Expression as it is Influenced by Proper 3′-End Processing

1. Luciferase and *Renilla* activity within the same sample is determined using the Promega Luciferase Assay System (Promega). Transfected cells are washed with PBS and lysed with the provided manufacture's lysis buffer (Promega).
2. The firefly luciferase activity presents after expression of pUbc-luc vectors is measured immediately following the addition of LAR II (Promega) to the prepared lysates.
3. The addition of the Stop & Glo® Reagent (Promega) quenches the luciferase activity and allows for the measurement of *Renilla* luciferase activity, which corresponds to the expression of the phRL vector.
4. The luciferase activity is normalized to the internal control values, and the calculated values are presented as the mean ± SD. To confirm the findings, at least three independent transfection experiments should be performed (*see* **Note 12**). Representative data is shown in **Fig. 4**.

4. Notes

1. Use care when handling acrylamide, which is a neurotoxin.
2. AP is a bleach. Be careful when handling.
3. All tissue culture work is done under sterile conditions in a laminar flow hood.
4. TRIZOL contains phenol so use care when handling.
5. DNA and proteins can be subsequently separated from the phenol extraction step if desired.
6. Be sure to wear gloves, a lab coat, and work behind a Plexiglas shield when working with radioactivity.
7. Remove any bubbles in the bottom buffer chamber of the electrophoresis apparatus as this may cause uneven current flow through the gel leading to uneven migration of RNA fragments.
8. Urea can get stuck in the wells of the gel and affect fragment migration. Therefore, it is advisable to rinse wells with 1× TBE immediately before sample loading.
9. All parameters can be adjusted within an RNase protection experiment if these conditions are not yielding optimal results. This includes the dilution factor, the

incubation temperature, and time of incubation. In addition, it may even be appropriate to use RNase T1 alone, which is also included with the RPA III™ Kit.
10. RNA is vulnerable to digestion by various RNases during any of these procedures. Therefore, it is vital that all experimental steps involving RNA are prepared with RNase-free reagents and equipment. In addition, RNases are very abundant on the skin making clean gloves a necessity throughout the procedures.
11. When using RNase, use filter tip pipet tips in all steps containing the RNase. If RNases are retained in the pipet, all RNA will be degraded.
12. Multiple repetitions of each experiment must be performed to calculate standard deviations and to ensure confidence in the results.

Acknowledgments

The authors thank Songchun Liang, Sarah Darmon, Bin Tian, and Pat O'Connor for critical reading of the manuscript. C.S.L. was supported by NIH 5R03AR052038-03 and NSF award MCB-0462195.

References

1. Jacobson A, Peltz SW. 1996. Interrelationships of the pathways of mRNA decay and translation in eukaryotic cells. *Annu Rev Biochem* 65: 693–739.
2. Sachs AB, Sarnow P, Hentze MW. 1997. Starting at the beginning, middle and end: translation initiation in eukaryotes *Cell* 89: 831–838.
3. Wickens M, Anderson P, Jackson RJ. 1997. Life and death in the cytoplasm: messages from the 3′ end. *Curr Opin Genet Dev* 7: 220–232.
4. Edmonds M. 2002. A history of polyA sequences: from formation to factors to function. *Prog Nucleic Acid Res Mol Biol* 71: 285–389.
5. Zhao J, Hyman L, Moore C. 1999. Formation of mRNA 3′ ends in eukaryotes: mechanism, regulation and interrelationships with other steps in mRNA synthesis. *Microbiol Mol Biol Rev* 63: 405–445.
6. Hu J, Lutz CS, Wilusz J, Tian B. 2005. Bioinformatic identification of candidate *cis*-regulatory elements involved in human mRNA polyadenylation. *RNA* 11: 1485–1493.
7. Tian B, Hu J, Zhang H, Lutz C.S. 2005. A large-scale analysis of mRNA polyadenylation of human and mouse genes. *Nucleic Acids Res* 33: 201–212.
8. Natalizio BJ, Muniz LC, Arhin GK, Wilusz J, Lutz CS. 2002. Upstream elements present in the 3′ untranslated region of collagen genes influence the processing efficiency of overlapping polyadenylation signals. *J Biol Chem* 277: 42733–42740.
9. Hall-Pogar T, Zhang H, Tian B, and Lutz, CS. 2005. Alternative polyadenylation of cyclooxygenase-2. *Nucleic Acids Res* 33: 2565–2579.

13

Monitoring the Temporal and Spatial Distribution of RNA in Living Yeast Cells

Roy M. Long and Carl R. Urbinati

Summary

RNA localization is a cellular process to spatially restrict translation of specific proteins to defined regions within or between cells. Most localized mRNAs contain *cis*-acting localization elements in the 3′-untranslated region (UTR), which are sufficient for localization of an mRNA to a particular region of the cell. The *cis*-acting localization elements serve as assembly sites for *trans*-acting factors which function to sort the mRNA to the correct sub-cellular destination. Although fluorescent in situ hybridization (FISH) has been widely used to study mRNA localization, FISH has a weakness in that it is a static assay, as FISH requires that cells be fixed before hybridization. Consequently, FISH is not ideally suited for investigating dynamic mRNA localization processes. This limitation of FISH has been overcome by the development of techniques that allow the visualization of mRNA in living cells. Here, we present a protocol that tethers green fluorescent protein (GFP) to an mRNA of interest, allowing for the visualization of dynamic mRNA localization processes in living cells.

Key Words: 3′-UTR; fluorescence microscopy; GFP; MS2; RNA localization; yeast.

1. Introduction

Widely utilized by eukaryotic organisms, RNA localization is a mechanism to spatially restrict translation of specific proteins to distinct regions within and between cells. RNA localization substrates contain *cis*-acting elements in the 3′-UTR, which are sufficient for localization of the mRNA within the cell. While a number of distinct mechanisms have been identified, the most common mechanism for RNA localization appears to be directed transport. In this

From: *Methods in Molecular Biology, Vol. 419: Post-Transcriptional Gene Regulation*
Edited by: J. Wilusz © Humana Press, Totowa, NJ

pathway, the *cis*-acting localization element is identified by an RNA-binding protein, which interfaces with a molecular motor that functions to directly transport the mRNA to the site of localization. The yeast *S. cerevisiae* provides a comprehensive understanding of mRNA localization through directed transport. In yeast, *ASH1* mRNA is transported to the tip of the bud resulting in the asymmetric sorting of the transcriptional repressor Ash1p to daughter cell nuclei *(1)*. The RNA-binding protein, She2p, directly interacts with the *ASH1* mRNA *cis*-acting localization elements, and She2p physically associates with She3p *(1)*. In addition to interacting with She2p, She3p simultaneously associates with the myosin motor protein, Myo4p *(1)*. The formation of the Myo4p–She3p–She2p complex interfaces *ASH1* mRNA with the molecular motor which functions to directly transport *ASH1* mRNA on the polarized actin cytoskeleton to the tip of the bud *(1)*. In higher eukaryotic organisms, directed transport of mRNA localization substrates is not confined to the actin cytoskeleton, because microtubules and molecular motors specific for microtubules are also involved in mRNA localization *(2,3)*.

FISH has been an extremely valuable approach for screening mutants that deleteriously affect mRNA localization, and numerous FISH protocols are readily available *(4,5)*. However, fixation of cells for FISH prevents analysis of dynamic processes required for mRNA localization. To overcome this limitation of FISH, a number of different approaches have been developed to study mRNA localization in living cells. Fluorescently labeled RNA localization substrates have been microinjected into cells and the RNA localization pathway for these exogenous substrates monitored by fluorescence microscopy *(6,7)*. Alternatively, endogenous localized mRNAs can be monitored by in vivo hybridization with fluorescently labeled nucleic acid probes *(8,9)*. Neither of these approaches is compatible with analysis of mRNA localization in living yeast cells, because a protocol for introducing fluorescent nucleic acid probes into living yeast cells has not yet been developed. However, a method was developed to analyze *ASH1* mRNA localization in living yeast cells (*see* **Fig. 1A**). This assay has two components: green fluorescent protein (GFP) fused to the bacteria phage MS2 coat protein (MS2cp), which specifically binds with high affinity to a short RNA MS2 stem-loop structure and the target RNA containing the MS2 stem-loop structures *(10,11)*. When both components are simultaneously expressed in a yeast cell, GFP is recruited to the target RNA through the MS2cp–MS2 stem-loop interaction, allowing visualization of the target mRNA in living yeast cells by fluorescence microscopy (*see* **Fig. 1B**) *(10,11)*. Another version of this approach utilizes the RNA-binding protein U1A fused to GFP along with the target mRNA containing the binding site for U1A *(12)*. Following development of this approach for analysis of mRNA localization in living yeast,

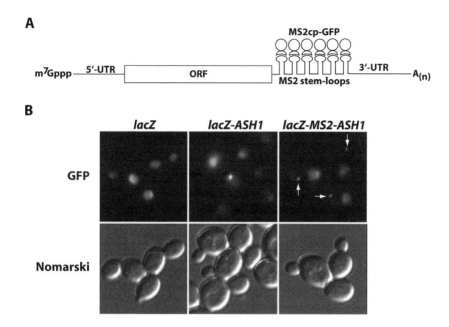

Fig. 1. Visualization of mRNA in living yeast cells by using green fluorescent protein (GFP)-MS2 tethered to an mRNA of interest. (**A**) Schematic representation of the GFP-MS2 system for visualizing an mRNA of interest in living yeast cells. When GFP-MS2 coat protein (MS2cp) is simultaneously expressed with an mRNA containing MS2 stem-loop structures in the 3′-UTR, GFP-MS2cp associates with the RNA, allowing visualization of the RNA in living yeast cells. (**B**) Visualization of localized mRNA in living yeast. Yeast were transformed with plasmids expressing either *lacZ* mRNA, *lacZ-ASH1* mRNA or *lacZ-MS2-ASH1* mRNA along with a plasmid expressing GFP-MS2cp. GFP particles containing localized mRNA are highlighted by the arrows. Since the GFP-MS2cp also contains a nuclear localization signal (NLS), unbound GFP-MS2cp is transported into the nucleus. Directing unbound GFP-MS2cp to the nucleus reduces background fluorescence in the cytoplasm.

this technology has been extended to investigate RNA degradation and nuclear transport in yeast as well as mRNA localization in higher eukaryotes *(13–17)*. Consequently, although the protocol described here is specific for analysis in yeast, with minimal modifications the approach is generally applicable to higher eukaryotic systems.

2. Materials

1. Plasmids pG14-MS2cp-GFP (*see* **Fig. 2A**), pSL-MS2$_6$ (**Fig. 2B**), and YEplac195 *(11,18)*.
2. Wild-type yeast strain W303-1a and/or mutant yeast strains (*see* **Note 1**).

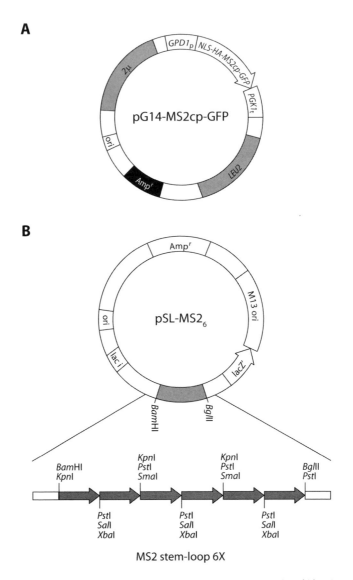

Fig. 2. Plasmid maps for pG14-MS2cp-GFP and pSL-MS2₆. (**A**) Plasmid pG14-MS2cp-GFP is a yeast multicopy plasmid containing the *LEU2* selectable marker and expressing MS2cp fused to GFP. The GFP-MS2cp fusion protein also contains a nuclear localization sequence (NLS) as well as the influenza virus hemagglutinin (HA) epitope. The NLS targets unbound cytoplasmic GFP-MS2cp to the nucleus, reducing cytoplasmic background fluorescence from GFP-MS2cp. (**B**) Plasmid pSL-MS2₆ is an *E. coli* vector containing the DNA encoding six copies of the MS2 RNA stem-loop

3. Liquid and solid YEPD media (*see* **Notes 2** and **3**).
4. Liquid and solid synthetic leucine and uracil drop-out medium (*see* **Notes 4** and **5**).
5. Sterile TE: 10 mM Tris–HCl pH 7.5, 1 mM EDTA.
6. Sterile lithium acetate/TE: 100 mM lithium acetate, 10 mM Tris–HCl pH 7.5, 1 mM EDTA.
7. 10 mg/ml denatured, sonicated sheared salmon sperm DNA (store at –20°C).
8. Sterile PEG/lithium acetate/TE: 40% polyethylene glycol (MW 3350), 100 mM lithium acetate, 10 mM Tris–HCl pH 7.5, 1 mM EDTA.
9. Sterile dimethyl sulphoxide (DMSO).
10. Glass slides containing agar slabs (*see* **Note 6**).
11. 1 mg/ml DAPI: 4´,6-diamidino-2-phenylindole dihydrochloride (store at –20°C and protect from light).
12. Epifluorescence microscope equipped with filter cubes for detection of GFP (excitation wavelength 488 nm/emission wavelength 500–530 nm) and DAPI for staining chromosomal DNA (excitation wavelength 360 nm/emission wavelength 435–485 nm), Nomarski optics, 60× PlanApo 1.4 N.A. objective as well as a CCD camera and imaging software.

3. Methods

3.1. Cloning MS2 Stem-Loop Sequences into RNA of Interest

1. A DNA fragment encoding six MS2 stem-loop structures is liberated from plasmid pSL-MS2$_6$ (*see* **Fig. 2B**) by simultaneous digestion with BamHI and BglII restriction endonucleases.
2. The gene corresponding to the mRNA of interest can be inserted into the multi-cloning site of plasmid YEplac195 which contains the yeast selectable marker *URA3*.
3. After the gene of interest is cloned into the YEplac195 vector, the BamHI/BglII DNA fragment encoding the six MS2 stem-loop structures can be inserted into the 3´-UTR of the gene of interest (*see* **Note 7**).

◀ ————————————————————

Fig. 2. *(Continued)* structure. The DNA for the six MS2 stem-loop structures can be liberated from pSL-MS2$_6$ by simultaneous digestion with the restriction enzymes BamHI and BglII. The BamHI site corresponds to the 5'end of the stem-loop structures, while BglII site corresponds to the 3' end of the stem-loop structures. If more than six MS2 stem-loop structures are required, the BamHI-BglII MS2$_6$ fragment can be inserted into pSL-MS2$_6$ digested with BglII, subsequently generating pSL-MS2$_{12}$. Twelve copies of the MS2 stem-loop structure can be liberated from pSL-MS2$_{12}$ by simultaneous digestion with BamHI and BglII.

3.2. Yeast Transformation

1. Inoculate 10 ml of YEPD with a single yeast colony and grow overnight at 30°C with shaking at 250 rpm.
2. Dilute the overnight culture into 50 ml of YEPD (OD_{600} of 0.5) and grow at 30°C with shaking at 250 rpm for an additional 2–4 h.
3. Harvest cells by centrifugation at 3500 × g for 5 min at room temperature.
4. Decant supernatant and resuspend cells in 40 ml TE.
5. Collect cells by centrifugation at 3500 × g for 5 min at room temperature.
6. Decant supernatant and resuspend cells in 2 ml lithium acetate/TE.
7. Incubate cells for 10 min at room temperature.
8. In a 1.5-ml microfuge tube, combine 100 µl of cells resuspended in lithium acetate/TE with 10 µl of 10 mg/ml denatured, sonicated sheared salmon sperm DNA and 1 µg of each plasmid (pG14-MS2cp-GFP and plasmid expressing mRNA-MS2 construct) being transformed. Vortex well to mix components.
9. Add 700 µl of PEG/lithium acetate/TE vortex vigorously and incubate at 30°C for 30 min.
10. Add 88 µl of DMSO, vortex to mix and incubate at 42°C for 7 min.
11. Collect cells by centrifugation at 20,000 × g for 30 s at room temperature.
12. Remove supernatant and resuspend cells in 1.0 ml TE.
13. Harvest cells by centrifugation at 20,000 × g for 30 s at room temperature.
14. Remove supernatant and resuspend cells in 200 µl of TE.
15. Spread cell suspension onto solid synthetic leucine and uracil drop-out medium and incubate plates at 30°C (*see* **Note 8**).

3.3. Propagation and Processing of Yeast Cells for Microscopic Analysis

1. Inoculate a 50-ml culture of liquid synthetic leucine and uracil drop-out medium with a single colony of yeast transformed with pG14-MS2cp-GFP and a plasmid expressing the mRNA of interest fused to the MS2 stem loops and grow the culture overnight at 30°C and 250 rpm.
2. When the culture reaches mid-log phase (OD_{600} 0.400), remove 5 ml of the culture to a sterile flask and add 50 µl of 1 mg/ml DAPI to the 5-ml culture (protect from light).
3. To stain chromosomal DNA, incubate the 5-ml culture at 30°C for 1 h at 250 rpm (protect from light).
4. Remove 1 ml of the culture to a 1.5-ml microfuge tube and collect the cells by centrifugation at 20,000 × g for 30 s at room temperature.
5. Remove the supernatant and resuspend cells in 10–20 µl of liquid synthetic leucine and uracil drop-out medium.

6. Spot 15 µl of the cell suspension on a glass slide and place a number 1 coverslip on the sample (*see* **Note 6**).

3.4. Image Acquisition

1. Place slide on microscope stage and locate focal plane containing the cells (*see* **Note 9**).
2. Capture image for GFP signal making certain not to exceed dynamic range of CCD camera (*see* **Notes 10** and **11**).
3. Capture image for DAPI signal.
4. Capture image for Nomarski optics.
5. Save raw data for each image in a TIFF format.

3.5. Scaling Images

1. Open raw data in Adobe Photoshop.
2. Locate the pixel with the highest fluorescent intensity input value (Image > Adjustments > Curves) and assign this input value an output value of 255.
3. Locate a pixel corresponding to the background fluorescent intensity input value and assign this input value an output value of 0 (*see* **Notes 12** and **13**).
4. After scaling the image, convert the Adobe Photoshop file to an 8-bit image (Image > Mode >8 bits/channel).
5. Save the scaled image as a new file in order not to permanently alter the raw data corresponding to the original image.
6. Scaled images can be used to generate figures for publication.

4. Notes

1. The genotype for W303-1a is *Mata, ura3, leu2-3,112, his3-11, trp1-1, ade2-1, ho, can1-100*.
2. Liquid YEPD medium is prepared by combining 900 ml of sterile 1.1× YEP medium with 100 ml of sterile 20% glucose (D). 1.1× YEP medium is made by adding 10 gm of Bacto-yeast extract and 20 gm of Bacto-peptone to 900 ml of distilled H_2O, and the mixture autoclaved for 15 min at 250°F and 15 lb/sq in of pressure.
3. Solid YEPD medium is prepared by combining 450 ml of sterile 4.5% Bacto-agar and 450 ml of sterile 2.2× YEP medium with 100 ml of sterile 20% glucose.
4. Solid synthetic leucine and uracil drop-out medium is prepared by combining 450 ml of sterile 4.5% Bacto-agar, 450 ml of sterile nutrients (composed of 20 mg adenine sulfate, 20 mg L-tryptophan, 20 mg L-histidine HCl, 20 mg L-arginine HCl, 20 mg L-methionine, 30 mg L-tyrosine, 30 mg L-isoleucine, 30 mg L-lysine HCl, 50 mg L-phenylalanine, 100 mg L-glutamic acid, 100 mg L-aspartic acid,

150 mg L-valine, 200 mg L-threonine, 400 mg L-serine, and 6.7 gm Bacto-yeast nitrogen base without amino acids) with 100 ml of sterile 20% glucose.

5. Liquid synthetic leucine and uracil drop-out medium is prepared by combining 900 ml of sterile nutrients (composed of 20 mg adenine sulfate, 20 mg L-tryptophan, 20 mg L-histidine HCl, 20 mg L-arginine HCl, 20 mg L-methionine, 30 mg L-tyrosine, 30 mg L-isoleucine, 30 mg L-lysine HCl, 50 mg L-phenylalanine, 100 mg L-glutamic acid, 100 mg L-aspartic acid, 150 mg L-valine, 200 mg L-threonine, 400 mg L-serine, and 6.7 gm Bacto-yeast nitrogen base without amino acids) with 100 ml 20% sterile glucose.

6. If time-lapse images are required, cells should be analyzed on 2% agarose slabs composed of synthetic leucine and uracil drop-out medium. Agarose slabs are prepared by pipeting a drop of molten medium on a glass microscope slide and immediately placing another slide on top of the medium. After the medium has hardened, the top slide is removed by gently sliding it off to the side. The agar surface is subsequently trimmed to the size of a number 1 coverslip, and a sample of the cell suspension to be analyzed is placed on the agar slab. A coverslip is placed on the sample and sealed with molten petroleum jelly.

7. The insertion of the MS2 stem-loop structures should not interfere with normal functions for the 3′-UTR.

8. For wild-type yeast strains, transformed colonies will be visible after 3 days of incubation at 30°C.

9. If the signal corresponding to the RNA of interest is not readily observed over the background fluorescence, the intensity of the RNA signal can be increased by increasing the expression level of the mRNA through the use of a galactose-inducible yeast promoter contained on a multicopy plasmid or by increasing the number of MS2 stem-loop structures in the 3′-UTR of the mRNA which will result in more GFP-MS2 molecules residing on the mRNA.

10. A 12-bit camera has a dynamic range of 4096 (0–4095) values. Consequently, for a given pixel, a fluorescence intensity value of 4095 indicates that the dynamic range of the camera has been exceeded. When the dynamic range of the camera is exceeded, the exposure time needs to be reduced such that the fluorescent intensity value for the brightest pixel is less than 4095. After determining which sample is the brightest, choose an exposure time which results in a fluorescent intensity value of approximately 3500. This exposure time is then used to collect all GFP images within a given experiment.

11. Movies corresponding to RNA movement can be generated by setting up the CCD camera and imaging software accordingly.

12. After scaling the image, the values between 0 and 255 must remain a linear function.

13. If the intensities of various GFP fluorescent signals are to be compared, the images must be captured with the same exposure time and must be scaled identically in Adobe Photoshop.

Acknowledgments

We thank Sharon Landers for comments on this manuscript. We apologize to our colleagues whose work we were unable to cite. Studies on *ASH1* mRNA localization in the Long laboratory were supported by R01 GM060392, the Pew Scholars Program in the Biomedical Sciences, and the March of Dimes Birth Defects Foundation.

Reference

1. Gonsalvez, G. B., Urbinati, C. R., and Long, R. M. (2005) RNA localization in yeast: moving towards a mechanism. *Biol Cell* **97**, 75–86.
2. Lopez de Heredia, M., and Jansen, R. P. (2004) mRNA localization and the cytoskeleton. *Curr Opin Cell Biol* **16**, 80–85.
3. St Johnston, D. (2005) Moving messages: the intracellular localization of mRNAs. *Nat Rev Mol Cell Biol* **6**, 363–375.
4. Chartrand, P., Bertrand, E., Singer, R. H., and Long, R. M. (2000) Sensitive and high-resolution detection of RNA in situ. *Methods Enzymol* **318**, 493–506.
5. Cole, C. N., Heath, C. V., Hodge, C. A., Hammell, C. M., and Amberg, D. C. (2002) Analysis of RNA export. *Methods Enzymol* **351**, 568–587.
6. Bullock, S. L., and Ish-Horowicz, D. (2001) Conserved signals and machinery for RNA transport in Drosophila oogenesis and embryogenesis. *Nature* **414**, 611–616.
7. Glotzer, J. B., Saffrich, R., Glotzer, M., and Ephrussi, A. (1997) Cytoplasmic flows localize injected oskar RNA in Drosophila oocytes. *Curr Biol* **7**, 326–337.
8. Bratu, D. P., Cha, B. J., Mhlanga, M. M., Kramer, F. R., and Tyagi, S. (2003) Visualizing the distribution and transport of mRNAs in living cells. *Proc Natl Acad Sci USA* **100**, 13308–13313.
9. Politz, J. C., Tuft, R. A., and Pederson, T. (2003) Diffusion-based transport of nascent ribosomes in the nucleus. *Mol Biol Cell* **14**, 4805–4812.
10. Beach, D. L., Salmon, E. D., and Bloom, K. (1999) Localization and anchoring of mRNA in budding yeast. *Curr Biol* **9**, 569–578.
11. Bertrand, E., Chartrand, P., Schaefer, M., Shenoy, S. M., Singer, R. H., and Long, R. M. (1998) Localization of ASH1 mRNA particles in living yeast. *Mol Cell* **2**, 437–445.
12. Takizawa, P. A., and Vale, R. D. (2000) The myosin motor, Myo4p, binds Ash1 mRNA via the adapter protein, She3p. *Proc Natl Acad Sci USA* **97**, 5273–5278.
13. Boulon, S., Basyuk, E., Blanchard, J. M., Bertrand, E., and Verheggen, C. (2002) Intra-nuclear RNA trafficking: insights from live cell imaging. *Biochimie* **84**, 805–813.
14. Brodsky, A. S., and Silver, P. A. (2000) Pre-mRNA processing factors are required for nuclear export. *RNA* **6**, 1737–1749.
15. Chubb, J. R., Trcek, T., Shenoy, S. M., and Singer, R. H. (2006) Transcriptional pulsing of a developmental gene. *Curr Biol* **16**, 1018–1025.

16. Forrest, K. M., and Gavis, E. R. (2003) Live imaging of endogenous RNA reveals a diffusion and entrapment mechanism for nanos mRNA localization in Drosophila. *Curr Biol* **13**, 1159–1168.

17. Sheth, U., and Parker, R. (2006) Targeting of aberrant mRNAs to cytoplasmic processing bodies. *Cell* **125**, 1095–1109.

18. Gietz, R. D., and Sugino, A. (1988) New yeast-Escherichia coli shuttle vectors constructed with in vitro mutagenized yeast genes lacking six-base pair restriction sites. *Gene* **74**, 527–534.

14

Analysis of mRNA Partitioning Between the Cytosol and Endoplasmic Reticulum Compartments of Mammalian Cells

Samuel B. Stephens, Rebecca D. Dodd, Rachel S. Lerner, Brook M. Pyhtila, and Christopher V. Nicchitta

Summary

All eukaryotic cells display a dramatic partitioning of mRNAs between the cytosol and endoplasmic reticulum (ER) compartments—mRNAs encoding secretory and integral membrane proteins are highly enriched on ER-bound ribosomes and mRNAs encoding cytoplasmic/nucleoplasmic proteins are enriched on cytosolic ribosomes. In current views, this partitioning phenomenon occurs through positive selection—mRNAs encoding signal sequence-bearing proteins are directed into the signal recognition particle pathway early in translation and trafficked as mRNA/ribosome/nascent polypeptide chain complexes to the ER. In the absence of an encoded signal sequence, mRNAs undergo continued translation on cytosolic ribosomes. Recent genome-wide analyses of mRNA partitioning between the cytosol and the ER compartments have identified subsets of mRNAs that are non-canonically partitioned to the ER—although lacking an encoded signal sequence, they are translated on ER-bound ribosomes. These findings suggest that multiple, and as yet unidentified, pathways exist for directing mRNA partitioning in the cell. In this contribution, we briefly review the literature describing the subcellular partitioning patterns of mRNAs and present a detailed methodology for studying this fundamental, yet poorly understood process.

Key Words: mRNA; ribosome; cytosol; endoplasmic reticulum; subcellular fractionation; isopycnic flotation.

From: *Methods in Molecular Biology, Vol. 419: Post-Transcriptional Gene Regulation*
Edited by: J. Wilusz © Humana Press, Totowa, NJ

1. Introduction

In eukaryotic cells, protein synthesis is compartmentalized; mRNAs encoding soluble, cytoplasmic proteins are translated on free, cytoplasmic ribosomes whereas mRNAs encoding secretory and integral membrane proteins are translated on endoplasmic reticulum (ER)-bound ribosomes. Extending from the pioneering studies of Palade and coworkers, which identified the ER as the site of translation for secretory and integral membrane protein-encoding mRNAs, a number of important questions were posed: What is the mechanism of ribosome binding to the ER? Are cytoplasmic and ER-bound ribosomes identical? How are secretory and integral membrane protein-encoding mRNAs trafficked to the ER?

A landmark series of highly influential experiments demonstrated that mRNA partitioning to the ER was a co-translational process requiring the synthesis of a signal sequence or other protein topogenic signal, such as a transmembrane domain *(1–5)*. In this pathway, a newly synthesized signal sequence is bound by the signal recognition particle (SRP) as it emerges from the ribosomal nascent chain exit site. The mRNA/ribosome/nascent chain complex (RNC) is then targeted to the ER membrane through direct interaction with an ER-localized SRP receptor. At the ER membrane, the RNC binds to the protein-conducting channel, with ongoing mRNA translation being directly coupled to protein translocation *(2,3,6–8)*. mRNAs lacking an encoded signal sequence, as is the case for cytosolic proteins, cannot enter the SRP pathway and thus remain cytosolic (default pathway). In this way, the partitioning of mRNAs to the ER operates through positive selection to yield the segregation of mRNAs between the two compartments on the basis of the compartmental fate of their translated products *(1–5)*.

The landmark studies noted above have provided remarkable insights into the machinery and mechanism of mRNA partitioning in the eukaryotic cell. Nonetheless, the experimental emphasis on in vitro reconstitution approaches left unresolved a very fundamental question: Does the SRP pathway direct mRNA partitioning in vivo? *(9)*. From the perspective of mRNA partitioning—and the prediction that mRNA localization to the ER requires "positive selection"—the available data suggest strongly that mRNA partitioning in the eukaryotic cell is a complex process that likely requires mechanisms other than, or in addition to, the SRP pathway. For example, genome-wide studies examining the mRNA population identities of cytosolic and ER-bound polysomes performed in yeast, fly, mouse, and human cell systems have identified an intriguing and wholly unexpected overlap in the composition of the two mRNA pools *(10–14)*. As expected, mRNAs encoding signal sequence-bearing (i.e., secretory and integral membrane) proteins were highly enriched in the membrane-bound polypore fraction. However, discrete subsets of mRNAs lacking signal sequences were also highly enriched in ER-bound polypore membrane fraction. In addition, somewhat heterogeneous distributions, or enrichments, of other subclasses of mRNA were observed. For example, the mRNA encoding the soluble glycolytic enzyme glyceraldehyde phosphate dehydrogenase (GAPDH) resided in both compartments, although it was enriched on cytosolic polysomes. In contrast, the mRNAs encoding the heat-shock protein, Hsp90, and the ER stress–response transcription factor XBP-1 (or Hac1 in yeast), which are both soluble cytosolic proteins, were enriched on ER-bound polyribosomes *(12,13,15)*. From these data, it appears that signal sequence-directed co-translational trafficking is not the only cellular mechanism for partitioning mRNAs to the ER. Drawing on biological precedent for mechanisms of mRNA localization, mRNA partitioning between the cytosol and the ER compartments likely utilizes multiple pathways, and presumably, such pathways utilize *cis*-localization elements *(9)*. Regardless of the mechanism, the data obtained in the noted genomic analyses are significant, for they re-open investigation into the basic mechanisms governing mRNA partitioning in the eukaryotic cell. In the sections that follow, methodologies for studying this important process are presented.

The methods presented here describe biochemical techniques for isolating highly enriched cytosol and total membrane-bound and ER membrane-bound mRNA fractions, to be used in RNA analyses. The first method utilizes a sequential detergent extraction technique to obtain cytosol and membrane-bound mRNA fractions and is based on the differential sensitivities of the plasma membrane and organelle membranes to β-sterol-binding detergents, such as digitonin and saponin. The second method uses a more rigorous approach of buoyant density centrifugation to isolate highly enriched ER membrane-derived vesicles. By either method, RNA can be isolated from the subcellular fractions and the partitioning patterns and ultimately mechanism of partitioning, studied.

2. Materials

As with all methods for studying RNAs, appropriate care must be used to avoid degradation of the RNAs of interest. To this end, all buffers are typically made in either 0.1% diethyl pyrocarbonate (DEPC)-treated water or are DEPC treated when applicable to eliminate trace sources of RNase activity (*see* **Note 1**). Glassware is typically baked at 300°C for 4 h. Additionally, plastic materials such as centrifuge tubes may be washed with 0.1% (w/v) sodium dodecyl sulfate (SDS) or, if more stringent standards are required, 4M guanidine thiocyanate (GT) buffer (*see* **Subheading 2.4.**) and rinsed with either DEPC-treated water or Milli-Q water. For additional discussions of techniques used in RNA preparations *see* **ref. 16**.

2.1. Cell Culture

1. Mammalian cell cultures (NIH 3T3, HEK293, Cos7, HeLa, Jurkat E6-1, J558L, etc.).
2. Phosphate-buffered saline (PBS) (Gibco/Invitrogen, Carlsbad, California, USA).
3. poly-L-lysine (Sigma-Aldrich; St. Louis, MO) (freeze in 0.1 mL aliquots) (*see* **Note 2**).

2.2. Sequential Detergent Extraction

1. Digitonin (Calbiochem; San Diego, CA) is prepared as a 1% (w/v) stock in DMSO (freeze in 0.2 mL aliquots).
2. RNase Out ribonuclease recombinant inhibitor 40 U/μL (Invitrogen, Carlsbad, California, USA) or equivalent.
3. Stock solutions: 4 M potassium acetate (KOAc); 2 M potassium 4-(2-hydroxyethyl)-1-piperazineethanesulfonate (KHEPES) pH 7.2; 1 M magnesium acetate ($Mg(OAc)_2$); 1M dithiothreitol (DTT) (freeze in 50 μL aliquots); 100 mM phenylmethylsulfonate fluoride (PMSF) (prepared in isopropanol; stable for approximately 2 weeks); 0.2 M ethyleneglycol bis (2-aminoethylether)-*N,N,N´,N´*-tetraacetic acid (EGTA) pH 8.0; 10% (v/v) Nonidet P-40 (NP-40); 10% (w/v) sodium deoxycholate (DOC).
4. Permeabilization buffer: 110 mM KOAc, 25 mM KHEPES pH 7.2, 2.5 mM $Mg(OAc)_2$, 1 mM EGTA, 0.015% digitonin, 1 mM DTT, 1 mM PMSF, 40 U/mL RNase Out (Invitrogen). Digitonin, DTT, PMSF, and RNase Out must be added fresh.
5. Wash buffer: 110 mM KOAc, 25 mM KHEPES pH 7.2, 2.5 mM $Mg(OAc)_2$, 1 mM EGTA, 0.004% digitonin, 1 mM DTT, 1 mM PMSF. Digitonin, DTT, and PMSF must be added fresh.
6. Lysis buffer: 400 mM KOAc, 25 mM KHEPES pH 7.2, 15 mM $Mg(OAc)_2$, 1% (v/v) NP-40, 0.5% (w/v) DOC, 1 mM DTT, 1 mM PMSF, 40 U/mL RNase Out. DTT, PMSF, and RNase Out must be added fresh (*see* **Note 3**).

2.3. Mechanical Homogenization and Isopycnic Flotation

1. 2 mL glass Dounce homogenizer (Kimble/Kontes, Vineland, New Jersey, USA) and B (tight-fitting) pestle.
2. Hypotonic lysis buffer: 10 mM KOAc, 10 mM KHEPES pH 7.2, 1.5 mM $Mg(OAc)_2$, 1 mM DTT, 1 mM PMSF, 40 U/mL RNase Out. DTT, PMSF, and RNase Out must be added fresh (*see* **Note 4**).
3. Isotonic buffer (IB): 150 mM KOAc, 25 mM KHEPES pH 7.2, 5 mM $Mg(OAc)_2$, 1 mM DTT, 1 mM PMSF, 10 U/mL RNase Out. DTT, PMSF, and RNase Out must be added fresh.

4. Sucrose solutions: 2.5 M sucrose in IB; 1.9 M sucrose in IB; 1.3 M sucrose in IB. DTT, PMSF, and RNase Out must be added fresh.
5. Beckman SW-55 ultra-clear centrifuge tubes and rotor (or equivalent).
6. Beckman TLA100.3 centrifuge tubes and rotor (or equivalent).

2.4. RNA Extraction

1. Stock solutions: 0.5 M ethylenediamine tetraacetic acid (EDTA) pH 8.0; 750 mM sodium citrate pH 7.0, 10% (w/v) *N*-lauryl sarcosine; 3 M sodium acetate (NaOAc) pH 5.2.
2. 0.1 M sodium citrate pH 4.5 buffer saturated phenol (Sigma-Aldrich).
3. GT buffer: 4 M GT, 25 mM sodium citrate pH 7.0, 0.5% *N*-lauryl sarcosine, 5 mM EDTA.
4. Isopropanol.
5. 70% (v/v) ethanol.
6. Chloroform.

2.5. Denaturing Agarose Gels

1. 10× 3-(*N*-morpholino)propanesulfonic acid (MOPS) RNA gel running buffer: 0.2 M MOPS, 80 mM NaOAc, 10 mM EDTA pH 7.4. This buffer is light sensitive.
2. Agarose (electrophoresis grade).
3. 37.5% (w/v) formaldehyde.
4. Formamide (deionized).
5. DNA/RNA gel electrophoresis apparatus.
6. 10 mg/mL ethidium bromide. This solution is light sensitive.

2.6. Northern Blotting

1. 20× SSC: 3 M NaCl, 0.3 M NaOAc pH 7.0.
2. 10 N NaOH.
3. Methylene blue stain: 0.02% (w/v) methylene blue in 0.2 M NaOAc pH 5.2.
4. Hybond XL (GE Healthcare; Little Chalfont, Buckinghamshire, UK).
5. Whatman 3 MM filter paper or equivalent.
6. Standard laboratory paper towels.
7. UV cross-linking device (e.g., UV Stratalinker 2400 from Stratagene/Agilent Technologies, Santa Clara, CA, USA).
8. Church–Gilbert hybridization buffer: 7% (w/v) SDS, 1 mM EDTA, 0.5 M sodium phosphate (0.36 M Na_2HPO_4, 0.14 M NaH_2PO_4). Dissolve and store at 37°C to prevent precipitation.
9. 10 mg/mL sheared salmon sperm DNA (freeze in 1 mL aliquots) (*see* **Note 5**).
10. Hybridization oven and glass tubes.
11. Phosphorimager plates, cassettes, and scanner (e.g., Typhoon 9400; GE Healthcare).
12. T4 Polynucleotide kinase (PNK) and corresponding buffer (New England Biolabs, Ipswich, MA, USA)

13. γ-[³²P]-dATP; end-labeling grade (>7000 Ci/mmol; 160 mCi/mL) (MP Biomedicals; Irvine, CA)
14. Sephadex G-50 NICK Columns (GE Healthcare) or equivalent.
15. TE buffer: 10 mM Tris pH 7.5, 1 mM EDTA.
16. Liquid scintillation fluid, counting tubes, and beta-counter.

3. Methods

Two methods are discussed here to separate mammalian cells into highly enriched soluble cytosol and total membrane-bound or ER membrane-bound fractions. These methods have reproducibly demonstrated the well-established enrichments of proteins within distinct organelle fractions, while uncovering non-conventional mRNA distributions. In this context, the integrity and enrichment of subcellular fractions can only be evaluated by their enzymic composition and not by their mRNA content. This point is illustrated in **Figs 1–3** examining the protein and mRNA distributions of the metabolic enzyme lactate dehydrogenase (LDH) and the heat-shock chaperone HSP90. Although mRNAs encoding HSP90 and LDH are clearly present in the ER membrane-derived fractions (*see* **Fig. 3**), the proteins are themselves exclusive to the cytosol fraction (*see* **Figs 1** and **2**).

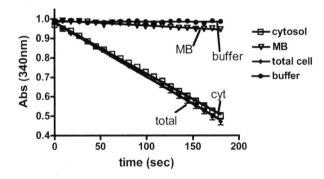

Fig. 1. Lactate dehydrogenase (LDH) activity is exclusively (>95%) recovered in the soluble cytosol fraction. Cytosol (cyt) and total membrane-bound fractions (MB) were isolated from HeLa cells through sequential detergent extraction. Total cell lysate was obtained from HeLa cells lysed in NP-40/DOC lysis buffer. LDH assay was performed according to the manufacturer's instructions (Sigma-Aldrich). Briefly, in a 1-mL cuvette, 10^4 cell equivalents were diluted in 0.5 mL of solution A containing NADH. Twenty microlitres of solution B-containing pyruvate was added and the absorbance at 340 nm monitored for 3 min sampling every 9 sec.

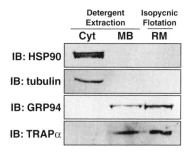

Fig. 2. sodium dodecyl sulfate–polyacrylamide gel electrophoresis (SDS–PAGE) and immunoblot analysis of cytosol, total membrane-bound and ER membrane-bound (RM) fractions. Cytosol (cyt), total membrane-bound (MB), and rough microsome (RM) fractions were isolated from HeLa cells through sequential detergent extraction or isopycnic flotation, respectively. 10 μg of protein derived from cytosol (cyt) or total membrane-bound (MB) fractions and 1 equivalent of HeLa RM (1 equivalent RM is defined as 1 μL of 50 A_{280} units per mL) were resolved on 12.5% Laemmeli SDS–PAGE minigels. Gels were transferred to polyvinylidene fluoride (PVDF) membranes in 50 mM CAPS pH 11, 0.075% (w/v) SDS, 20% methanol for 40 min using a semi-dry gel transfer apparatus at 100 mA per minigel. Primary antibodies were used at a 1:1000 dilution. Secondary antibodies (goat anti-rabbit-F(ab)$_2$-conjugated AlexaFluor-647, goat anti-mouse-F(ab)$_2$-conjugated AlexaFluor-555; Invitrogen) were used at a 1:3000 dilution. Images were collected using a Typhoon 9400 (GE Healthcare) and contrasted using Adobe Photoshop 7.0.

3.1. Subcellular Fractionation Through Sequential Detergent Extraction

Sequential digitonin-mediated detergent extraction provides a rapid and reliable means of separating soluble cytosolic and total membrane-bound subcellular fractions from mammalian cell cultures growing either as cell monolayers (*see* **Subheading 3.1.1.**) or in suspension (*see* **Subheading 3.1.2.**). This method takes advantage of the high enrichment of cholesterol in the plasma membrane relative to other membranous organelles by utilizing the selective cholesterol-binding activity of the detergent digitonin. In this respect, the plasma membrane is effectively permeabilized and the cytosolic contents released, whereas organelles such as the ER, Golgi, and nuclei remain virtually intact *(13, 15, 17, 18)*. Subsequent solubilization of the remaining cell membranes with lytic detergent conditions in the presence of magnesium yields a fraction enriched in organelle membrane and soluble lumenal contents,

A.

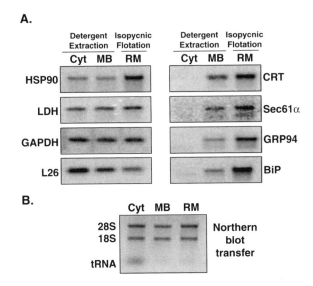

B.

Fig. 3. Northern blot analysis of cytosol and total membrane-bound and ER membrane-bound (RM) fractions. Two micrograms of RNA isolated from cytosol (cyt) and total membrane-bound (MB) (sequential detergent extraction), and rough microsome (RM) (isopycnic flotation) fractions obtained from HeLa cells was resolved on denaturing agarose gels and transferred to Hybond-XL by downward capillary transfer. **(B)** The ethidium bromide fluorescence of the transferred RNA was scanned using a Typhoon 9400 (GE Healthcare). **(A)** Membranes were hybridized with [^{32}P]-end-labeled antisense oligonucleotides corresponding to mRNAs encoding cytosolic proteins (mRNA$_{cyt}$) and mRNAs encoding secretory and integral membrane proteins (mRNA$_{sec}$). Prehybridization, hybridization, and washing steps were all performed in Church–Gilbert buffer at 50°C. Following overnight exposures to phosphorimager plates, images were collected using a Typhoon 9400 (GE Healthcare) and contrasted using Adobe Photoshop 7.0. mRNA$_{cyt}$: heat-shock protein 90 (HSP90), lactate dehydrogenase (LDH), glyceraldehyde phosphate dehydrogenase (GAPDH), and the large ribosomal subunit 26 (L26). mRNA$_{sec}$: calreticulin (CRT), Sec61α, GRP94, and BiP.

without disturbing the nuclear architecture. Fractions generated by this method have been rigorously evaluated for purity through immunoblot for marker proteins *(13, 15, 19)* (*see* **Fig. 2**), immunofluorescence microscopy *(15)*, enzymatic activities (*see* **Fig. 1**), and transmission electron microscopy *(13,17)*. This method is particularly advantageous as it allows direct comparisons of protein and RNA components present in cytosolic and/or membrane-bound fractions.

3.1.1. Sequential Detergent Extraction (Cell Monolayer)

The method described here uses volumes adjusted for a 75-cm^2 tissue culture flask; however, these volumes may be altered for other culture dish sizes (*see* **Notes 2, 6,** and **7**).

1. Aspirate media and wash cells with 5–10 mL ice-cold PBS to remove any excess culture media.
2. Gently coat the flask with 1 mL permeabilization buffer and slowly rock in a 4°C cold room or on ice for 5 min. At this point, loss of plasma membrane and soluble cytosolic contents can be visualized by using phase contrast microscopy.
3. Stand the flask upright and allow the buffer to drain to the bottom for 1 min. Transfer the soluble material (cytosol) to a 1.5-mL microcentrifuge tube and place in ice.
4. Gently wash the flask with 5 mL wash buffer and discard the buffer.
5. Coat the flask with 1 mL lysis buffer and rock in a 4°C cold room or on ice for 5 min. At this point, only nuclei and insoluble cytoskeletal components remain attached to the culture flask, which can be visualized by using phase contrast microscopy.
6. Stand the flask upright and allow the buffer to drain to the bottom for 1 min. Transfer the soluble material (membrane-bound) to a 1.5-mL microcentrifuge tube.
7. Clarify the soluble lysates at 7500× *g* for 10 min at 4°C and transfer the supernatants to new tubes.
8. At this point, fractions may be analyzed for protein (enzymatic activity, sodium dodecyl sulfate–polyacrylamide gel electrophoresis (SDS–PAGE), immunoblot) or RNA content (Northern, RT-PCR, cDNA microarray, ribonuclease protection assay) as desired. Additionally, the steps in this procedure can be tracked and evaluated in situ by immunofluorescence microscopy *(15)* (for polyribosome recovery, *see* **Note 8**).

3.1.2. Sequential Detergent Extraction (Cell Suspensions)

1. Aliquot 10^7 cells and centrifuge at 500 × *g* for 5 min (*see* **Note 9**).
2. Aspirate media, resuspend cells with 10 mL ice-cold PBS, and transfer to a 15-mL conical tube. Recover cells by centrifugation at 500 × *g* for 5 min.
3. Resuspend cells in 1 mL permeabilization buffer (10^7 cells/mL) and chill on ice for 5 min.
4. Centrifuge at 1000 × *g* for 5 min and transfer the supernatant (cytosol) to a 1.5-mL microcentrifuge tube.
5. Resuspend cells in 5 mL wash buffer (2 × 10^6 cells/mL) and recover by centrifugation at 1000 × *g* for 5 min. Remove the supernatant by aspiration.
6. Resuspend cells in 1 mL lysis buffer (10^7 cells/mL) and place on ice for 5 min to solubilize the remaining cellular membranes. Transfer the membrane-bound fraction to a 1.5-mL microcentrifuge tube.
7. Clarify the soluble lysates at 7500 × *g* for 10 min at 4°C and transfer the supernatants to new tubes.

8. At this point, fractions may be analyzed for protein (enzymatic activity, SDS–PAGE, immunoblot) or RNA content (Northern, RT-PCR, cDNA microarray, ribonuclease protection assay) as desired (for polyribosome recovery, *see* **Note 8**).

3.2. Mechanical Homogenization and Isopycnic Flotation

Mechanical homogenization followed by isopycnic flotation provides an alternative source of highly purified ER membranes relying on the distinct buoyant density of ribosome-studded rough microsomes (RM) that form during homogenization. In this method, ER membranes are separated from lighter cellular membranes such as Golgi and plasma membranes by density centrifugation using a discontinuous sucrose gradient. Although this technique is well established and considered by many to be the gold standard for obtaining ER membranes, it has several major drawbacks in cell cultures. First, this technique offers substantially lower yields of material compared with the sequential detergent extraction. Second, because of the degree of processing/handling, quantitative recovery of membranes in comparative experiments can be challenging. Last, but most important, this technique does not allow the recovery of an exclusive soluble cytosol fraction from cell cultures. The recovered soluble "cytosol" fraction is typically contaminated by organelle membrane and lumenal markers released during homogenization *(15)*. For this reason, the discussion of this method will focus only on isolation of membrane fractions.

1. Prepare approximately $1–3 \times 10^7$ tissue culture cells (*see* **Note 10**):

    a. For suspension cells, pellet at approximately 500× *g* at 4°C for 5 min.
    b. For cells in monolayer, remove the cells from the plate with trypsin/EDTA or by cell scraping into PBS (the latter technique is preferred). Recover by centrifugation at approximately 500 × *g* for 5 min.

2. Resuspend in 10 mL cold PBS and recover by centrifugation at 500 × *g* for 5 min. Remove the supernatant by aspiration.

3. Resuspend cells in 0.35 mL hypotonic buffer ($<10^8$ cells/mL) and gently homogenize on ice in a glass Dounce homogenizer using a tight-fitting (B) pestle with approximately 40 up–down strokes to rupture >90% of cells (*see* **Note 11**). The degree of cell disruption can be determined by examining a 1:100 (10^6 cells/mL) dilution under the microscope.

4. Dilute 0.30 mL of the homogenate (fivefold) with 1.3 mL 2.5 M sucrose IB (final sucrose concentration is approximately 2 M) and load 1.5 mL of diluted homogenate above a cushion of 0.25 mL 2.5 M sucrose IB in a SW-55 centrifuge tube. In this order, carefully layer 1.5 mL 1.9 M sucrose IB, 0.75 mL 1.3 M sucrose IB, 0.25 mL IB atop the homogenate.

5. Centrifuge at 290,000×g_{av} for 2.5 h at 4°C in a Beckman SW55 rotor (or equivalent).
6. Collect fractions by either of the following methods: (*see* **Note 12**)

   a. Manually puncture a hole in the bottom of a tube with a 16-gauge needle and dropwise collect fractions.
   b. Remove the RM layer in approximately 0.5 mL by using pipet, which bands below the 1.3 M/1.9 M sucrose interface (white to yellowish material).

7. To recover RM, dilute the RM fraction (fourfold) in 1.5 mL IB and centrifuge in a Beckman TLA100.3 rotor at 66,000×g_{av} for 20 min at 4°C (or equivalent). Remove the supernatant by aspiration.
8. Analyze as desired for protein or RNA content (for polyribosome recovery, *see* **Note 8**).

3.3. RNA Isolation

The method described here is one of many variations of RNA isolations that utilize the powerful denaturing agent, guanidinium thiocyanate, in conjunction with a mildly acidic phenol–chloroform extraction to recover RNA in an aqueous phase. Variations or alternative techniques will most certainly work as well.

1. Mix in a 2:1 ratio GT buffer and 0.1 M sodium citrate pH 4.5 buffer-saturated phenol. Add 7 µL concentrated BME (14.4 M) per mL GT buffer (final concentration approximately 0.1 M BME). This solution will be referred to as GT-phenol.
2. Add 2–3 vol GT–phenol mix to 1 vol of sample and mix by rocking or rotating end over end for 15 min at room temperature.
3. Add approximately 0.2 mL chloroform per 0.75 mL GT–phenol and thoroughly mix.
4. Separate phases by centrifugation at >3000 × g for 5–10 min or until phases are distinct.
5. Transfer the aqueous phase to a new tube and add 0.1 vol of 3 M NaOAc pH 5.2 and an equal volume of isopropanol. Let sample stand for 20 min to precipitate the RNA (*see* **Note 13**).
6. Recover the RNA precipitate by centrifugation at 13,000 × g for 10 min at 4°C and discard the supernatant.
7. Wash the RNA precipitate with a small volume of 70% ethanol and recover by centrifugation at 13,000 × g for 10 min at 4°C. (For samples recovered in 1.5 mL microcentrifuge tubes, this is typically 0.5 mL 70% ethanol). Discard the supernatant.
8. Allow the sample to air dry, or, alternatively, dry samples in a speedvac (take care not to overdry samples or they can be very difficult to resuspend afterwards).
9. Resuspend RNA pellets in 50–100 µL of DEPC-treated or nuclease-free water and determine the concentration of a 1:20-fold or 1:100-fold diluted aliquot by using UV absorbance at 260 nm (*see* **Notes 14** and **15**). An A_{260} of 1.0 equals 40 µg/mL RNA.

Adjust to desired concentration. A typical yield from 10^7 cells is approximately 50–100 μg. RNA may be stored frozen at –20°C for 1–2 years or in 70% ethanol at –80°C indefinitely.

3.4. Denaturing Agarose Gel Electrophoresis

RNA may be analyzed by any number of desired techniques such as Northern blot, dot blot, ribonuclease protection assay, RT-PCR, or cDNA microarray, to list a few. In this example, Northern blot analysis will be used to analyze mRNA content of subcellular fractions.

1. Thoroughly wash a DNA/RNA electrophoresis apparatus (gel tray, combs, buffer tank) with 0.1% (w/v) SDS and rinse with Milli-Q water.
2. For a 100 mL 1% agarose 3% formaldehyde gel, dissolve 1 g of electrophoresis grade agarose into 82 mL of DEPC-treated water in a baked Erlenmeyer flask using a microwave (at or near boiling). Allow to cool to <60°C and add 10 mL 10× MOPS RNA gel buffer and 8 mL of 37.5% formaldehyde. As formaldehyde is a rather potent volatile irritant, avoid inhalation. Pour gel into gel tray, set combs accordingly, and allow solidification to occur at room temperature or at 4°C.
3. Prepare RNA sample buffer: 99.6 μL formamide, 34.8 μL 37.5% formaldehyde, 15 μL 10× MOPS RNA gel running buffer and 0.5–1 μL 10 mg/mL ethidium bromide. Aliquot 1–10 μg RNA, and add RNA sample buffer to sample volume in a ratio of 11:4.
4. Heat sample at 70°C for 10–15 min, chill on ice for 2 min, and load onto the gel.
5. Electrophorese in 1× MOPS RNA gel running buffer (dilute 10× MOPS RNA gel running buffer with Milli-Q or DEPC-treated water) at 100–130V until desired migration and separation has occurred. Generally, 30–45 min of electrophoresis clearly resolves the 28S (approximately 5 kb) and the 18S (approximately 2 kb) rRNAs. Photograph or document if desired (*see* **Note 16**).

3.5. Northern Blotting

3.5.1. Membrane Transfer and Hybridization

This method of Northern blotting uses downward capillary flow to transfer RNA from an agarose gel to a solid membrane in a mildly alkaline saline solution (5× SSC + 10 mM NaOH). This is significantly faster (2 vs. 18 h) than upward transfer and tends to give more sharply resolved bands. Additionally, hybridization conditions described here use Church–Gilbert hybridization buffer. However, other buffers and corresponding conditions such as Denhardt's solution, or proprietary solutions such as ULTRA Hyb-Oligo Hybridization Buffer (Ambion/Applied Biosystems, Foster City, CA, USA) will work as well.

1. Precut 6 pieces of 3 MM filter paper and 1 piece of Hybond membrane to exact size of agarose gel. Cut two long strips of 3 MM filter paper of equivalent width

to the agarose gel and approximately three times the length to be used as a wick. Precut a 3- to 4-inch stack of laboratory paper towels slightly larger in size than the agarose gel.

2. Assemble stack of paper towels with 2 pieces of 3 MM filter paper on top.

3. Following electrophoresis, soak the agarose gel for 2–5 min in 5 × SSC + 10 mM NaOH.

4. Wet the Hybond membrane and 1 piece of filter paper in 5 × SSC + 10 mM NaOH and assemble on the bottom face of the gel with the membrane touching the gel. Carefully smooth out any air bubbles and place on top of the two dry filter papers above the paper towel stack, membrane side down.

5. Wet the three remaining pieces of filter paper in 5× SSC + 10 mM NaOH and place atop the agarose gel. Gently smooth out any air bubbles with a pipet using a slow rolling action.

6. Wet the two long strips of filter paper in 5 × SSC + 10 mM NaOH and place one end atop the agarose gel stack and the other in a buffer tank containing approximately 100 mL of 5 × SSC + 10 mM NaOH. This will be used as a wick for capillary transfer. Check that the wick is not touching the paper towels such that the buffer only transfers through the gel. Plastic wrap may be used to cover the stack.

7. Cover the stack with the gel-casting tray (<200 g), but do not add weight to compress the stack.

8. Transfer for 1–2 h undisturbed.

9. Carefully disassemble and crosslink the RNA to the Hybond membrane at 1200 mJ in a UV cross linker (Stratalinker; Stratagene) or equivalent. Following crosslinking, the RNA is no longer susceptible to RNase-mediated degradation.

10. If desired, check transfer by staining with methylene blue solution for 2 min and rinse with deionized water until rRNA bands are visible. It is not necessary to remove methylene blue stain completely before proceeding. Alternatively, the ethidium bromide fluorescence of the rRNA is generally detectable using a fluorescent gel imager (Typhoon 9400, GE Healthcare).

11. Prehybridize membrane at 50–60°C in 1× Church–Gilbert buffer (10–15 mL of buffer is usually sufficient depending on size of membrane and hybridization tube) with 100 µg/mL salmon sperm DNA, denatured at 95°C for 10 min just before use. Rotate the membrane in a hybridization oven for 30 min–1 h.

12. Add 1×10^7 cpm [^{32}P]-labeled DNA probe (*see* **Subheading 3.5.2.**) per 10 mL hybridization buffer containing heat-denatured 100 µg/mL salmon sperm DNA. Hybridize overnight (16–18 h) at 50–60°C (*see* **Note 17**).

13. Wash with 20–40 mL of 0.1× Church–Gilbert buffer for 30 min at hybridization temperature. Repeat. Temperature of washes may be dropped 5–10° if signal is low.

14. Wrap membrane in clear plastic wrap and add a small volume of 0.1× Church–Gilbert buffer to prevent the membrane from drying out (*see* **Note 18**). Expose to

a phosphorimager plate for 2–18 h and scan with a Phosphorimager Plate Scanner (Typhoon 9400, GE Healthcare). Signals are generally detectable after 2 h whereas longer exposures will improve detection.

15. Membranes may be stripped by the addition of 0.5% boiling SDS and agitating for 30 min–1 hr (until room temperature). Verify removal of hybridized radioactive probe by using phosphoimage analysis.

3.5.2. [^{32}P]-End Labeled Oligonucleotide Probe Synthesis

As there are a wide range of methods for generating [^{32}P]-labeled DNA probes (PCR, random-primed labeling, end-labeling of oligonucleotides, etc.), it is far beyond the scope of this chapter to discuss the variations and merits of these different techniques. We typically use end-labeled oligonucleotides with great success.

1. Design antisense oligonucleotide probes using the following guidelines: Probes should be 26–45 nucleotides in length, have a 78–90°C T_m value, and possess 46–62% GC content.
2. Assemble the kinase reaction on ice as follows (final volume 20 μL): 10 pmol of oligo, 2 μL 10× T4 PNK buffer, 160 μCi (1 μL) γ–[^{32}P]-dATP, 1 μL T4 PNK, H$_2$O up to 20 μL. Allow reaction to proceed at 37°C for 1 h.
3. Stop the enzymatic reaction by heating at 95°C for 2 min (optional).
4. Purify the [^{32}P]-labeled probe by column chromatography using NICK Sephadex G-50 columns (or equivalent) according to the manufacturer's instructions as follows: decant the storage buffer and wash the resin with 3 mL TE buffer. Add the reaction mix to the resin and wash with 0.4 mL TE buffer discarding the flowthrough. Elute the end-labeled probe with 0.4 mL TE buffer.
5. Determine the specific activity of 1 μL of purified probe through liquid scintillation. A good labeling reaction will generate probes at a specific activity of 5×10^6 cpm/pmol or more.
6. Use probe at 1×10^7 cpm per 10 mL hybridization buffer.

4. Notes

1. To inactivate RNases by DEPC alkylation, add DEPC to a final concentration of 0.1% (v/v), mix thoroughly, and incubate at 37°C overnight. Autoclave to remove unreacted DEPC for 15 min. Compounds containing free amino groups such as Tris should not be DEPC-treated, whereas others such as MOPS or sucrose should not be autoclaved. Thus, take care to properly prepare buffers containing these agents using DEPC-treated water rather than DEPC treating.
2. Some tissue culture cells such as HEK293 tend to weakly adhere to culture flasks and are thus prone to detachment during the permeabilization and wash steps of the sequential detergent extraction (*see* **Subheading 3.1.**). This can be remedied by pre-coating flasks with poly-L-lysine before adding cell cultures.

3. A number of alternative detergents can be used in place of the classic 1% NP-40/0.5% DOC mixture such as substituting Triton X-100 or Nikkol for NP-40 with similar results. The choice of detergent in some ways depends on the desired output. For example, membrane solubilization with milder detergents such as 2% digitonin or *n*-dodecyl-beta-D-maltoside (Calbiochem) can preserve the ER ribosome–translocon (Sec61/TRAP/ribophorins) interactions, whereas a NP-40/DOC cocktail will disrupt these complexes *(19)*. However, in addition to the higher cost of digitonin and *n*-dodecyl-beta-D-maltoside, these detergents are somewhat weaker, will require longer incubation times, and generally yield slightly less material than NP-40/DOC.

4. The nuclei of NIH 3T3 cells tend to be more sensitive to low magnesium levels relative to other cell lines tested during hypotonic lysis. This can lead to rupture of the nuclear membranes and cause contamination of RM fractions with DNA. This can be easily remedied by adjusting the magnesium concentration of the hypotonic lysis solution to 5 mM $Mg(OAc)_2$.

5. Lyophilized salmon sperm DNA (Sigma-Aldrich) is resuspended at 10 mg/mL in deionized water by slowly stirring at room temperature for several hours. Shear DNA by progressively passing through a range of 16–21 gauge syringe needles (5–10 times each) or until a workable viscosity is achieved. The mixture is then aliquoted (1 mL) and heated at 95°C for 10 min. Store at –20°C.

6. It is essential that cell permeabilization occur on the monolayer rather than in suspension. For reasons that are not entirely clear, release of supramolecular structures such as polyribosome-containing mRNAs is prevented when adherent cells are permeabilized in suspension, perhaps because of cytoskeletal exclusion. For cells that typically grow in suspension, this does not seem to be the case.

7. The volume used for the permeabilization and lysis buffers should be sufficient to coat the entire surface of the cell monolayer. Typical volumes for a given culture flask size are 1.5 mL for a 175-cm^2 flask, 1 mL for a 75-cm^2 flask, 0.6–1 mL for a 10-cm^2 dish, and 0.3–0.5 mL for a 4-cm^2 dish.

8. To recover polyribosomes, pre-treat cell cultures with 0.2 mM cycloheximide for 5 min at 37°C before solubilization and add 0.2 mM cycloheximide fresh to all buffers. Following detergent solubilization, layer the clarified lysate over a 1.0-M sucrose cushion containing 400 mM KOAc, 25 mM KHEPES pH 7.2, 15 mM $Mg(OAc)_2$, 1 mM DTT, 1 mM PMSF, 10 U/mL RNase Out (2:1 or 3:1 lysate : cushion). Centrifuge at $230,000 \times g_{av}$ for 30–40 min in a Beckman TLA100.2 or 100.3 rotor at 4°C. Ribosomal pellets are clear and glassy. Alternatively, resolve polyribosomes by velocity sedimentation through 15–50% linear sucrose gradients at $151,000 \times g$ for 3 h in a Beckman SW41 rotor at 4°C. Dropwise collect fractions manually by tube puncture or using an automated Isco Gradient Fractionator with a continuous UV flow cell *(9)*.

9. For larger quantities of suspension cells, simply adjust the permeabilization, wash, and lysis buffer volumes accordingly (10^7 cells/mL represents the upper limit for these conditions).

10. 3×10^7 cells represent the upper limit of material for a single SW55 centrifuge tube. If more material is desired, simply increase the amounts of cells and buffer to accommodate multiple centrifuge tubes.

11. More or less strokes may be necessary to fully rupture cells. Too many strokes can over homogenize cells leading to mixing of organeller fractions, whereas too few may not yield enough lysed cells.

12. Collecting samples from the top using a pipet allows the extraction of bands of material with specific buoyancies such as RM. Alternatively, analysis of the entire gradient would be best achieved by drop-wise collection of fractions (tube puncture). Insoluble material can plug the drain hole preventing collection by this method. If this occurs, either load less homogenate or centrifuge the homogenate before diluting in sucrose to remove nuclei and other insoluble debris as follows: centrifuge samples at $1000 \times g$ for 5 min following homogenization. Wash the pellet with 1/2 vol of homogenization buffer and repeat centrifugation. Combine the supernatants and proceed. This step generally reduces the final yield of all membrane fractions.

13. If working with small amounts of material, samples may be placed on ice or at $-20°C$ for 2 h overnight to facilitate precipitation. Additionally, 5 µg of glycogen may be added as carrier.

14. Trace amounts of DEPC inhibit most DNA/RNA polymerases; thus, nuclease-free water (Ambion) may be preferred if these applications are to follow.

15. It may be necessary to heat RNA samples to 70°C for 5–10 min to aid in resuspension.

16. MOPS running buffer can be reused many (>5) times until either RNase contamination is suspected or flies are floating in it.

17. Stringency using Church–Gilbert hybridization buffer is temperature dependent and must be empirically determined based upon the probe (length, T_m, etc.).

18. Do not allow the membrane to dry while probe is still bound, as this will permanently fix it to the membrane, preventing stripping.

Acknowledgments

The authors thank Matthew Potter and Robert Seiser for their significant contributions to the establishment of the methods described here as well as other past and present members of the Nicchitta laboratory for their insightful comments. This work was supported by NIH grant GM077382 to C. V. N. and American Heart Association predoctoral fellowship 0515333U to S. B. S.

Reference

1. Palade, G. (1975) Intracellular aspects of the process of protein synthesis. *Science* **189**(4200): p. 347–358.
2. Blobel, G. and B. Dobberstein (1975) Transfer of proteins across membranes. I. Presence of proteolytically processed and unprocessed nascent immunoglobulin light chains on membrane-bound ribosomes of murine myeloma. *J Cell Biol* **67**(3): p. 835–851.
3. Blobel, G. and B. Dobberstein (1975) Transfer of proteins across membranes. II. Reconstitution of functional rough microsomes from heterologous components. *J Cell Biol* **67**(3): p. 852–862.
4. Blobel, G., et al. (1979) Translocation of proteins across membranes: the signal hypothesis and beyond. *Symp Soc Exp Biol* **33**: p. 9–36.
5. Lingappa, V.R. and G. Blobel (1980) Early events in the biosynthesis of secretory and membrane proteins: the signal hypothesis. *Recent Prog Horm Res* **36**: p. 451–475.
6. Gilmore, R. and G. Blobel (1983) Transient involvement of signal recognition particle and its receptor in the microsomal membrane prior to protein translocation. *Cell* **35**(3 Pt 2): p. 677–685.
7. Walter, P., I. Ibrahimi, and G. Blobel (1981) Translocation of proteins across the endoplasmic reticulum. I. Signal recognition protein (SRP) binds to in-vitro-assembled polysomes synthesizing secretory protein. *J Cell Biol* **91**(2): p. 545–550.
8. Walter, P. and A.E. Johnson (1994) Signal sequence recognition and protein targeting to the endoplasmic reticulum membrane. *Annu Rev Cell Biol* **10**: p. 87–119.
9. Nicchitta, C.V., et al. (2005) Pathways for compartmentalizing protein synthesis in eukaryotic cells: the template-partitioning model. *Biochem Cell Biol* **83**(6): p. 687–695.
10. Mechler, B. and T.H. Rabbitts (1981) Membrane-bound ribosomes of myeloma cells. IV. mRNA complexity of free and membrane-bound polysomes. *J Cell Biol* **88**(1): p. 29–36.
11. Mueckler, M.M. and H.C. Pitot (1981) Structure and function of rat liver polysome populations. I. Complexity, frequency distribution, and degree of uniqueness of free and membrane-bound polysomal polyadenylate-containing RNA populations. *J Cell Biol* **90**(2): p. 495–506.
12. Diehn, M., et al. (2000) Large-scale identification of secreted and membrane-associated gene products using DNA microarrays. *Nat Genet* **25**(1): p. 58–62.
13. Lerner, R.S., et al. (2003) Partitioning and translation of mRNAs encoding soluble proteins on membrane-bound ribosomes. *RNA* **9**(9): p. 1123–1137.
14. Kopczynski, C.C., et al. (1998) A high throughput screen to identify secreted and transmembrane proteins involved in Drosophila embryogenesis. *Proc Natl Acad Sci USA* **95**(17): p. 9973–9978.

15. Stephens, S.B., et al. (2005) Stable ribosome binding to the endoplasmic reticulum enables compartment-specific regulation of mRNA translation. *Mol Biol Cell* **16**(12): p. 5819–5831.

16. Sambrook, J., E.F. Fritsch, and T. Maniatis (2001) *Molecular Cloning: A laboratory manual.* 3rd ed. Cold Spring Harbor, New York, USA: Cold Spring Harbor Press.

17. Seiser, R.M. and C.V. Nicchitta (2000) The fate of membrane-bound ribosomes following the termination of protein synthesis. *J Biol Chem* **275**(43): p. 33820–33827.

18. Adam, S.A., R.S. Marr, and L. Gerace (1990) Nuclear protein import in permeabilized mammalian cells requires soluble cytoplasmic factors. *J Cell Biol* **111**(3): p. 807–816.

19. Potter, M.D. and C.V. Nicchitta (2002) Endoplasmic reticulum-bound ribosomes reside in stable association with the translocon following termination of protein synthesis. *J Biol Chem* **277**(26): p. 23314–23320.

15

In Vivo and In Vitro Analysis of Poly(A) Length Effects on mRNA Translation

Jing Peng, Elizabeth L. Murray, and Daniel R. Schoenberg

Summary

Regulating gene expression at the translational level controls a wide variety of biological events such as development, long-term memory, stress response, transport and storage of certain nutrients, and viral infection. Protein synthesis at steady-state level can be directly measured with Western blot or using an easy-to-detect reporter such as luciferase. However, these methods do not measure the association of mRNA with ribosomes, which is more meaningful in understanding the mechanism and dynamics of translation. This chapter describes the use of sucrose density gradients for analysis of polysome profiles. RNA or protein samples extracted from gradient fractions are commonly used for further analysis of their association with translating ribosomes. We also describe an in vitro translation system prepared from HeLa S3 cell cytoplasmic extract that shows dependency on the mRNA cap and length of the poly(A) length tail, both features of translation in vivo. This is particularly useful to study the *cis*- and *trans*-acting factors involved in translational control. Lastly, we describe a method for transfecting cells with an in vitro prepared RNA to study the impact of poly(A) length on translation. This approach is particularly useful for characterizing *cis*-acting elements that work in conjunction with poly(A) in regulating translation.

Key Words: Translation; sucrose density gradient; polysome; HeLa cytoplasmic extract; in vitro translation; poly(A) tail.

1. Introduction

Initiation is the primary regulated step in the translation of most eukaryotic mRNAs *(1)*. Two features common to most mRNAs, the 5′ cap and the 3′ poly(A) tail play principal roles in translation initiation *(2,3)*. Much of our current view of the steps involved in initiation derives from the circular

From: *Methods in Molecular Biology, Vol. 419: Post-Transcriptional Gene Regulation*
Edited by: J. Wilusz © Humana Press, Totowa, NJ

polysome model *(4)*, in which eIF4G brings together the 5′ cap bound by eIF4E and the 3′ poly(A) tail bound by poly(A)-binding protein (PABP). Initiation commences with the binding of eIF3 and the met-tRNA-bound 43S ribosome subunit to this complex. Most vertebrate mRNAs exit the nucleus with a poly(A) tail of approximately 200 residues *(5)*, and this undergoes progressive shortening in the cytoplasm until it reaches a limit of 10–15 residues *(6)*, following which it is no longer associated with eIF4E or PABP *(7)* and instead becomes associated with Dcp1p, Dcp2p, and the cytoplasmic Lsm1p-7p proteins in discrete sites of mRNA degradation termed processing bodies, or P-bodies *(7,8)*. It was known for some time that decreasing poly(A) tail length is associated with reduced translation of maternal mRNAs, and recently, a similar process of deadenylation is linked to micro RNA-mediated repression of target mRNAs. Depending on the mRNA poly(A) shortening has been linked to reduced translation or to enhanced degradation. Thus, one must evaluate a particular sequence element in the context of differing length poly(A) tails to understand its function in controlling translation.

Physical measurements determined that 12 adenosines constitute the minimum length for binding by poly(A)-binding protein (PABP) *(9)*. On longer adenosine tracts, PABP binds approximately every 25 residues. The notion that poly(A) is a length-dependent enhancer of translation is supported by experiments performed in vitro using cell extracts *(2,10–12)* and in electroporated or RNA-transfected cells *(12–14)*. Tethering PABP to mRNA using MS2 fusions can stimulate translation initiation, indicating that the impact of increasing poly(A) length on translation initiation results from the increased availability of PABP bound to the mRNA 3′ end. More recent work with PABP-depleted extracts proved that PABP is an initiation factor that works by enhancing the formation of both the 48S initiation complex and the 60S subunit joining *(15)*.

Here, we describe the overall approach used in our laboratory to analyze the impact of poly(A) tail length on translation both in vivo and in vitro. Ribosome binding is the hallmark of translation in vivo, and the first half of this chapter describes the use of linear sucrose density gradients to study the efficiency of ribosome loading onto a particular mRNA. This classical approach to separating translating ribosome-bound complexes has proven to be particularly powerful for studying translating mRNP complexes when combined with immunoprecipitation, microarrays, and proteomics. Although gradient analysis has its place, it is of limited usefulness in evaluating the impact of features such as poly(A) tail length on translational regulation by a particular *cis*-acting element. For this one must turn to in vitro approaches, and we describe here the preparation of an in vitro translation system derived from cytoplasmic extract of HeLa S3 cells that is dependent on the presence of 5′ cap for initiation and shows increasing translation with increasing length of poly(A). We also describe an approach we used successfully to examine the translational efficiency of capped transcripts with varying length poly(A) tails introduced into cells.

2. Materials

2.1. Preparation of Sucrose Density Gradients

1. Dual chamber gradient maker (Model SG50, Hoefer) fitted with a small spin bar that just fits within the diameter of one chamber.
2. 1 M NaOH, to treat the gradient maker (*see* **Note 1**).
3. Diethylpyrocarbonate (DEPC)-treated double-distilled water or water processed through a polishing apparatus such as those sold by Millipore, Billerica, MA or Barnstead, Dubuge, IA.
4. High- and low-concentration sucrose buffers. Start with 15 and 40% sucrose buffers: 15 or 40% w/v sucrose, 20 mM HEPES-KOH (*N*-2-hydroxyethylpiperazine-*N*′-2-ethanesulfonic acid) (pH 7.5), 100 mM KCl, 10 mM $MgCl_2$, 100 μg/ml cycloheximide, 1 mM phenylmethylsulfonylfluoride (PMSF), 25 μl/ml protease inhibitor cocktail (Sigma, St. Louis, MO optional), 10 U/ml RNaseOUT™ (Invitrogen, Carlsbad, CA), 1 mM dithiothreitol (DTT). Prepare fresh in DEPC-treated deionized water (*see* **Note 2**).
4. Peristaltic pump. We use an LKB Bromma 2132 Microperpex peristaltic pump, but comparable units will suffice.
5. Centrifuge tubes. We use Beckman centrifuge tubes 9/16 × 3½ inches (14 × 89 mm) P.A. part number 331372, but comparable tubes will suffice.
6. Glass capillary tubes.
7. Magnetic stirring plate.

2.2. Sample Preparation for Sucrose Density Gradient Analysis

1. To insure a sufficient amount of material for application to gradients and to have sufficient material for subsequent analysis of protein and RNA grow enough cells to fill a 150-mm tissue culture dish (*see* **Note 3**).
2. Cycloheximide, dissolved in 70% ethanol at 100 mg/ml. Store at −20°C for up to 2 weeks.
3. Phosphate-buffered saline (PBS), sterile, ice-cold.
4. Lysis buffer: 20 mM HEPES-KOH (pH 7.5), 100 mM KCl, 10 mM $MgCl_2$, 0.25% NP-40 (Nonidet P-40), 100 μg/ml cycloheximide, 1 mM PMSF, 25 μl/ml protease inhibitor cocktail (Sigma), 100 U/ml RNaseOUT™ (Invitrogen), 1 mM DTT. Prepare fresh in DEPC-treated double-distilled water (*see* **Note 4**).
5. Sterile cell scrapers.
6. 1-ml syringes.
7. Needles, gauge to vary with size of cells to be lysed.

8. Ice buckets, filled. Prepare enough so each plate can sit on ice while cells are prepared.
9. Refrigerated microcentrifuge.
10. Ultracentrifuge capable of 225,004 × *g*. We use a Sorvall UltraPro 80 ultracentrifuge.
11. Swinging bucket rotor, we use a Sorvall TH-641 swinging bucket rotor. Alternatives, such as a Beckman SW41, can also be used.

2.3. Sample Collection for Sucrose Density Gradient Analysis

1. Glass capillary tubes.
2. Peristaltic pump.
3. UV monitor. We use an LKB 2138 Uvicord S UV monitor. This can be substituted with various comparable alternative models.
4. Fraction collector. We use an LKB 2112 Redirac fraction collector. This can be substituted with various comparable alternative models.
5. Chart recorder. We use an LKB 2210 two-channel recorder. This can be substituted with various alternative models.

2.4. Preparation of HeLa Cytoplasmic Extract for In Vitro Translation

1. Complete medium for HeLa S3 cells: Dulbecco's Modified Eagle's Medium (DMEM, Invitrogen) supplemented with 4 mM L-glutamine (Invitrogen) and 10% fetal bovine serum (FBS, Invitrogen).
2. PBS (Amresco, Solon, OH).
3. Hypotonic MC buffer: 10 mM HEPES/KOH, pH 7.4, 10 mM potassium acetate, 0.5 mM magnesium acetate, 5 mM DTT, 1 mM PMSF, 25 μl/ml protease inhibitor cocktail (Sigma). Prepare fresh.
4. Spinner flasks that are suitable for growing cells in suspension. We use 1 L ProCulture Spinner Flasks (Corning, Corning, NY).
5. Magnetic stirring plate for spinner cell culture.
6. Dounce homogenizer with a tight-fitting pestle.
7. Refrigerated microcentrifuge.

2.5. Preparation of m⁷GpppG-Capped Transcripts with Varying Length Poly(A) Tails

1. mMESSAGE mMACHINE™ High Yield Capped RNA Transcription SP6 Kit (Ambion, Austin, TX).
2. Plasmid template that has luciferase coding sequence downstream of an SP6 promoter and varying length poly(A) at the 3′ end (*see* **Note 5**).

2.6. Preparation of ApppG-Capped Transcripts

1. MEGAscript® transcription SP6 kit (Ambion).

2. ApppG, 20 mM (Ambion).
3. Plasmid template that has luciferase coding sequence downstream of a SP6 promoter and varying length poly(A) at the 3´ end (*see* **Note 5**).

2.7. In vitro translation

1. HeLa cytoplasmic extract.
2. In vitro transcripts diluted in DEPC-treated water at the concentration of 15 fmol/µl (*see* **Note 6**).
3. Translation buffer (5×): 30 mM HEPES-KOH, pH 7.4, 330 mM KCl, 5 mM MgCl$_2$, 5 mM ATP, 0.25 mM GTP, 125 µM bovine liver tRNA, 50 µM amino acid mix (Promega), 21 mM β-mercaptoethanol, 50 mM creatine phosphate, 125 ng/µl creatine phosphokinase (optional), 1.2 mM spermidine. Store aliquots for single use at –80°C. These can be used for up to 1 year. (*see* **Note 7**).
4. RNaseOut™ (Invitrogen).
5. Cycloheximide, 100 mg/ml in 70% ethanol.
6. Firefly luciferase assay reagent (Promega, Madison, WI).
7. Luminometer for luciferase activity assay. We use the Sirius Luminometer (Berthold Detection Systems, Oak Ridge, TN). This can be substituted with various comparable alternative models.

2.8. Cell Transfection with Capped and Polyadenylated RNA

1. For LM(tk-) cells used in our laboratory, complete medium consists of DMEM (Invitrogen) supplemented with 4 mM L-glutamine (Invitrogen) and 10% FBS (Invitrogen).
2. OPTI-MEM® medium (Invitrogen).
3. Lipofectamine 2000™ (Invitrogen).
4. Dual-Luciferase® Reporter Assay System (Promega).

3. Methods
3.1. Preparation of Sucrose Density Gradients

1. Treat the gradient maker with 1 M NaOH to remove RNase contamination. Soak at least for 20 min and then rinse thoroughly with deionized water. While the gradient maker is soaking, prepare sucrose buffers and place on ice.
2. In a cold room, prepare the pump and gradient maker. Disconnect the pump from other equipment, attach a capillary tube to one end of the tubing and place the other end of the tubing over a container to catch waste. Insert the capillary tube into DEPC-treated double-distilled water and turn on the pump to run water through the tubing for a few (2–5) minutes to clean it. The pump should run at 1 ml/min. Remove tubing from capillary tube and run water out of the tubing.
3. Next, attach the gradient maker to one end of the pump tubing. The gradient maker has two cylindrical chambers connected by a small channel controlled by

a stopcock. One chamber connects to an outlet through a second channel; this is also controlled by a stopcock. Connect the gradient maker to the tubing on the pump through this outlet.

4. Place the gradient maker on the stir plate and place the stir bar in the chamber closest to the outlet. Center the gradient maker so that the stir bar turns smoothly in its chamber.

5. Close all stopcocks, add several millilitres of DEPC-treated double-distilled water to the chamber furthest from the outlet, turn on the magnetic stirring plate, turn on the pump, and open the stopcocks to wash and check the system.

6. When the water has drained from gradient maker and air has entered the tubing, turn off the pump and the magnetic stirring plate and close the stopcocks. Add the higher percentage sucrose buffer to chamber farthest from the tubing connection; add the lower percentage sucrose buffer to the chamber closest to tubing connection. Use an equal volume of each buffer. For example, to pour a 11-ml gradient, put 5.5 ml high-concentration sucrose buffer in the chamber farthest from the outlet, and 5.5 ml low-concentration buffer in the chamber closest to the outlet.

7. Turn on the magnetic stirring plate to the lowest speed that allows for mixing. Turn on the peristaltic pump and open the stopcock to begin mixing of the sucrose solutions. The lighter sucrose solution should immediately enter the tubing from the gradient maker, and the chamber with the spin bar should show a swirling pattern caused by the mixing of the heavier sucrose solution with the lighter one.

8. Insert one end of a clean capillary tube into the tubing before the sucrose buffer reaches the farthest end of tube and place the glass capillary into the bottom of a centrifuge tube. Mixed sucrose buffers should enter the centrifuge tube at approximately 1 ml/min. Care should be taken at this step to avoid introducing bubbles (*see* **Note 8**).

9. Gradients are finished when most of each sucrose solution is gone from the gradient maker—i.e., before bubbles enter the gradient in the centrifuge tube. Stop the peristaltic pump and carefully remove the glass capillary from gradient (*see* **Note 9**).

10. Gradients are stored at 4°C. This can be done in the rotor bucket, and gradients can be prepared the evening before use. To insure reproducibility between each gradient, the gradient maker is rinsed with DEPC-treated double-distilled water before the next gradient is prepared (*see* **Note 10**).

3.2. Sample Preparation for Sucrose Density Gradient Analysis

1. Turn on ultracentrifuge to allow the machine to cool to 4°C before use. Pre-cool the rotor in a cold room, refrigerator, or place it into the cooled ultracentrifuge.

2. Cycloheximide stops elongating ribosomes on mRNAs and is useful for stabilizing polysome complexes. If you choose to use this reagent, add it to cells at a final concentration of 100 µg/ml within 15 min of cell harvest.

3. Prepare the lysis buffer and keep it on ice. Move materials for sample preparation to cold room. Prepare a sufficient number of ice buckets to accommodate all of

the plates of cells. Keeping everything cold is important for assay quality and reproducibility.

4. Remove cells from the CO_2 incubator and place the plates on ice in a cold room. Wash plates twice with ice-cold PBS and tilt each plate to drain remaining PBS.

5. Add 1 ml ice-cold PBS and remove adherent cells from the surface using a cell scraper. The suspended cells are removed to a 1.5-ml microfuge tube.

6. Cells are collected by centrifuging at 1000 × g for 2 min, 4°C, after which the PBS is removed leaving the cell pellet.

7. Add 0.5 ml lysis buffer to each cell pellet and pipet gently to resuspend.

8. Lyse cells by passage through syringe (*see* **Note 11**). The size of the needle and the number of passages is dependent on cell type (*see* **Note 12**). Monitor lysis through microscopy (*see* **Note 13**).

9. Pellet nuclei and cell debris by centrifuging at 12,000 × g for 5 min, 4°C. Carefully load the supernatant (cytoplasmic extract) onto the previously prepared sucrose gradients.

10. Use lysis buffer to adjust all of the centrifuge tube samples to the same weight for balancing the rotor.

11. Add small drop of Spinkote® grease to screw threads on lids of buckets and screw lids on buckets.

12. Add all of the buckets to the pre-cooled rotor and carefully place onto the spindle in the ultracentrifuge.

13. The rotor is centrifuged at 225,000 × g for 2.5 h at 4°C. We do not use the brake to slow the rotor, and it takes about 75 min for rotor to stop spinning at end of run without the brake.

3.3. Sample Collection From Sucrose Density Gradients

1. The UV monitor, chart recorder and fraction collector are kept in a cold room or refrigerated chromatography cabinet. The UV monitor is turned on when the centrifuge run of the sucrose gradients is over but before rotor stops spinning to allow sufficient time for the lamp to warm up and stabilize (30–60 min before use).

2. The individual gradient fractions are collected into microfuge tubes. These are labeled and placed into the fraction collector to cool. We typically use 11-ml gradients, and the added sample volume and void volume within the tubing requires 24 tubes per gradient for collecting 0.5 ml fractions.

3. When the centrifuge run is finished, the rotor is carefully removed from the ultra-centrifuge, the buckets are removed from the rotor and placed into their holder, and taken to the cold room where forceps are used to remove individual tubes from the buckets.

4. Prepare the collection apparatus by connecting the output of the UV monitor to a chart recorder and the flow cell within the monitor to the fraction collector. Connect pump tubing to the UV monitor inlet tubing, allowing the outlet from UV monitor tubing to hang over a container to collect waste. Reverse the pump direction and

rinse tubing once with DEPC-treated double-distilled water until you introduce a bubble into the line.

5. Insert a capillary tube into the end of tubing and place this into the 15% sucrose buffer solution. Turn on the pump and fill the line with the 15% sucrose solution. At this point the pump is turned off. Adjust the baseline on the chart recorder to zero before adjusting zero absorbance on the UV monitor. Remove the capillary tube from tubing and turn on the pump until you introduce an air bubble into line.

6. A new capillary tube is placed on the end of the tubing, and this is immobilized vertically on a ring stand using a small clamp. The clamp on the ring stand is loosened, and the capillary is slowly lowered to the bottom of the centrifuge tube taking care not to disturb the gradient.

7. At this point, the pump is turned on and fractions are pumped from the bottom of the tube through the UV monitor and into the fraction collector until all of the solution has been removed. When air enters the line, the pump is turned off. A typical profile for absorbance at 254 nm is shown in **Fig. 1**. In this particular experiment, 48 0.25 ml fractions were collected, but as few as twenty-four 0.5 ml fractions provide sufficient separation for most applications. Because these are collected from the bottom of the tube, the numbering in this figure starts with the most rapidly sedimenting fractions. The tubing and flow-cell within the UV monitor are rinsed with DEPC-treated double-distilled water after each gradient, and the fractionation process outlined above is repeated.

8. When all of the gradients have been collected, the apparatus is cleaned by rinsing with DEPC-treated double-distilled water, taking care to clean up any residual

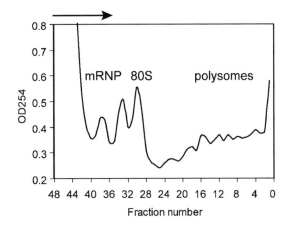

Fig. 1. Absorbance profile of RNP and polysome complexes separated on a sucrose density gradient. A cytoplasmic extract of LM(tk-) murine fibroblasts was separated on a 15–40% sucrose density gradient. Shown is the absorbance profile at 254 nm as the indicated fractions pass through a UV monitor.

sucrose that may have spilled. The individual fractions may be processed immediately or stored at –80°C until ready for further analysis (*see* **Note 14**).

3.4. Preparation of HeLa Cytoplasmic Extract for In Vitro Translation

1. HeLa S3 cells are maintained in complete medium in plastic cultureware in a humidified 37°C incubator supplemented with 5% CO_2. To grow HeLa S3 cells in suspension, trypsinized cells are counted and seeded into a 250-ml spinner flask at a concentration of 4–8 × 10^4cells/ml. When the cell density reaches 4–6 × 10^5 cells/ml, the entire 250-ml culture is inoculated into a 1-L spinner flask containing 750 ml of medium. The cells are ready for harvest when the cell density reaches 4–6 × 10^5 cells/ml (*see* **Note 15**).
2. Harvest the cells by centrifugation at 200 × *g* for 10 min at 4°C in a Sorvall HS-4 rotor.
3. Wash the cell pellet three times using 10 vol of ice-cold PBS (*see* **Note 16**).
4. Resuspend the cell pellet with an equal volume of hypotonic MC buffer.
5. Allow the cells to swell on ice for 5–10 min.
6. Lyse the cells with 25–30 strokes of a Dounce homogenizer using a tight-fitting (A) pestle (*see* **Note 17**).
7. Remove the cell debris and nuclei by centrifugation at 15,000 × *g* for 20 min at 4°C in a refrigerated microcentrifuge (*see* **Note 18**).
8. Collect the supernatant and store in small aliquots at –80°C (*see* **Note 19**).

3.5. Preparation of m⁷Gpppg-Capped Transcripts with Varying Length Poly(A) Tails

1. Set up a 20 µl in vitro transcription reaction on ice following the standard instruction from the mMESSAGE mMACHINE® kit. The reaction mixture includes 1 µg of linear plasmid DNA template, 10 µl 2× NTP mix, 2 µl 10× Reaction Buffer, and 2 µl SP6 RNA polymerase (*see* **Note 20**).
2. Incubate the reaction mixture at 37°C overnight (*see* **Note 21**).
3. Add 1 µl DNase I and incubate for 15 min at 37°C to remove the plasmid DNA template.
4. Add 30 µl water and 25 µl Lithium Chloride Precipitation Solution and incubate at –20°C for at least 30 min (*see* **Note 22**).
5. Centrifuge at 15,000 × *g* for 15 min at 4°C to pellet the transcript.
6. Carefully remove the supernatant and wash the pellet once with 1 ml 70% ethanol.
7. Resuspend the transcript in 30 µl water and measure the concentration with a spectrophotometer (*see* **Note 23**).

3.6. Preparation of ApppG-Capped Transcripts

1. Set up a 20-µl in vitro transcription reaction on ice following the standard instruction from the MEGAscript® SP6 kit. The reaction mixture includes 1 µg of linear plasmid DNA template, 2 µl 10× Reaction Buffer, and 2 µl each ATP, CTP, and

UTP Solution, 0.4 µl GTP solution, 4 µl 20 mM 5′ApppG, and 2 µl SP6 Enzyme Mix (*see* **Note 20**).

2. Incubate the reaction mixture at 37°C overnight (*see* **Note 21**).
3. Add 1 µl DNase I and incubate for 15 min at 37°C to remove the plasmid DNA template.
4. Add 30 µl water and 25 µl Lithium Chloride Precipitation Solution and incubate at –20°C for at least 30 min (*see* **Note 22**).
5. Centrifuge at 15,000 × *g* for 15 min at 4°C to pellet the transcript.
6. Carefully remove the supernatant and wash the pellet once with 1 ml 70% ethanol.
7. Resuspend the transcript in 30 µl water and measure the concentration with a spectrophotometer (*see* **Note 23**).

3.7. In vitro Translation

1. Assemble the reaction mixture on ice as follows: 2 µl transcript (30 fmol), 2 µl water, 2 µl 5× translation buffer, 4 µl HeLa cytoplasmic extract, and 1 µl (4 units) RnaseOut™. If desired, you may include 6 fmol of a control transcript such as m^7GpppG-capped Renilla luciferase transcript with a 78 nucleotide poly(A) tail. This control transcript is used to normalize results obtained with firefly luciferase transcripts with varying length poly(A) tails (*see* **Fig. 2**). If you choose to do this, reduce the amount of test firefly luciferase RNA construct in the reaction from 30 to 24 fmol.
2. Incubate the reaction mixture at 37°C for 60–90 min.
3. Stop the translation reaction by adding 1 µl of 100 µg/µl cycloheximide.
4. Dilute 2 µl of the reaction mixture with 8 µl 1× PBS, add 50 µl luciferase assay reagent from the Dual-Luciferase® Reporter Assay kit, and measure firefly and Renilla luciferase activity with a luminometer. Results showing the impact of the cap and increasing length poly(A) on translation in vitro *(12)* are presented in **Fig. 2**. The first two samples consist of luciferase RNA with a non-functional ApppG cap and either no poly(A) or a 98 residue poly(A) tail. The next five samples show the translation of capped firefly luciferase mRNA with 0, 14, 20, 54, and 98 residue long poly(A) tails. In this figure, firefly luciferase translation is normalized to an internal control of Renilla luciferase mRNA (*see* Northern blot in **Fig. 2B**), and the fold increase is determined by the amount of activity relative to capped luciferase mRNA with no added poly(A) (cap A0).

3.8. Translation of Luciferase mRNA with Varying Length Poly(A) Tails in RNA-Transfected Cells

1. LM(tk-) cells are grown in complete medium in a humidified 37°C incubator supplemented with 5% CO_2. The day before transfection, $1.6 × 10^4$ cells in 100 µl complete medium are seeded into each well of a 96-well culture dish.

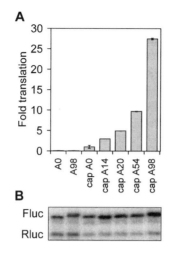

Fig. 2. In vitro translation of firefly luciferase mRNAs with increasing poly(A) tail lengths in HeLa extract. (**A**)Triplicate samples of firefly luciferase mRNA with a non-functional ApppG cap and 0 or 98 nucleotide poly(A) tails (A0, A98), or m^{7}GpppG-capped firefly luciferase mRNA with poly(A) tails of 0, 14, 20, 54, or 98 nucleotides were translated in HeLa cytoplasmic extracts together with a fixed amount of m^{7}GpppG-capped Renilla luciferase mRNA. The relative light units of each luciferase reporter were determined in a luminometer and results with each sample of firefly luciferase were normalized to the internal Renilla luciferase control. The value for m^{7}GpppG-capped firefly luciferase mRNA with no poly(A) (cap A0) was arbitrarily set to one to quantify the fold increase in translation observed with increasing length poly(A) on the reporter mRNA. Shown are the mean ± standard deviation for the triplicate determinations. (**B**) Each of the samples in (**A**) were pooled and RNA extracted after translation in vitro was analyzed using Northern blot. The blot was hybridized with a mixed probe for firefly (Fluc) and Renilla (Rluc) luciferase mRNA.

2. Dilute 0.4 µl Lipofectamine 2000® in 10 µl OPTI-MEM® for each well of cells to be transfected. This can be prepared in a master mix. Incubate the mixture for 5–10 min at room temperature.

3. During the incubation, dilute 160 ng (~240 pmol) firefly luciferase transcript and 40 ng (~120 pmol) m^{7}GpppG-capped Renilla luciferase transcript with 78-nt poly(A) tail in 10 µl OPTI-MEM® for each well of cells to be transfected. (*see* **Note 24**).

4. Combine the diluted transfection reagent from step 2 and diluted RNA from step 3 and incubate the mixture for 20 min at room temperature.

5. Add the transfection mixture to each well of cells and return the cells to the cell culture incubator.

6. Harvest the cells 60–90 min after transfection (*see* **Note 25**). Wash each well of cells twice with 100 µl PBS, then lyse the cells with 1× Passive Lysis Buffer (diluted

Peng et al.

A

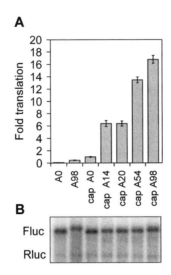

B

Fig. 3. Translation of firefly luciferase mRNA with increasing poly(A) tail lengths in RNA-transfected LM(tk-) cells. (**A**) Triplicate cultures of LM(tk-) cells were transfected with the indicated firefly luciferase transcripts with varying length poly(A) tails together with a Renilla luciferase RNA control as in **Fig. 2**. Firefly luciferase activity was normalized to that of co-transfected Renilla luciferase, and the results are plotted as in **Fig. 2**. (**B**) RNA extracted from the transfected cells was analyzed using Northern blot as in **Fig. 2** using a mixed probe for firefly (Fluc) and Renilla (Rluc) luciferase mRNA.

in water from the 5× buffer provided by Dual-Luciferase Reporter Assay kit) by incubation at room temperature for 15 min (*see* **Note 26**).

7. Take 10 µl of the cell lysate for use in firefly and Renilla luciferase activity assays. As outlined above, use 50 µl luciferase activity assay reagent from the kit and assess results with a luminometer.

8. An example of the data obtained using this approach is shown in **Fig. 3**. As seen with in vitro translation using HeLa extract, there is little translation of ApppG-capped RNA regardless of poly(A) tail length and translation of capped luciferase mRNA clearly increases with increasing poly(A) length. A Northern blot of RNA recovered from RNA-transfected cells is an essential control for this experiment to insure that any changes observed in translation are not due to differences in mRNA decay.

4. Notes

1. All equipment and buffers should be prepared in a manner that is consistent with preventing RNase contamination and protein degradation. Plastic tubes and tips should be free of DNase and RNase. Solutions should be made with DEPC-treated

deionized water and equipment should be soaked in 1 M NaOH. Sample preparation should be performed on ice in the cold room.

2. The volume of sucrose buffers to prepare depends on the volume of the gradients, the number of gradients, and the volume contained in the tubing connected to the pump. Each gradient is composed of equal volumes of the high- and low-concentration sucrose buffers. Sufficient amount of the lower concentration sucrose buffer should be prepared to have enough for use in filling the tubing between each gradient and zeroing the UV monitor.

3. High-quality cell cultures are essential to the quality of the resulting polysome profile. Cells should be in log-phase growth at the time of harvest. While they should not be crowded on the plate, it is recommended that they are near confluency to have enough material (protein or RNA) for subsequent studies to be performed with the collected gradient fractions.

4. Although only 0.5 ml lysis buffer is required per sample, prepare extra for use in balancing samples before centrifugation. This should be prepared immediately before use.

5. The plasmids need to have the gene of interest downstream of a common bacteriophage promoter (T3, T7, or SP6) for transcription in vitro. We use luciferase (either firefly or Renilla) because it is very easy to measure the activity after the in vitro translation reaction. To analyze the relationship between translation efficiency and poly(A) tail length of the mRNA, we have prepared a series of plasmids that contain 0, 14, 20, 54, and 98 adenosines downstream of the luciferase coding sequence. It is important to design the templates so that they are linearized by digestion with a restriction enzyme that cuts at the precise end of the poly(A) stretch. The plasmids developed in our laboratory have an SstII site at this location. In addition, we commonly use a capped Renilla luciferase control transcript with a 78 residue poly(A) tail as an internal control in each reaction.

6. The amount of transcript for the translation reaction needs to be titrated. The amount of protein produced will increase with increasing amounts of input RNA; however, after a certain point, the translation reaction becomes less dependent on poly(A) length. For the luciferase RNA used in our laboratory, 30 fmol of transcript in a 10-µl reaction produces enough protein and also shows good poly(A) length dependency.

7. The conditions of the translation reaction need to be optimized, especially magnesium and potassium concentrations. While we have provided here concentrations that work reproducibly in our hands, subtle differences in sample preparation can affect this.

8. Both chambers of the gradient maker should drain equally. If not, check that there is no obstruction in the connection between the two chambers. Care should be taken to avoid getting bubbles between the chambers as this is a common source of this problem. In addition, the entire apparatus should be checked for leaks before pouring the gradients.

9. Bubbles cause mixing of the gradient and care should be taken to prevent this. In the event that bubbles occur in a gradient, it is best to pour this out and prepare a new gradient.

10. Sucrose gradients may be prepared the night before use, but no longer than that. If prepared too far in advance, there is enough diffusion that the separation quality is reduced.

11. Alternately, cells may be lysed with a Dounce homogenizer.

12. The gauge needle required at this step depends on size of cells to be lysed. A higher gauge needle (i.e., smaller diameter) is used for smaller cells. We use a 28½-gauge needle for LM(tk-) cells (mouse fibroblast) and a 25-gauge needle for Cos-1 cells (monkey).

13. Although some clearing of the cell suspension may be evident after lysis by passing through the syringe, it is best to monitor this under the microscope. Lysed cells should appear as nuclei clinging to strips of outer membrane and debris should be obvious in surrounding medium. To prepare a wet mount, place about 20 μl of lysed cells on a microscope slide, cover with a coverslip, and view under phase contrast at 40× magnification. If more than 10% of the cells remain intact, the suspension should be passed twice more through the syringe and needle and then examined once again under the microscope. Although monitoring the cell lysis is important, it is also important to stress that one should strive to minimize the amount of time spent handling these extracts before application to the gradient.

14. RNA and proteins can be recovered from each fraction. We use 2 ml Trizol® (Invitrogen) to extract RNA and 10% trichloroacetic acid (TCA) to precipitate protein from each 0.5 ml fraction. If a protein is particularly abundant, gradient fractions can be processed by several spins through a centrifugal microdialyzer (such as a Microcon®) to remove excess sucrose.

15. The HeLa S3 cells in our laboratory are HeLa S3 (tet off) from Clontech/BD Biosciences, San Jose, CA. The doubling time of this cell line is about 24 h under the conditions described above. If cell density is less than 10^4/ml, the cells tend to grow more slowly but they are still healthy. It may be necessary to collect cells by centrifugation when replacing the medium every 3–4 days. However, it is critical to harvest the cells at the optimal density for best translation activity.

16. The cell pellet is about 3.5 ml for every litre of cells grown at density of 6×10^5/ml.

17. It is recommended that cell lysis is monitored as in **Note 13** by using phase contrast microscopy. Ideally one will see lysed cell debris and intact nuclei at this step. It is important to minimize nuclear lysis as these nuclear contaminants inhibit translation activity.

18. If the supernatant is still very cloudy, repeat the centrifugation step. The final extract should be almost clear with a light yellowish color. Try to remove the white lipid layer on top of the supernatant, but the extract will still work with a little bit of lipid.

19. In our experience, extracts retain essentially full translational activity through two cycles of freezing/thawing. We normally store fresh extract in 1 ml aliquots. Before use, a tube is placed on ice to thaw and any extract that is not used can be re-frozen in 100 µl aliquots and used one more time.

20. If the gene of interest is downstream of a different promoter (T3 or T7), use the corresponding RNA polymerase instead.

21. SP6 RNA polymerase is slower than T3 and T7, so the reaction takes longer for maximum production of the transcript. We normally use overnight (~16 h) incubation. If the T3 or T7 RNA polymerases are used, a 2-h incubation is sufficient.

22. It is critical to remove the unincorporated m^{7}GpppG cap analog from the transcript, because it will inhibit the translation reaction in later step. LiCl precipitation is a simple and fast way to recover the transcription product while leaving the unincorporated nucleotides, including m^{7}GpppG cap analog, in the supernatant. Other methods such as spin column chromatography can also be used. However, phenol : chloroform extraction and ethanol (or isopropanol) precipitation is not recommended because the incorporated nucleotides are not removed completely.

23. It is recommended that each transcript be analyzed by using denaturing gel electrophoresis for quality control purposes. In addition, this gel can be used as a purification step to insure the quality of the transcript if necessary.

24. We normally transfect each firefly luciferase transcript in triplicate.

25. The transcripts transfected in the cells start to degrade 2 h after transfection. To assess the impact of a particular sequence element on translation, it is important to harvest the cells before this. As an additional control, it is important to extract RNA and run a Northern blot to insure that RNA decay has not occurred or that any sequence element within the reporter RNA has not preferentially activated decay.

26. It is optional to remove the cell debris by centrifugation before measuring luciferase activities.

Acknowledgments

This work was supported by PHS grant R01 GM38277 and GM55407 to D.R.S. E.L.M. was supported in part by PHS grant T32 CA09338, and support for core facilities was provided by PHS center grant P30 CA16058 from the National Cancer Institute to The Ohio State University Comprehensive Cancer Center.

References

1. Arava, Y., Wang, Y., Storey, J. D., Liu, C. L., Brown, P. O., and Herschlag, D. (2003) Genome-wide analysis of mRNA translation profiles in Saccharomyces cerevisiae. *Proc. Natl. Acad. Sci. USA* **100,** 3889–3894.

2. Raught, B., Gingras, A. C., and Sonenberg, N. (2000) Regulation of ribosomal recruitment in eukaryotes, in *Translational Control of Gene Expression*

(Sonenberg, N., Hershey, J. W. B., Mathews, M. B., and Ab, G., eds), Cold Spring Harbor Laboratory Press, Cold Spring Harbor, NY, pp. 245–293.

3. Mangus, D. A., Evans, M. C., and Jacobson, A. (2003) Poly(A) binding proteins: multifunctional scaffolds for the post-transcriptional control of gene expression. *Genome Biol.* **4,** 223.

4. Sachs, A. B. (2000) Physical and functional interactions between the mRNA cap structure and the poly(A) tail, in *Translational Control of Gene Expression* (Sonenberg, N., Hershey, J. W. B., Mathews, M. B., and Ab, G., eds), Cold Spring Harbor Laboratory Press, Cold Spring Harbor, NY, pp. 447–465.

5. Wahle, E. and Keller, W. (1996) The biochemistry of polyadenylation. *Trends Biochem. Sci.* **21,** 247–250.

6. Tharun, S. and Parker, R. (1997) Mechanisms of mRNA turnover in eukaryotic cells, in *mRNA Metabolism and Post-transcriptional Gene Regulation* (Harford, J. and Morris, D. R., eds). Wiley, New York, pp. 181–200.

7. Tharun, S. and Parker, R. (2001) Targeting an mRNA for decapping: displacement of translation factors and association of the Lsm1p-7p complex on deadenylated yeast mRNAs. *Mol. Cell* **8,** 1075–1083.

8. Sheth, U. and Parker, R. (2003) Decapping and decay of messenger RNA occur in cytoplasmic processing bodies. *Science* **300,** 805–808.

9. Sachs, A. B. (1987) A single domain of yeast poly(A)-binding protein is necessary and sufficient for RNA binding and cell viability. *Mol. Cell. Biol.* **7,** 3268–3276.

10. Preiss, T., Muckenthaler, M., and Hentze, M. W. (1998) Poly(A)-tail-promoted translation in yeast: implications for translational. *RNA* **4,** 1321–1331.

11. Bergamini, G., Preiss, T., and Hentze, M. W. (2000) Picornavirus IRESes and the poly(A) tail jointly promote cap-independent translation in a mammalian cell-free system. *RNA* **6,** 1781–1790.

12. Peng, J. and Schoenberg, D. R. (2005) mRNA with a <20 nt poly(A) tail imparted by the poly(A)-limiting element is translated as efficiently *in vivo* as long poly(A) mRNA. RNA 11, 1131–1140.

13. Gallie, D. R. (1991) The cap and poly(A) tail function synergistically to regulate mRNA translational efficiency. *Genes Dev.* **5,** 2108–2116.

14. Grudzien, E., Kalek, M., Jemiely, J., Darzynkiewicz, and Rhoads, R. E. (2006) Differential inhibition of mRNA degradation pathways by novel cap analogs *J. Biol. Chem.* **281,** 1857–1867.

15. Kahvejian, A., Svitkin, Y. V., Sukarieh, R., M'Boutchou, M. N., and Sonenberg, N. (2005) Mammalian poly(A)-binding protein is a eukaryotic translation initiation factor, which acts via multiple mechanisms. *Genes Dev.* **19,** 104–113.

16

A Ribosomal Density-Mapping Procedure to Explore Ribosome Positions Along Translating mRNAs

Naama Eldad and Yoav Arava

Summary

The number and distribution of ribosomes on a transcript provide useful information in ascertaining the efficiency of translation. Herein we describe a direct method to determine the association of ribosomes with specific regions of an mRNA. The method, termed Ribosome Density Mapping (RDM), includes cleavage of ribosomes-associated mRNAs with RNase H and complementary oligodeoxynucleotide followed by separation of the cleavage products on a sucrose gradient. The gradient is then fractionated and the sedimentation position of each mRNA fragment is determined by northern analysis. Although developed for yeast mRNAs, RDM is likely to be applicable to various other systems.

Key Words: Translation; mRNA; ribosome density; polysome; RNase H cleavage.

1. Introduction

Ribosomal association with an mRNA provides important information regarding various aspects of translation. The number of ribosomes associated with an mRNA is usually related to the amount of proteins that it synthesizes, and changes in ribosomal association are linked in many cases to changes in the amount of synthesized protein. Thus, in cases where direct measurements of protein levels are unavailable, assays of ribosomal association may be used as a proxy.

In studies related to translation per se, the number of ribosomes on an mRNA serves as a convenient output for the sum of events that occur during its

From: *Methods in Molecular Biology, Vol. 419: Post-Transcriptional Gene Regulation*
Edited by: J. Wilusz © Humana Press, Totowa, NJ

translation. These events include ribosome binding (i.e., initiation), ribosome progression along the mRNA (elongation), and ribosome dissociation (termination). It is therefore possible to gain insights into various aspects of translation by examining changes in ribosomal association following the disturbance of various events.

However, exploring translation by examining the overall number of ribosomes on an mRNA is limited as it only reports the balance between all underlying steps, and information regarding a specific affected step is not available. This limitation is critical because there might be many scenarios in which different processes will lead to the same number of bound ribosomes. For example, slow ribosomal dissociation from the mRNA might lead to an increase in ribosome number, exactly as will happen if the rate of ribosome binding is high or if ribosomes pause at specific sites during their progression along the mRNA. On the other hand, a decrease in the number of ribosomes on an mRNA might be due to either lower rates of ribosome binding or limited processivity (i.e., frequent dissociation of ribosomes during elongation).

Information regarding the distribution of ribosomes along an mRNA, in addition to their number, can help in generating a more complete picture regarding the translation process. Many molecular events are expected to change ribosomal density along the mRNA; for example, limited processivity will be characterized by higher density at the 5′ end compared with the 3′ end, and slower termination might yield a local increase in density at the termination region. Moreover, many regulatory events, such as pausing of ribosomes at the 5′ UTR or at certain positions along the ORF, will result in a characteristic distribution pattern. Thus, methods that assess the *distribution* of ribosomes along an mRNA provide important information about the individual translation stages and their variation among different mRNAs and under different conditions.

Determination of ribosome distribution along an mRNA can be done in vitro by a "toe printing" procedure in which a radiolabeled primer is used to synthesize cDNA until the bound ribosome blocks the extension process. This method is used routinely to study ribosomal association at the initiation and termination sites *(1)*. Pause sites along the mRNA can be detected by micro-coccal digestion of mRNA associated with ribosomes, followed by annealing to cDNA and a primer extension reaction that will terminate at the position of the pause site; this assay relies on the accumulation of ribosomes at a specific position on the mRNA and was used to identify pause sites in vitro on preprolactin mRNA induced by the signal recognition particle *(2,3)*. Finally, ribosomal position can sometimes be inferred indirectly from the results of mutagenesis experiments, e.g., *(4–8)*.

The ribosome density-mapping (RDM) procedure *(9)* described below is a general and direct method for identifying ribosomal association with different regions of mRNAs in vivo.

1.1. General Concept

The main idea behind RDM is the determination of the number of ribosomes on fragments of an mRNA of interest rather than the full transcript. It is based on the well-established procedure of isolating mRNAs associated with ribosomes by velocity sedimentation in a sucrose gradient, with an additional step whereby the mRNA of interest is cleaved at a specific site by RNase H and antisense oligodeoxynucleotide (ODN). Four general steps are involved in the procedure (*see* **Fig. 1**): (1) Separation of all cellular mRNAs according to their ribosomal association in a sucrose gradient; (2) Collection of a specific fraction and cleavage of the mRNA of interest at specific sites by RNase H and ODN; (3) Separation of the cleavage products according to their ribosomal association in a sucrose gradient; and (4) Determination of the sedimentation position of the cleavage products (i.e., their number of ribosomes) by northern analysis. Each of these steps is described in **Subheading 3**.

Each of the steps involved in RDM has been utilized separately in many organisms and for the study of various mRNAs. Thus, although the protocol presented herein is for the analysis of yeast cells grown at optimal conditions, it can be easily modified to other experimental systems.

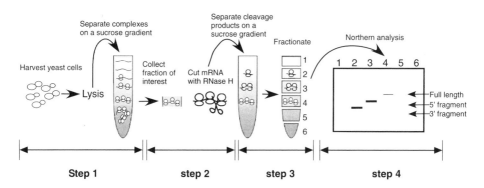

Fig. 1. Schematic presentation of the ribosome density-mapping (RDM) steps. See details of each step in **Subheading 3**.

2. Materials

1. Cyclohexamide (CHX): for stock solution, dissolve CHX in water to 10 mg/ml (vortex well) and store at –20°C. CHX is highly toxic.
2. Heparin: for stock solution, dissolve heparin in water to 10 mg/ml and store at –20°C.
3. RNase inhibitor: ribonuclease inhibitor (porcine) (Takara Cat. 2311A, 40 U/µl) or RNasin (Promega Cat. N2511, 40 U/µl).
4. RNase NEB H: New England Biolabs (Cat. M0297, 5 U/µl).
5. Lysis buffer: 20 mM Tris–HCl pH 7.4, 140 mM KCl, 1.5 mM MgCl$_2$, 0.5 mM dithiothreitol (DTT), 1% Triton X-100, 0.1 mg/ml CHX, 1 mg/ml heparin.
6. Gradient (with heparin): 11 ml 10–50% sucrose in a buffer: 20 mM Tris–HCl pH 7.4, 140 mM KCl, 5 mM MgCl$_2$, 0.5 mM DTT, 0.1 mg/ml CHX, 1 mg/ml heparin.
7. Gradient (without heparin): 11 ml 10–50% sucrose in a buffer: 20 mM Tris–HCl pH 7.4, 140 mM KCl, 5 mM MgCl$_2$, 0.5 mM DTT, 0.1 mg/ml CHX.
8. Antisense ODN: approximately 20 bases long and approximately 50% GC content. Oligonucleotides should be desalted before use.
9. 5× RNase H buffer: 0.1 M Tris–HCl pH 7.4, 0.5 M KCl, 0.1 M MgCl$_2$, 0.5 mM DTT, 2.5 mg/ml CHX.
10. Lysis minus detergent (LMD) buffer: 20 mM Tris–HCl pH 7.4, 140 mM KCl, 1.5 mM MgCl$_2$, 0.5 mM DTT, 0.1 mg/ml CHX, 1 mg/ml heparin.
11. 8 M guanidinium HCl (GuHCl): for stock solution dissolve GuHCl in water. Heat to allow a complete dissolving.
12. 10× 3-(*N*-morpholino)propanesulfonic acid (MOPS) buffer: 0.4 M MOPS, 0.1 M NaAcetate, 0.01 M ethylenediaminetetraacetic acid (EDTA). Bring to pH 7.0 with acetic acid.
13. RNA-loading buffer: for 15 mL stock mix 10 mL of 100% formamide, 3 mL of 37% formaldehyde, 2 mL of 10× MOPS buffer and 25 µL of 10 mg/ml ethidium bromide. Keep at –20°C.
14. 6× Loading dye: 50% glycerol, 0.25% bromophenol blue, and 1 mM EDTA.
15. RNA marker: Century Plus from Ambion. Keep at –20°C. No dyes (e.g., bromophenol blue or xylene cyanol) are added to its loading buffer as these usually shade some of the bands.
16. Hybridization solution: 0.4 M Na$_2$HPO$_4$/NaH$_2$PO$_4$ pH 7.2, 6% Sodium Dodecyl Sulfate (SDS), 1 mM EDTA.
17. Hybridization wash 1: 40 mM Na$_2$HPO$_4$/NaH$_2$PO$_4$ pH 7.2, 5% SDS, 1 mM EDTA.
18. Hybridization wash 2: 40 mM Na$_2$HPO$_4$/NaH$_2$PO$_4$ pH 7.2, 1% SDS, 1 mM EDTA.
19. SM 24 rotor 13 mL tubes (95 × 16.8 mm polypropylene from Sarstedt cat no. 55.518).
20. Glass beads 0.45–0.55 mm diameter. Cool to 4°C before use.
21. Bead Beater: Mini BeadBeater-8 from BioSpec (Bartlesville, OK, USA).

22. Collection system: The gradients are separated and collected using a collection system that includes
   a) ISCO UA-6® Absorbance detector with a 254-nm type 11 filter.
   b) ISCO Retriever II fraction collector.
   c) Brandel syringe pump (SYR-101).
   d) Brandel model number 184 fractionator with 5-mm flow cell. A complete description of the components and manuals can be obtained from the suppliers: Isco, Inc., Lincoln, NE, USA:http://www.isco.com/, and Brandel, Gaithersburg, MD, USA:http://www.brandel.com/.

3. Methods

3.1. Separation of mRNAs According to Their Ribosomal Association

The protocol presented here is aimed at determining ribosomal density as similar as possible to the situation within the cell. The first step is the isolation of mRNAs associated with ribosomes from cells in the presence of cycloheximide and in a cold environment, conditions that are expected to minimize changes in the ribosomal status. Cells can be lysed according to *(10)* or by using the following protocol that provides better yields. Other protocols should work as well, yet we have not tested them in RDM. Following lysis, the extract is separated by velocity sedimentation in a sucrose gradient, which leads to separation of mRNA molecules according to the number of ribosomes they are associated with (*see* **Fig. 2A**) *(10)*. This separation step allows collection of a specific fraction—the fraction into which the mRNA of interest sediment. There are two main reasons for separating the extract on a gradient rather than performing the next steps on the entire lysate: (1) Isolation of a subset of mRNAs with a specific number of bound ribosomes simplifies the interpretation of the results; (2) Large amounts of heparin are used during the lysis to inhibit RNase activity (we have not yet found any comparable alternative to heparin). This heparin does not sediment into the gradient, and therefore, the isolated fraction is clean of heparin and can be used in the following enzymatic step.

1. Grow 50 ml of yeast cells to OD_{600} 0.5–0.8 in YPD (*see* **Note 1**).
2. Add CHX to a final concentration of 0.1 mg/ml, transfer to 50 ml conical tube on ice, and immediately spin down the cells at 4000 rpm (3220 × *g*) for 4 min at 4°C.
3. Discard the supernatant, resuspend the cell pellet in 4-ml ice-cold lysis buffer and spin cells again as in **step 2** (*see* **Note 2**).
4. Wash again as in **step 3**.
5. Resuspend cells with 350 µl of lysis buffer, transfer to a micro tube with a screw cap and add ice-cold glass beads to cover the cells and lysis buffer.
6. Break cells in a bead beater by two pulses of 1.5 min at maximum level.

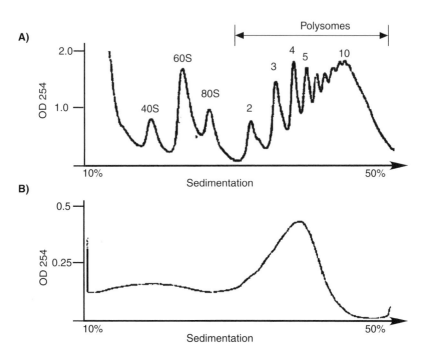

Fig. 2. Representative OD$_{254}$ profiles obtained from **step 1 (A)** and **step 3 (B)** of ribosome density-mapping (RDM). **(A)** Cells grown in rich medium were harvested and lysate was separated on 10–50% linear sucrose gradient. The gradient was fractionated, while the OD$_{254}$ was measured continuously using the ISCO UA-6 system. Most of the OD$_{254}$ signal is obtained from the rRNA present in the different complexes, and the peaks represent the sedimentation position of the ribosomal subunits (40S and 60S) and mRNAs associated with one ribosome (80S), 2, 3, or more ribosomes. **(B)** The fraction that contains 4 and 5 ribosomes was collected and treated with RNase H and ODN. The cleaved sample was separated on a 10–50% sucrose gradient, and fractions were collected while measuring its OD$_{254}$. As only a specific fraction was collected, the profile contains only a single peak at the appropriate position. Note that different detector sensitivities were used in **(A)** and **(B)**, and therefore their OD$_{254}$ scales (y-axis) are different.

7. To recover the lysate, perforate the bottom of the tube with a heated needle and place it on top of a 15-ml conical tube. As the diameter of the screw-cap tube is smaller than the diameter of the 15 ml conical tube, an adaptor is needed. The cylinder of a 5-ml syringe is well suited for this purpose. The assembly containing the punctured lysate-containing tube, the adaptor, and the 15-ml conical tube is spun at 4,000 rpm (3220 × g) for 1 min at 4°C.

8. Transfer the resulting lysate from the 15-ml tube into a new ice-cold micro tube. Usually there is a small pellet that should be resuspended with the supernatant and also be transferred to the micro tube.
9. Centrifuge for 5 min at 9500 rpm ($8300 \times g$) at 4°C in a microcentrifuge and transfer the supernatant into a new ice-cold micro tube.
10. Bring to a final volume of 1 mL with lysis buffer and carefully load the lysate on 10–50% sucrose gradient without heparin (*see* **Note 3**).
11. Separate complexes by ultra centrifugation using SW41 rotor at 35,000 rpm for 2.25 h at 4°C. The separation times can be adjusted according to the desired resolution.

3.2. RNase H Cleavage

Following centrifugation, the fraction of interest is collected for cleavage by RNase H and ODN. This is usually the fraction that contains the majority of transcripts of the gene of interest (i.e., its peak fraction) but could be any other fraction that contains sufficient amounts of the mRNA of interest. The collection can be done manually or with an automated fraction collector. We use the UA-6 system from ISCO that provides continuous OD_{254} readings of the sample (*see* **Fig. 2A**). The OD_{254} indicates the sedimentation position of each polysomal complex (i.e., 1, 2, 3 ribosomes) and therefore allows accurate collection of complexes of interest.

1. Collect the fraction of interest (620 µl) into an ice-cold micro tube (*see* **Note 4**). Immediately add 70 µl of 0.1 M DTT (final concentration 10 mM) and 7 µl of 40 u/µl RNase Inhibitor (final concentration 0.4 u/µl).
2. Transfer 600 µl of the above to another micro tube containing 10 µl of ODN (10 pmole/µl) complementary to the mRNA region to be cleaved (*see* **Notes 5–7**). The remainder (70–100 µl) serves as a control sample (uncut) and will be subjected to the following incubations (**steps 3–6**).
3. Put the sample into a 500 ml beaker containing water at 37°C and allow cooling for 20 min. This supposedly enables annealing of the ODN to its target sequence (*see* **Notes 5–7**).
4. Add 150 µl 5× RNase H buffer and RNase H to a final concentration of 16.5 u/ml.
5. Incubate the samples at 37°C for 30 min.
6. Stop the reaction by bringing the volume of the ODN sample to 1 ml with LMD buffer. This will also dilute the sucrose in the sample to allow overlaying on a 10–50% sucrose gradient. Set aside 1/10 of the reaction as a pre-gradient control (cut) (*see* **Note 8**).

3.3. Separation of the Cleavage Products

1. Overlay the sample on a sucrose gradient containing heparin and centrifuge as in **Subheading 3.1., step 11** (*see* **Note 9**).

2. After centrifugation, collect multiple fractions from the gradient. The number of fractions to be collected depends on the desired resolution; a reasonable number to start with is 18 (fraction vol. ~0.6 ml). Using Retriever II and a Brandel syringe pump, this number of fractions is achieved by a fraction time of 0.9 min/fraction and a syringe speed of 0.75 ml/min.

3. Collect fractions into SM-24 rotor 13-ml tubes containing 1.5 vol. of 8 M GuHCl (final concentration of 5.3 M).

4. Add 2.5 sample volumes of 100% ethanol. Mix well and incubate at –20°C for at least 1 h.

5. Centrifuge samples for 20 min at 12,000 rpm (~18,000 × g) at 4°C. Carefully discard the supernatant.

6. Wash with 500 μl of ice-cold 80% ethanol, centrifuge as in **step 4** and carefully discard the supernatant. Note that the pellet might be unstable.

7. Resuspend with 400 μl TE, transfer to micro tube, and precipitate again by adding 40 μl of 3 M NaAc pH 5.3 and 1 ml of 100% ethanol. Mix well and incubate at –20°C for at least 1 h.

8. Centrifuge samples for 20 min at 14,000 rpm (18,000 g) at 4°C. Discard the supernatant.

9. Wash by adding 100 μl ice cold 80% EtOH. Centrifuge as in **step 7** and discard the supernatant.

10. Dry the pellet and resuspend in 10 μl of sterile water. Keep at –20°C. Take about half of the sample for northern analysis. This amount might vary according to the expression level of the tested mRNA.

3.4. Determination of the Sedimentation Position of Cleavage Products

The fraction into which the cleavage products sediment is determined using standard northern analysis. The protocol presented here is based on *(11)* with small modifications. It is relatively simple and provides reliable results when analyzing abundant mRNAs [>5 copies/cell *(12)*] that yield cleavage fragments exceeding approximately 200 nts in length (for shorter fragments, an RNase protection assay can be used). From our experience, the best northern results are obtained when analyzing mRNAs expressed from their genomic locus. We received a slight decrease in the quality of the northern analysis when analyzing genes expressed from a plasmid with their natural 5´ and 3´ regions, and a strong decrease in quality when heterologous genes were used (e.g., Green Flourescent Protein (GFP) or luciferase reporters). The decrease in quality is characterized by the appearance of multiple bands or smears that hinder the detection of cleavage product.

1. Prepare 1.2–2% agarose gel (depending on the expected sizes) in 1× MOPS and formaldehyde (1.3% final concentration). Dissolve the agarose first in water, let cool to 65°C, and then add the MOPS and formaldehyde.
2. Mix 5 µl of each RNA sample with 7.5 µl of RNA-loading buffer.
3. Incubate at 55°C for 15 min to open RNA secondary structures and add 2.5 µL of 6× loading dye.
4. Load the uncut and cut controls on the gel together with RNA size marker and a sample (2–10 µg) of untreated RNA (e.g., RNA extracted by hot phenol–SDS method). Run the gel in 1× MOPS buffer to obtain the best resolution of expected cleavage products (*see* **Notes 10** and **11**).
5. Blot the RNA from the gel to a nylon membrane.

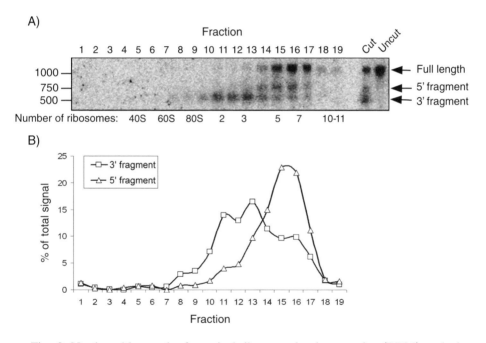

Fig. 3. Northern blot result of a typical ribosome density-mapping (RDM) analysis. (**A**) An abundant yeast mRNA (YGR086C) was cleaved at the last third of its coding region, and cleavage products were separated by sucrose gradient into 19 fractions. Fractions were run on 1.6% agarose gel, blotted to a nylon membrane, and hybridized to a radioactive probe recognizing the entire coding region. Arrows indicate the migration positions of the 5′ and 3′ cleavage products, as well as some uncut full-length mRNA. The lane marked "uncut" represents the uncut control (i.e., a sample from **Subheading 3.2., step 2**) and the lane marked "cut" represents the cut control (i.e., a sample from **Subheading 3.2., step 6**). (**B**) Quantitation results of the 3′ and 5′ fragments (*see* **Note 12**).

6. Cross-link the RNA to the membrane using a UV cross-linker or a dry oven.
7. The quality of RNA and transfer can be evaluated by methylene blue staining. This is done by immersing the membrane in 5% acetic acid for 5 min and then in 0.1% methylene blue in 5% acetic acid for 5 min. The membrane is then washed in water and distinct bands of the 25S (3400 nts), 18S (1800 nts), 5.8S (160 nts), 5S (120 nts) rRNA, as well as the RNA marker bands, should appear.
8. Prepare by random priming a radioactive probe to hybridize with the membrane. Templates for probes are usually PCR products that correspond to the entire transcript and are at least 100 nts long.
9. Hybridize the probe with the membrane at 57°C for at least 6 h in hybridization buffer.
10. Following hybridization, wash the membrane twice for 15 min (each wash) in hybridization wash 1 and twice for 15 min in hybridization wash 2 at 57°C.
11. Expose to a phosphor-imager screen or film. Three bands should appear: two that are similar in size to the cleavage products (5′ and 3′ to the RNase H cleavage site) and a longer band similar in size to the full-length transcript and represents the remainder of an uncut mRNA (*see* **Fig. 3**).

4. Notes

1. Do not use larger amounts of cells as this leads to a massive degradation of RNA in later steps, probably because of overload with RNases.
2. It is important to wash with excess of lysis buffer to remove YPD leftovers. The quick and complete removal of dextrose helps in blocking further initiation events.
3. For gradient preparation instructions, see *(10)*.
4. Do not collect too much of the polysomal fraction because this fraction also contains a significant amount of RNases. The overload of RNases cannot be competed by the RNase inhibitor and therefore might lead to massive degradation.
5. RNase H can cleave RNA and DNA hetero duplexes as short as 4 bp *(13)*. Thus, it is recommended verifying that the designed ODN has minimal complementarity to other sequences. This is of particular importance for regions that are rich in repeating sequences (such as A : U in the UTRs).
6. We used many ODNs ranging in length from 18 to 25 nts and having 40–60% GC content. These ODNs varied in efficiency, yet we are unable to directly link the differences in efficiency to their length or GC content. This is probably because many additional factors, such as structure of the cleavage site or presence of a ribosome, affect cleavage efficiency. It is therefore recommended to design several ODNs for the desired region and compare their efficiency in a test reaction that includes polysomal RNA and the ODNs. We perform the test reactions on isolated polysomal fraction and not on clean RNA (e.g., RNA isolated by the hot phenol or Tri-reagent methods), as many ODNs perform differently within these populations.
7. Multiple ODNs can be used in the same reaction, complementary either to the same transcript or to different transcripts. If several ODNs to the same transcript are used, note that partial products are also expected (i.e., cleavage by only one

ODN). Thus, make sure that the full cleavage products and the various partial cleavage products are of distinct sizes. In cases where no good separation in size exists, it is possible to design probes for northern analysis that would be specific to only some of the products.

8. As Mg^{2+} ions are needed for ribosome stability, EDTA treatment is a common control for ribosomal association. For such control, a gradient without $MgCl_2$ and supplemented with 20 mM EDTA is used during the second centrifugation. In addition, in **Subheading 3.2., step 6**, an LMD buffer supplemented with 30 mM EDTA and without $MgCl_2$ is used.

9. Because only the fraction of interest is collected for RNase H cleavage, the OD_{254} of the second centrifugation includes only a single peak of the collected fraction (*see* **Fig. 2B**). It is therefore impossible to use this OD_{254} profile to determine the number of ribosomes on the mRNA cleavage products. There are ways of overcoming this limitation: (1) Use of an external profile—It is possible to use the OD_{254} profile of whole-cells extract (e.g., **Fig. 2A**) separated on a similar gradient and centrifuged under exactly the same conditions. Since the sedimentation procedure is highly reproducible, the position of peaks in this gradient can be used as a guide for the sedimentation of complexes in the fractions of the RDM analysis; (2) Spike-in extract—another option is to set aside a sample (~1/10 of the extract) of the whole-cells extract before loading the first gradient (*see* **Subheading 3.1., step 10**) and spike it into the products of the RNase H cleavage before loading on the second gradient (*see* **Subheading 3.3., step 1**). The spiked aliquot will result in an OD_{254} profile that contains multiple peaks, which can be used to determine the sedimentation position of the various complexes.

10. Correct identification of the cleavage products can be done according to size. It is important to use the appropriate percent of gel for the desired resolution, run an RNA size marker (as DNA markers migrate differently), resuspend the RNA marker in the same buffer as the samples (because variation in ionic conditions might change its migration), and include the uncut and cut controls.

11. MOPS buffer has a weak buffering capacity; thus, it is advised to circulate the buffer between chambers during the running time. Increased pH in the upper chamber might lead to RNA degradation.

12. In a real experiment, the sedimentation of fragments is not restricted to a single fraction (as depicted schematically in **Fig. 1**) but distributed to several adjacent fractions. The distribution might represent a true biological variation (i.e., association of fragments with various numbers of ribosomes) or might be caused by technical aspects (i.e., mixing the sample during loading on the gradient).

References

1. Sachs, M.S., Wang, Z., Gaba, A., Fang, P., Belk, J., Ganesan, R., Amrani, N. and Jacobson, A. (2002) Toeprint analysis of the positioning of translation apparatus components at initiation and termination codons of fungal mRNAs. *Methods* **26,** 105–114.

2. Wolin, S.L. and Walter, P. (1988) Ribosome pausing and stacking during translation of a eukaryotic mRNA. *EMBO J* **7**, 3559–3569.

3. Wolin, S.L. and Walter, P. (1989) Signal recognition particle mediates a transient elongation arrest of preprolactin in reticulocyte lysate. *J Cell Biol* **109**, 2617–2622.

4. Mueller, P.P. and Hinnebusch, A.G. (1986) Multiple upstream AUG codons mediate translational control of GCN4. *Cell*, **45**, 201–207.

5. Abastado, J.P., Miller, P.F., Jackson, B.M. and Hinnebusch, A.G. (1991) Suppression of ribosomal reinitiation at upstream open reading frames in amino acid-starved cells forms the basis for GCN4 translational control. *Mol Cell Biol* **11**, 486–496.

6. Werner, M., Feller, A., Messenguy, F. and Pierard, A. (1987) The leader peptide of yeast gene CPA1 is essential for the translational repression of its expression. *Cell* **49**, 805–813.

7. Sagliocco, F.A., Vega Laso, M.R., Zhu, D., Tuite, M.F., McCarthy, J.E. and Brown, A.J. (1993) The influence of 5'-secondary structures upon ribosome binding to mRNA during translation in yeast. *J Biol Chem* **268**, 26522–26530.

8. Yaman, I., Fernandez, J., Liu, H., Caprara, M., Komar, A.A., Koromilas, A.E., Zhou, L., Snider, M.D., Scheuner, D., Kaufman, R.J. *et al.* (2003) The zipper model of translational control: a small upstream ORF is the switch that controls structural remodeling of an mRNA leader. *Cell* **113**, 519–531.

9. Arava, Y., Boas, F.E., Brown, P.O. and Herschlag, D. (2005) Dissecting eukaryotic translation and its control by ribosome density mapping. *Nucleic Acids Res* **33**, 2421–2432.

10. Arava, Y. (2003) Isolation of polysomal RNA for microarray analysis. *Methods Mol Biol* **224**, 79–87.

11. Kuhn, K.M., DeRisi, J.L., Brown, P.O. and Sarnow, P. (2001) Global and specific translational regulation in the genomic response of Saccharomyces cerevisiae to a rapid transfer from a fermentable to a nonfermentable carbon source. *Mol Cell Biol* **21**, 916–927.

12. Wang, Y., Liu, C.L., Storey, J.D., Tibshirani, R.J., Herschlag, D. and Brown, P.O. (2002) Precision and functional specificity in mRNA decay. *Proc Natl Acad Sci USA* **99**, 5860–5865.

13. Donis-Keller, H. (1979) Site specific enzymatic cleavage of RNA. *Nucleic Acids Res* **7**, 179–192.

17

Identification of Changes in Gene Expression by Quantitation of mRNA Levels

Wendy M. Olivas

Summary

Sequence elements within mRNA-untranslated regions and their binding partners are key controllers of mRNA stability. Changes in mRNA stability can often be detected by changes in steady-state mRNA abundance, or a more careful analysis of mRNA half-lives can be performed following transcriptional repression. This chapter presents methods to isolate RNA from both yeast and mammalian cells for either steady-state or half-life analyses. In addition, two reliable methods to quantitate mRNA levels, northern blot analysis and real-time PCR, are outlined and compared.

Key Words: Steady-state RNA levels; RNA quantitation; mRNA degradation; mRNA decay; northern blot; real-time PCR; qPCR; transcriptional shut off.

1. Introduction

Each mRNA has an intrinsic lifespan that is dependent at least in part on the presence or absence of specific sequence elements in its untranslated regions. Specifically, the 3´ untranslated region (3´ UTR) often contains regulatory motifs that either enhance or inhibit the degradation of the mRNA, typically through interactions with proteins or regulatory RNAs *(1–3)*. It is the balance between rates of transcription and degradation that determines the steady-state abundance of an mRNA. Thus, changes in the rate of degradation of an mRNA can usually be detected by changes in its steady-state abundance. However, an *x*-fold change in an mRNA's degradation rate is not necessarily reflected as

From: *Methods in Molecular Biology, Vol. 419: Post-Transcriptional Gene Regulation*
Edited by: J. Wilusz © Humana Press, Totowa, NJ

an *x*-fold change in an mRNA's steady-state abundance. Moreover, there are many instances when changes in degradation rates cannot be identified simply through analysis of steady-state levels because of compensatory changes in transcription rates.

To evaluate the role of 3´ UTR regulatory elements and their binding partners in mRNA degradation, transcriptional shut-off assays are typically performed. There are several different approaches that have been used in diverse organisms to inhibit transcription such that degradation of the existing pool of mRNA can be monitored over time. The half-life of an mRNA is measured as the time required for 50% of the initial mRNA pool at the time of transcriptional inhibition to degrade. For global transcriptional repression and decay analysis of endogenous mRNAs, chemical inhibitors have been used such as actinomycin D in mammalian cells *(4,5)* and thiolutin and 1,10-phenanthroline in *Saccharomyces cerevisiae (6)*. In addition, the *rpb1-1* allele in *S. cerevisiae*, a temperature-sensitive mutant of the RNA polymerase II catalytic subunit, has been used extensively for mRNA decay studies *(7,8)*. However, all of these inhibitors cause general perturbations to the cell and induce a stress response, which may itself alter the decay profiles of certain mRNAs *(6,9,10)*.

Alternatively, vectors with inducible/repressible promoters have been designed such that specific genes or just 3´ UTR regions of interest can be cloned and the stability of the resulting transcripts evaluated in cells after removal/addition of the promoter-specific inducer/repressor. Although this method is limited to analysis of the cloned sequence versus any endogenous mRNA, most promoter-repressor systems are not stressful to the cell nor affect other components of the mRNA decay machinery. Such vector systems include the Tet on/off constructs controlled by doxycycline addition, which have been used in both mammalian and yeast cells *(11,12)*, the serum-inducible *c-fos* promoter constructs used in mammalian cells *(13)*, and the GAL upstream activating sequence (UAS) constructs used in yeast that can be rapidly induced by galactose and repressed by dextrose *(14)*.

Regardless of whether steady-state mRNA abundance or mRNA levels following transcription repression is being analyzed, a method of quantitating these mRNA levels after cellular extraction must be chosen. The two most common and reliable methods are northern blot analysis and real-time/quantitative PCR (qPCR). The most important information that can be derived from both these methods is not the absolute quantity of transcripts, but the comparison of relative differences in mRNA levels between samples. The advantages of northern blot analysis are that this method is simple and inexpensive, it requires only extraction of total RNA without need to further process or amplify the RNA, it allows for easy analysis of both RNA integrity and RNA loading, a single northern blot can be probed multiple times for different RNAs, and it can be used to monitor different phases of degradation such as the reduction of poly(A) tail lengths during deadenylation *(15)*. The downsides to this method are that it can be difficult to detect low-abundance mRNAs, especially if cell sample sizes are small, and radioactive probes are usually required for mRNA detection. In contrast, qPCR can be used with small sample sizes and low-abundance mRNAs, and no radioactivity is involved. However, qPCR involves multiple processing steps of the RNA (reverse transcription and qPCR) that require critical optimization for accurate results.

With qPCR, relative mRNA levels can be assessed individually with one cDNA amplified per reaction tube, or multiple cDNAs can be amplified concurrently in a single multiplex reaction. Individual amplification is easier to optimize, and detection of the resulting PCR product by incorporation of SYBR Green dye is fairly inexpensive. However, separate reactions are needed for each gene of interest and each internal reference gene, which uses more sample and increases the opportunity for error. In multiplex reactions, each cDNA to be amplified requires both a set of PCR primers and an internal fluorophore-labeled probe, and it is critical that each primer/probe set is both specific and efficient with no primer interactions. In addition, all multiplex reactions must also be tested in individual reactions to make sure that one reaction does not skew the amplification of another reaction when multiplexed, especially when monitoring targets of widely different expression levels.

This chapter describes methods to isolate and quantitate RNA levels from both yeast and mammalian cell cultures. For the purposes of this chapter, the northern blot method is presented for analysis of yeast RNA, and single reaction SYBR Green qPCR is presented for analysis of mammalian RNA, although the methods can be used interchangeably once total RNA is isolated from the respective cell culture.

2. Materials

2.1. Harvesting of Yeast Cells and Isolation of Total Yeast RNA

1. Diethyl pyrocarbonate (DEPC)-treated H_2O.
2. LET solution: 25 mM Tris–HCl, pH 8.0, 100 mM LiCl, 20 mM ethylenediaminete-traacetic acid (EDTA), DEPC-treated.
3. Phenol equilibrated with LET.
4. Phenol : chloroform (1:1) equilibrated with LET.
5. Chloroform.
6. 3 M NaOAc, DEPC-treated.

7. 100% ethanol (Sigma, St. Louis, MO, USA), store in –20°C.
8. 70% ethanol made with DEPC-treated H_2O, store in –20°C.
9. Acid-washed glass beads, 425–600 μm (Sigma G8772).

2.2. Northern Agarose Gel Preparation, Electrophoresis, and Blotting

1. $10\times$ 3-(N-morpholino)propanesulfonic acid (MOPS) buffer: Dissolve 41.8 g MOPS in 800 mL H_2O, pH 7.0 with NaOH, add 16.6 mL $3M$ NaOAc and 20 mL 0.5 M EDTA, add H_2O to 1 L.
2. 1.25% agarose gel: Mix 1.25 g agarose, 10 mL $10\times$ MOPS buffer and 72 mL H_2O; microwave until fully dissolved, swirling frequently; cool to approximately 70°C and then add 18 mL of 37% formaldehyde; pour gel immediately (*see* **Note 1**).
3. Gel running buffer: $1\times$ MOPS buffer.
4. Northern dye mix: 1mM EDTA, 0.12% bromophenol blue, 0.12% xylene cyanol, 50% glycerol. Components should be diluted from DEPC-treated stock solutions.
5. Gel loading buffer: 500 μL deionized formamide, 375 μL 37% formaldehyde, 110 μL Northern dye mix, 50 μL $10\times$ MOPS.
6. 10 mg/mL ethidium bromide.
7. $10\times$ SSC transfer buffer: 1.5 M NaCl, 0.15 M Na citrate, pH 7.

2.3. Preparation of Radioactive Oligonucleotide Probe

1. $10\times$ T4 polynucleotide kinase buffer (New England Biolabs, Ipswich, MA, USA).
2. 100 ng/μL oligonucleotide.
3. γ-(^{32}P) ATP (6000 Ci/mmol).
4. T4 Polynucleotide kinase (New England Biolabs).
5. 2% Blue dextran.
6. 2% Phenol red.
7. Sephadex G-25 slurry: In a 50-mL tube, add approximately 40 mL of H_2O to approximately 2 g of dry Sephadex G-25, let swell overnight and store at 4°C.
8. Sephadex G-25 spin column: Place a small cotton plug in the bottom of a 1-mL syringe barrel and place barrel in a 15-mL conical tube. Using a pasture pipette, fill syringe with Sephadex G-25 gel slurry. Allow the water to drain, and add more Sephadex G-25 slurry to fill syringe. Centrifuge syringe column in 15 mL tube for 2 min at 2500 × *g* to pack column. Remove all water from bottom of 15-mL tube and then replace column into 15-mL tube.

2.4. Hybridization of Oligonucleotide Probe to Northern Blot

1. 0.1% SDS/0.1× SSC solution.
2. $100\times$ Denhardts solution: Mix 20 g Ficoll, 20 g polyvinylpyrrolidone and 20 g bovine serum albumin in H_2O (warmed to 65°C) to 1 L final volume. Filter through 0.45 μm using prefilter (may only get ~250 mL through each filter before clogging). Store at –201191C.

3. Oligo hybridization buffer: 6× SSC, 10× Denhardts solution, 0.1% SDS; store at 4°C.
4. 0.1% SDS/6× SSC solution.

2.5. Isolation of Total RNA from Mammalian Cells

1. Trizol reagent (Invitrogen, Carlsbad, CA, USA).
2. Chloroform.
3. Isopropanol.
4. DEPC-treated H_2O.
5. 70% ethanol made with DEPC-treated H_2O, store at –20°C.
6. DNase I buffer (Fermentas).
7. RNase-free DNase I (1 U/μL; Fermentas, Hanover, MD, USA)
8. Phenol : chloroform : isoamyl alcohol (25:25:1) equilibrated with 10 m*M* Tris–HCl, pH 8.0.
9. 10 *M* NH₄OAc, DEPC-treated.
10. 100% ethanol, store at –20°C.

2.6. Preparation of cDNA

1. Random hexamer or oligo(dT) primer, 500 ng/μL.
2. DEPC-treated H_2O.
3. 5× Reverse transcriptase buffer (Promega, Madison, WI, USA).
4. Reverse transcriptase (ImProm-II, Promega).
5. 10 m*M* dNTPs.
6. 25 m*M* MgCl₂.
7. RNasin (40 U/μL, Promega).

2.7. Validation of Primer Specificity and Efficiency

1. iTaq SYBR Green Supermix (Bio-Rad, Hercules, CA, USA): 2× mix contains 0.4 m*M* each of dATP, dCTP, and dGTP, 0.8 m*M* dUTP, 50 U/ml iTaq DNA polymerase, 6 m*M* Mg^{2+}, SYBR Green I, ROX reference dye, stabilizers.
2. Upstream primer, 2.5 μ*M*.
3. Downstream primer, 2.5 μ*M*.

3. Methods

3.1. Growth and Harvesting of Yeast Cells for Steady-State RNA Analysis

1. Grow cells to an OD_{600} of 0.4 in 100 mL of an appropriate medium (*see* **Note 2**).
2. Pellet the yeast cells from each 50 mL of culture in a separate conical tube by centrifugation for 2 min at 2500 × *g*. Decant off the supernatant (*see* **Note 3**).
3. Resuspend each cell pellet in 1 mL DEPC-treated H_2O, transfer cell solution to 2-mL tube, centrifuge 15 s at 2000 × *g*, and pipet off supernatant. Place cell

pellet immediately in dry ice. The pellet can be stored at –20°C up to 1 week until ready to isolate RNA.

3.2. Transcriptional Shut-off Assays for Yeast mRNA Half-life Analysis

1. Grow cells to an OD_{600} of 0.4 in 200 mL of an appropriate medium. For yeast cells containing the *rpb1-1* allele for global transcriptional repression, the culture is grown in rich media (YEPD) at 24°C. For yeast cells containing the gene of interest under the control of the GAL UAS, the culture is grown in selective media with 2% galactose (*see* **Note 4**).
2. Pellet the yeast cells from each 50 mL of culture in a separate conical tube by centrifugation for 2 min at 2500 × *g*. Decant off the supernatant.
3. For global transcriptional repression through the *rpb1-1* allele, combine the cell pellets from all four tubes by resuspension of one pellet in 10 mL of 24°C YEPD, then transferring the cell solution to the second tube to resuspend that pellet, and continuing until all four cell pellets are combined into one tube of cell solution. Working quickly, add 11 mL of YEPD warmed to 55°C and transfer to a flask pre-warmed in a 37°C water bath. This will rapidly alter the temperature of the solution to 37°C to shut off transcription. Remove a 2-mL sample of the cell solution to a 2-mL tube, centrifuge 15 s at 2000 × *g*, and pipet off supernatant. Place cell pellet immediately in dry ice. This sample represents the zero minute time point.
4. For specific transcriptional repression of a gene under the control of the GAL UAS, combine the cell pellets from all four tubes by resuspension of one pellet in 10 mL of 30°C selective media lacking a sugar carbon source, then transferring the cell solution to the second tube to resuspend that pellet, and continuing until all four cell pellets are combined into one tube of cell solution. Working quickly, add 11 mL of 30°C selective media containing 8% dextrose and transfer to a flask pre-warmed in a 30°C water bath. The resulting 4% dextrose will rapidly repress the GAL UAS to shut off transcription. Remove a 2-mL sample of the cell solution to a 2-mL tube, centrifuge 15 s at 2500 × *g*, and pipet off supernatant. Place cell pellet immediately in dry ice. This sample represents the zero minute time point.
5. At designated time points after transcriptional repression, remove 2 mL aliquots of the cell solution, centrifuge 15 s at 2000 × *g*, and pipet off supernatant. Place cell pellets immediately in dry ice. Typical time points used are 2, 4, 6, 8, 10, 15, 20, 30, and 60 min after transcriptional repression.

3.3. Isolation of Total Yeast RNA

1. Resuspend frozen cell pellet (in 2-mL tube) in 150 µL LET.
2. Add 150 µL phenol equilibrated with LET.
3. Add 0.5 mL of glass beads.
4. Vortex for 5 min at top speed to disrupt cells (*see* **Note 5**).
5. Add 250 µL DEPC-treated H_2O and 250 µL phenol : chloroform equilibrated with LET.

6. Vortex briefly to mix then centrifuge 2 min at 21,000 × *g*.

7. Remove top aqueous phase to 1.5-ml tube containing 400 µL phenol : chloroform equilibrated with LET (*see* **Note 6**). Vortex briefly to mix and then centrifuge 2 min at 21,000 × *g*.

8. Remove top aqueous phase to new 1.5-ml tube containing 400 µL of chloroform (*see* **Note 7**). Vortex briefly to mix then centrifuge 2 min at 21,000 × *g*.

9. Remove top aqueous phase to new 1.5-ml tube containing 40 µL of DEPC-treated 3 *M* NaOAc. Add 1 mL cold 100% ethanol. Place at –20°C for at least 30 min to aid precipitation of RNA (*see* **Note 8**).

10. Pellet the RNA by centrifugation for 10 min at 21,000 × *g*. Decant off supernatant. Wash pellet by adding 200 µL cold 70% ethanol and inverting tube several times. Centrifuge 5 min at 21,000 × *g*. Decant off supernatant. Dry pellet in SpeedVac for no more than 5 min or air dry (*see* **Note 9**). Resuspend pellet in 100 µL DEPC-treated H_2O. The RNA can be stored at –20°C.

11. Determine the yield of RNA by measuring OD_{260} (*see* **Note 10**). 1.0 OD_{260} = 40 µg/mL RNA.

3.4. Northern Agarose Gel Preparation, Electrophoresis, and Blotting

1. Pour gel of appropriate percent agarose. For resolution of mRNAs between 300 and 2000 nucleotides, a 1.25% agarose gel works well.

2. Once gel is solidified, add 1× MOPS gel running buffer to electrophoresis chamber.

3. Aliquot 20–40 µg of total RNA into 1.5-mL tube. Make sure to use the same amount of total RNA per sample to accurately compare levels of specific mRNAs.

4. Dry down RNA in SpeedVac, but do not over-dry (*see* **Note 9**).

5. Resuspend RNA in 4 µL DEPC-treated H_2O. Add 15 µL Gel loading buffer (*see* **Note 11**).

6. Heat denature the RNA at 100°C for 3 min, quick spin, and then chill on ice.

7. Pipet entirety of each RNA sample into gel wells.

8. Electrophorese at 70 V with buffer recirculation to avoid a pH gradient (*see* **Note 12**).

9. To assess the integrity of the resolved RNA as well as to visually compare the amount of RNA loaded per lane, stain the gel with ethidium bromide following electrophoresis. The ribosomal RNAs (28S and 18S) should be seen as distinct bands in equal intensities between samples (*see* **Fig. 1**). For staining, soak gel in 200 mL H_2O containing 50 µL of 10 mg/mL ethidium bromide for 8 min with gentle shaking. Remove excess ethidium bromide by washing gel 3 × 5 min in 200 mL H_2O for 5 min each with gentle shaking. Image gel under UV light.

10. Transfer RNA from gel to nylon membrane (Nytran SuperCharge, Whatman, Florham Park, NJ, USA) by capillary blotting using 10× SSC transfer buffer. UV crosslink blot.

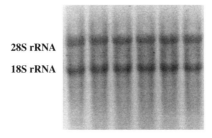

28S rRNA

18S rRNA

Fig. 1. Ethidium bromide staining of Northern agarose gel. Total RNA from different sample preparations was electrophoresed on a 1.25% agarose gel and ethidium bromide stained to visualize the rRNA bands for analysis of RNA integrity and equal loading.

3.5. Preparation of Radioactive Oligonucleotide Probe

1. Assemble the following components in the order listed in a 1.5-mL tube:
   13 µL H$_2$O
   2 µL 10× T4 Polynucleotide Kinase buffer
   2 µL 100 ng/µL oligonucleotide
   2 µL γ-(^{32}P) ATP (6000 Ci/mmol; *see* **Note 13**)
   1 µL T4 Polynucleotide Kinase
2. Incubate the reaction at 37°C for 1 h.
3. Add 65 µL H$_2$O, 10 µL of 2% blue dextran, and 5 µL of 2% phenol red.
4. Pipet the reaction onto the top of a Sephadex G-25 spin column.
5. Centrifuge the column for 2 min at 2500 × *g* and collect the flow-through containing the purified oligonucleotide probe (*see* **Note 14**).
6. Using a liquid scintillation counter, measure the counts per minute (cpm) of 1 µL of purified probe (*see* **Note 15**).

3.6. Hybridization of Oligonucleotide Probe to Northern Blot

1. Wash blot for 1 h at 65°C in 100 mL of 0.1% SDS/0.1× SSC solution with gentle shaking (*see* **Note 16**).
2. Drain off wash solution.
3. Pre-hybridize blot for 1 h at 42°C in 50 mL Oligo Hybridization buffer with gentle shaking.
4. Drain off pre-hybridization solution.
5. Add radioactive probe to 5 mL of new Oligo Hybridization buffer pre-warmed to 42°C and filter through a 0.45-µm syringe filter (*see* **Note 17**).
6. Combine filtered probe with another 40 mL of pre-warmed Oligo Hybridization buffer and pour onto blot.
7. Hybridize probe to blot overnight at 42°C with gentle shaking (*see* **Note 18**).

8. Wash blot 3 × 5 min at room temperature in 100 mL 0.1% SDS/6× SSC solution with gentle shaking.
9. Wash blot at approximately 10°C below estimated oligonucleotide Tm for 20 min with gentle shaking. For most probes, this can be done at 50°C.
10. Wrap blot in Saran Wrap for exposure to a phosphorimager screen. Blots can be stored in Saran Wrap at –20°C. **Figure 2A** shows an example of a northern blot analysis of a transcriptional shut off assay to monitor mRNA decay.
11. To strip an oligonucleotide probe off a blot, add 100 mL of boiling 0.1% SDS/0.1× SSC to the blot in a plastic container with lid slightly opened. Gently shake for 10 min at room temperature then drain off. Repeat with another 100 mL of boiling 0.1% SDS/0.1× SSC. Blot is now ready for subsequent probe hybridizations starting at the pre-hybridization (*see* **Subheading 3.6, step 3**).

3.7. Quantitation of RNA Levels From Northern Blot Analysis

1. Hybridize blot as described above to a stably expressed reference RNA (*see* **Note 19**). The *scRI* RNA, an RNA polymerase III transcript *(16)*, is commonly used to normalize loading of yeast RNA as its expression is not altered by changes in RNA polymerase II activity. Expose Saran-wrapped blot to a phosphorimager screen. **Figure 2B** shows hybridization of the *scRI* control RNA to the same blot shown in **Fig. 2A**.
2. Quantify the radioactive intensity of each RNA band (including both the RNA of interest and the normalization control) using software such as ImageQuant (*see* **Note 20**).
3. Calculate the relative RNA levels of the control RNA in each lane by dividing the radioactive volume of each band by the volume of the band in the first lane, thus setting the relative intensity of the first lane to 1.0.

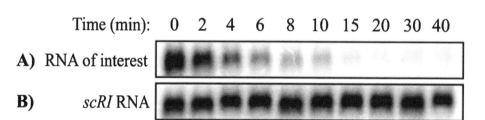

Fig. 2. Northern blot analysis of mRNA decay. Total RNA was isolated from *rpb1-1* yeast at the indicated times after transcriptional repression. RNA was electrophoresed on a 1.25% agarose gel and blotted to nylon membrane. (**A**) The blot was hybridized first with a radioactive probe to specifically recognize the mRNA of interest. (**B**) After stripping the first probe off the northern blot, the blot was reprobed for *scRI* RNA, the loading control.

4. To normalize the RNA levels of the RNA of interest, divide the radioactive volume of each mRNA band by the above-calculated relative intensity of the respective control RNA band. To analyze transcriptional shut off assays, the RNA remaining at each time point is calculated by dividing the normalized RNA intensity of each band by the normalized RNA intensity of the band at the zero minute time point (representing the original pool of RNA). The RNA remaining at each time point is graphed (y-axis) versus time (x-axis) to determine the time at which 50% of the RNA is degraded (the half-life).

3.8. Isolation of Total RNA from Mammalian Cells

1. Culture cells in an appropriate medium. For steady-state mRNA analysis, suspend cells in Trizol Reagent for lysis by repetitive pipetting (*see* **Note 21**). The Trizol can be added directly to a culture dish following removal of the medium, or cells can be detached by standard trypsin/scraping methods and pelleted by centrifugation before resuspending in Trizol. Use 1 ml of Trizol per 10 cm^2 if cells are still on the culture dish or 1 ml of Trizol per 5×10^6 cells if pelleted.
2. Incubate the lysis reaction for 5 min at room temperature. At this point the sample can be stored at –80°C.
3. Add 0.2 mL of chloroform per mL of Trizol.
4. Vortex 15 s and centrifuge at 12,000 $\times$ g for 10 min at 4°C.
5. Transfer the top aqueous phase to a new 1.5-mL tube and add 0.5 mL isopropanol per mL of Trizol reagent used in original lysis step (*see* **Note 7**).
6. Vortex briefly to mix, centrifuge at 12,000 $\times$ g for 10 min at 4°C, and decant off supernatant.
7. Wash the now-visible RNA pellet in 1 ml of cold 70% ethanol by inverting, followed by centrifugation at 12,000 $\times$ g for 5 min at 4°C.
8. Pipet off the supernatant and allow RNA pellet to air dry (*see* **Note 9**).
9. Resuspend RNA pellet in 160 µL DEPC-treated H_2O.
10. Add 20 µL DNase I buffer and 20 µL of RNase-free DNase (*see* **Note 22**).
11. Incubate reaction for 20 min at 37°C.
12. Add 200 µL phenol/chloroform equilibrated in TE, vortex briefly to mix, and centrifuge 2 min at 21,000 $\times$ g.
13. Transfer top aqueous phase to new 1.5-mL tube and add 66 µL of DEPC-treated 10 M NH_4OAc and 500 µL cold 100% ethanol.
14. Vortex briefly to mix, centrifuge 10 min at 21,000 $\times$ g, and decant off supernatant.
15. Wash the RNA pellet in 250 µL of cold 70% ethanol by inverting, followed by centrifugation at 21,000 $\times$ g for 5 min.
16. Pipet off the supernatant and allow RNA pellet to air dry (*see* **Note 9**).
17. Resuspend RNA pellet in 20 µL DEPC-treated H_2O.
18. Determine the yield of RNA by measuring OD_{260}. 1.0 OD_{260} = 40 µg/mL RNA.

3.9. Preparation of cDNA

1. Reverse transcription of RNA into cDNA can be primed by random hexamers or oligo(dT) (*see* **Note 23**). For annealing reaction, mix 1–2 μg of DNase-treated RNA with 500 ng of either random hexamer or oligo(dT) primer. Add DEPC-treated H$_2$O to total volume of 5 μL. The same amount of RNA should be used per sample.
2. Incubate at 70°C for 5 min and then quench on ice.
3. In a separate tube, mix the following components for the RT reaction:

   a. 4 μL 5× Reverse Transcriptase buffer
   b. 5.6 μL DEPC-treated H$_2$O
   c. 1 μL 10 m*M* dNTPs
   d. 2.4 μL 25 m*M* MgCl$_2$
   e. 1 μL RNasin
   f. 1 μL Reverse Transcriptase

4. Combine the RT reaction with the annealing reaction and incubate first for 5 min at 25°C, then 1 h at 42°C, and then 5 min at 70°C.
5. The cDNA is now ready for use in PCR applications.

3.10. Design of Primers for qPCR

1. Design PCR primers to amplify a product between 100 and 200 bp, with 150 bp as an ideal product size (*see* **Note 24**).
2. Ideally, primers should have the following properties:

   a. Annealing temperatures between 58 and 60°C and within 1°C of each other.
   b. GC content of 40–60%.
   c. Length of approximately 20 nt.
   d. No complementarity, strong secondary structure, or runs of multiple bases.

3.11. Validation of Primer Specificity and Efficiency

1. Validate primers by preparing reactions with several different dilutions of cDNA (*see* **Note 25**).
2. Prepare six 10-fold dilutions of cDNA for the templates (1×, 0.1×, 0.01×, 0.001×, 0.0001×, and 0.00001×).
3. Each standardization reaction is composed of the following:

   a. 12.5 μL iQ SYBR Green Supermix (*see* **Note 26**).
   b. 1 μL Upstream primer (2.5 μ*M*).
   c. 1 μL Downstream primer (2.5 μ*M*).
   d. 2 μL template cDNA.
   e. 8.5 μL H$_2$O.

4. Set up a master mix containing all of the above reaction components except the template cDNA. The master mix should be sufficient for 21 reactions. Aliquot 75.9 µL of master mix into each of six tubes (*see* **Note 27**). Set up reactions on ice.
5. Add 6.6 µL of each template cDNA dilution to one of the six tubes of master mix.
6. To analyze each reaction in triplicate, aliquot 25 µL of each template mix into each of three wells of a PCR plate.
7. Prepare a single reaction with H_2O instead of cDNA template for use as a negative control—combine 2 µL H_2O with 23 µL of remaining master mix.
8. Following PCR, calculate the primer efficiency and specificity using software such as MyIQ (Bio-Rad).

3.12. Preparation of qPCR Reactions

1. Set up master mixes for all reactions and test each template sample in triplicate as described for the standardization reactions above.
2. At least one and up to three reference genes should be amplified for each template sample to normalize quantitations *(17)*. Appropriate reference genes such as GAPDH or β-actin should be stably expressed across samples. Ideally, the reference gene should be of similar abundance to the gene of interest. Ribosomal RNA can be used as a reference, but because of its high abundance, it is not ideal for normalization of low-level transcripts.
3. Relative RNA abundance is calculated from differences in sample C_T values, the cycle number at which the amplified target reaches a significant threshold. Although a detailed discussion of the various quantitation methods is beyond the scope of this chapter, the reader is directed to *(18–20)*.

4. Notes

1. The gel apparatus used should be dedicated to RNA agarose gels to avoid problems associated with contamination. Gel equipment can be made ready for RNA use by soaking in 3% H_2O_2, followed by an ethanol rinse. The gels should be poured and electrophoresed in a fume hood because of the formaldehyde.
2. To accurately compare RNA levels between different strains/conditions, it is imperative to harvest each culture at the same optical density (OD). For many genes, it is best to compare RNA levels from cells harvested during exponential growth. However, RNA levels from genes such as those involved in the stress response may be better compared in cells harvested at higher ODs.
3. It is easy to determine whether a yeast culture is contaminated with bacteria, as the bacteria will not pellet well during this centrifugation and the supernatant will be cloudy.
4. Yeast cultures are typically grown at 30°C in media containing 2% dextrose unless otherwise noted. Standard protocols for culturing yeast are available *(21,22)*.
5. Use of a multi-tube vortexer is advantageous for this step. Be sure the lids of the tubes are tight to avoid phenol leakage.

6. When removing the aqueous phase, do not get too close to the phenol layer. It is better to lose a little yield than to try and get the entire aqueous layer and contaminate the RNA. It should be possible to remove 400 µL of the aqueous phase.

7. Dispose of all residual phenol properly.

8. RNA is quite stable in ethanol and can be stored safely at –20°C if needed after this step.

9. Do not over-dry RNA, as it is very difficult to get back into solution. Resuspension can be aided by heating up to 65°C for a few minutes.

10. A 250-fold dilution of the RNA works well for spectrophotometer readings at OD_{260}. A 50-mL cell culture harvested at OD_{600} of 0.4 typically yields approximately 400 µg of RNA.

11. RNA is more easily resuspended in H_2O rather than directly into the gel loading buffer. Resuspension of the RNA is key to making sure equal amounts of RNA get loaded onto the gel.

12. In a 1.25% agarose gel, the bromophenol blue dye migrates similar to a 500 nucleotide RNA. Typically, gels are electrophoresed 4 h at 70 V.

13. The use of unpurified γ-(^{32}P) ATP at 6000 Ci/mmol (PerkinElmer, Waltham, MA, USA, NEG035C) allows the creation of probes with high specific activity.

14. If the column works properly, the blue dextran and oligonucleotide probe should completely move through column into liquid captured at bottom of 15-mL tube, while the phenol red and unincorporated nucleotides should remain within the column.

15. For a probe with high specific activity, which is especially important for detection of low-abundance mRNAs, the probe should be at least 10^6 cpm/µL.

16. All washes and hybridizations can be easily performed in sealed plastic containers such as 10-cup square Rubbermaid containers.

17. Filtering the probe helps reduce the spotty background sometimes seen with oligonucleotide probes.

18. Hybridization should be performed at least 15°C below the melting temperature (Tm) of the oligonucleotide probe. The Tm of an oligonucleotide can be roughly estimated under the hybridization conditions as Tm (°C) = 4(G + C content) + 2(A + T content). For most oligonucleotide probes, hybridization can be performed at 42°C.

19. It is best to always probe blots for the RNA(s) of interest first before probing for the normalization control RNA, as most control RNAs are highly abundant and their probes are sometimes difficult to completely strip off the blot.

20. To quantitate RNA bands using ImageQuant software, draw a box tight around one RNA band in the blot image. Copy and paste the box to each subsequent RNA band to ensure each box (and the number of pixels quantitated within each box) is the exact same size. To account for background noise of the blot, copy a box

onto a background region of the blot and subtract the radioactive volume within this box from all other RNA band volumes.

21. For half-life analysis of mRNA using actinomycin D as a transcriptional inhibitor, culture cells in the appropriate medium and then harvest an aliquot of cells representing the steady-state pool of RNA at time zero. Add actinomycin D to the remaining cells to a concentration of 3 μg/mL and then harvest aliquots of cells at increasing time points after treatment, such as 2, 4, 6, 8, 12, and 24 h. Each harvested cell aliquot is then subjected to Trizol disruption and subsequent RNA isolation.

22. It is important to DNase-treat the RNA to remove contaminating DNA that could contribute to subsequent PCR products.

23. Oligo(dT) will only prime mRNAs, prohibiting the use of rRNA as a normalization control in subsequent qPCR reactions. In addition, oligo(dT) priming introduces a 3′ end bias such that segments closer to the 3′ end are copied more efficiently and become over-represented, whereas mRNAs with short poly(A) tails become under-represented. Therefore, random hexamers are a better primer alternative.

24. Many programs are available to aid in the selection of primers for use in both SYBR-green and Taqman-type methods, including the publicly available Primer 3 program (http://frodo.wi.mit.edu/cgi-bin/primer3/primer3_www.cgi).

25. Primer validation results should give a PCR efficiency of 80–120% and a correlation coefficient of >0.98. In addition, the melting curve should be a single peak at a temperature of 80–90°C. Multiple peaks are indicative of primer-dimers or non-specific amplification. Primers not meeting these specifications should not be used.

26. Avoid repeated freeze/thaw cycles with the iQ SYBR Green Supermix.

27. It is important to set up reactions using a master mix to minimize the effect of pipetting error. Also, be sure that pipets are adequately calibrated.

Acknowledgments

I thank Carol Wilusz for technical assistance with the mammalian RNA and qPCR protocols, and Randi Ulbricht and Florencia Lopez Leban for assistance with manuscript preparation. This work was supported by a grant from the National Institutes of Health (GM63759).

References

1. Derrigo, M., Cestelli, A., Savettieri, G., and Di Liegro, I. (2000) RNA-protein interactions in the control of stability and localization of messenger RNA. *Int. J. Mol. Med.* **5,** 111–123.

2. Grzybowska, E. A., Wilczynska, A., and Siedlecki, J. A. (2001) Regulatory functions of 3′ UTRs. *Biochem. Biophys. Res. Commun.* **288,** 291–295.

3. Bartel, D. P. (2004) MicroRNAs: genomics, biogenesis, mechanism, and function. *Cell* **116,** 281–297.

4. Sobell, H. M. (1985) Actinomycin and DNA transcription. *Proc. Natl. Acad. Sci. USA* **82**, 528–5331.

5. Atasoy, U., Curry, S. L., Lopez de Silanes, I., Shyu, A. B., Casolaro, V., Gorospe, M., and Stellato, C. (2003) Regulation of eotaxin gene expression by TNF-α and IL-4 through mRNA stabilization: involvement of the RNA-binding protein HuR. *J. Immunol.* **171**, 4369–4378.

6. Grigull, J., Mnaimneh, S., Pootoolal, J., Robinson, M. D., and Hughes, T. R. (2004) Genome-wide analysis of mRNA stability using transcription inhibitors and microarrays reveals posttranscriptional control of ribosome biogenesis factors. *Mol. Cell. Biol.* **24**, 5534–5547.

7. Nonet, M., Scafe, C., Sexton, J., and Young, R. (1987) Eukaryotic RNA polymerase conditional mutant that rapidly ceases mRNA synthesis. *Mol. Cell. Biol.* **7**, 1602–1611.

8. Herrick, D., Parker, R., and Jacobson, A. (1990) Identification and comparison of stable and unstable mRNAs in *Saccharomyces cerevisiae*. *Mol. Cell. Biol.* **10**, 2269–2284.

9. Herruer, M. S., Mager, W. H., Raue, H. A., Vreken, P., Wilms, E., and Planta, R. J. (1988) Mild temperature shock affects transcription of yeast ribosomal protein genes as well as the stability of their mRNAs. *Nucleic Acids Res.* **16**, 7917–7929.

10. Adams, C. C. and Gross, D. S. (1991) The yeast heat shock response is induced by conversion of cells to spheroplasts and by potent transcriptional inhibitors. *J. Bacteriol.* **173**, 7429–7435.

11. Gossen, M. and Bujard, H. (1992) Tight control of gene expression in mammalian cells by tetracycline-responsive promoters. *Proc. Natl. Acad. Sci. USA* **89**, 5547–5551.

12. Gari, E., Piedrafita, L., Aldea, M., and Herrero, E. (1997) A set of vectors with a tetracycline-regulatable promoter system for modulated gene expression in *Saccharomyces cerevisiae*. *Yeast* **13**, 837–848.

13. Loflin, P. T., Chen, C. Y., Xu, N., and Shyu, A. B. (1999) Transcriptional pulsing approaches for analysis of mRNA turnover in mammalian cells. *Methods* **17**, 11–20.

14. Heaton, B., Decker, C., Muhlrad, D., Donahue, J., Jacobson, A., and Parker, R. (1992) Analysis of chimeric mRNAs identifies multiple regions within the *STE3* mRNA which promote rapid mRNA decay. *Nucleic Acids Res.* **20**, 5365–5373.

15. Decker, C. J. and Parker, R. (1993) A turnover pathway for both stable and unstable mRNAs in yeast: evidence for a requirement for deadenylation. *Genes Dev.* **7**, 1632–1643.

16. Felici, F., Cesareni, G., and Hughes, J. M. X. (1989) The most abundant small cytoplasmic RNA of *Saccharomyces cerevisiae* has an important function required for normal cell growth. *Mol. Cell. Biol.* **9**, 3260–3268.

17. Vandesompele, J., De Preter, K., Pattyn, F., Poppe, B., Van Roy, N., De Paepe, A., and Speleman, F. (2002) Accurate normalization of real-time quantitative RT-PCR

data by geometric averaging of multiple internal control genes. *Genome Biol.* **3,** research 0034.1–0034.11.

18. Wong, M. L. and Medrano, J. F. (2005) Real-time PCR for mRNA quantitation. *BioTechniques* **39,** 75–85.
19. Leclerc, G. J., Leclerc, G. M., and Barredo, J. C. (2002) Real-time RT-PCR analysis of mRNA decay: half-life of Beta-actin mRNA in human leukemia CCRF-CEM and Nalm-6 cell lines. *Cancer Cell Int.* **2,** 1.
20. http://www.gene-quantification.info/
21. Guthrie, C. and Fink, G., eds. (2002) Guide to Yeast Genetics and Molecular and Cell Biology, Part B. *Methods Enzymol.* **350.**
22. Amberg, D. C., Burke, D. J., and Strathern, J. N. (2005) *Methods in Yeast Genetics,* Cold Spring Harbor Laboratory Press, Cold Spring Harbor, NY.

18

Application of the Invader® RNA Assay to the Polarity of Vertebrate mRNA Decay

Elizabeth L. Murray and Daniel R. Schoenberg

Summary

The inability of structural elements within a reporter mRNA to impede processive decay by the major 5′ and 3′ exonucleases has been a major obstacle to understanding mechanisms of vertebrate mRNA decay. We present here a new approach to this problem focused on quantifying the decay of individual portions of a reporter mRNA. Our approach entails two parts. The first involves the use of a regulated promoter, such as one controlled by tetracycline (tet), to allow reporter gene transcription to be turned off when needed. Cells stably expressing the tet repressor protein are transiently or stably transfected with tet-regulated beta-globin genes in which the sequence element under study is cloned into the 3′-UTR. The second involves the quantification of beta-globin mRNA using the Invader RNA assay, a sensitive and quantitative approach that relies on signal amplification instead of target amplification. Because the Invader RNA assay does not depend on downstream primer binding, the use of multiple probes across the reporter beta-globin mRNA allows for quantification of the decay of individual portions of the mRNA independent of events acting at other sites.

Key Words: mRNA decay; Invader RNA assay; invasive cleavage assay; transcriptional pulse; transcriptional turn-off.

1. Introduction

Measuring the decay rate of a particular mRNA fundamentally involves two steps: turning off reporter gene transcription and quantitation of the decrease over time of its encoded mRNA. In the past, transcriptional inhibitors such as actinomycin D and 5,6-dichloro-1β- D-ribofuranosylbenzimidazole (DRB) have

From: *Methods in Molecular Biology, Vol. 419: Post-Transcriptional Gene Regulation*
Edited by: J. Wilusz © Humana Press, Totowa, NJ

been used to stop transcription; more current methods use regulated systems whereby the transcription of the gene of interest can be specifically controlled. Traditionally, mRNA detection has relied on northern blotting, ribonuclease protection assay (RPA), and reverse transcription–PCR (RT–PCR). The method that we use combines the inducible tetracycline (tet)-repressor system and a novel method for mRNA quantification, the invasive cleavage assay, or Invader RNA assay.

Various methods to measure mRNA decay have been described. One approach is to radioactively label mRNAs and measure the approach to steady-state or decay. Other approaches involve using transcriptional inhibitors such as actinomycin D or DRB to stop transcription and then measure the decay of the mRNA transcript *(1)*. Actinomycin D interferes with transcription by inter-calating with DNA and DRB inhibits transcription by inhibiting pTEFb. Both transcriptional inhibitors can have significant effects on normal cell physiology leading to possible artifacts. Most current studies use a regulated promoter such as the serum-inducible c-fos promoter or a promoter flanked upstream by multiple tet-operator elements to produce either a pulse of transcription or allow for decay from steady-state by turning off transcription *(2)*. These methods have the advantages of specifically controlling expression of a single gene with minimal effects on normal cell physiology.

Using this system, there are several ways to measure reporter mRNA half-life. Transcriptional pulsing has been previously described in detail *(2)*. Briefly, this approach involves inducing reporter mRNA expression for a defined period, for example by removing tet from the medium, followed by its re-addition to repress transcription. Provided that the pulse length is shorter than the half-life of the mRNA, this results in production of a homogeneous population of reporter mRNAs. In situations where the mRNA half-life is very short or the amount of mRNA induced during a pulse of transcription is insufficient for subsequent quantitation, this system can be used to study decay from steady state by maintaining cells in medium lacking tet and following decay after its addition.

Methods commonly used to quantify mRNA decay include northern blotting, RPA, and RT–PCR with radiolabeled primers or real-time PCR. The advantage of northern blotting is that it tracks decay of the full-length mRNA and as such can provide evidence for deadenylation and in rare cases for endonuclease cleavage. RPA is more sensitive and better for quantifying changes in less-abundant mRNAs than northern blotting, and RT-PCR and real-time PCR offer even greater sensitivity and flexibility when dealing with limited amounts of RNA. PCR-based methods require a reverse transcription step that by definition depends on the integrity of a downstream primer binding site, in addition to which PCR can be subject to target amplification artifacts. The Invader RNA assay (Third Wave Technologies, Inc.; Madison, WI; www.twt.com) offers a new way to quantify mRNA decay *(3)*. This assay is applicable to small quantities of input RNA, quantitative, highly sensitive, and free from potential artifacts associated with PCR target amplification.

The Invader RNA assay consists of two isothermal reactions run in the same well of a microtiter plate. In the primary reaction, three oligonucleotides (a stacking oligonucleotide, a primary probe oligonucleotide, and an Invader oligonucleotide) bind adjacent to one another at a specific locus on the target mRNA (*see* **Fig. 1**). The primary probe oligonucleotide consists of two portions, one which binds the mRNA and a second which is non-complementary to the target mRNA and forms a 5´ flap. The Invader oligonucleotide binds the target

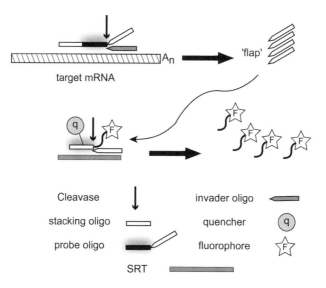

Fig. 1. Outline of the Invader RNA assay. The steps involved in the Invader RNA assay are shown. The stacking oligomer abuts the probe oligonucleotide on the target mRNA, and the invader oligonucleotide overlaps the junction of the probe oligonucleotide and the flap by one nucleotide. Cleavase enzyme cuts within the distorted structure formed by this complex to generate flaps in direct proportion to the concentration of target mRNA. These serve as the invader in a second reaction in which an oligonucleotide bearing a fluorescent tag and a quencher are separated by the cleavage site in a complex assembled on a generic second reaction template (SRT). Fluorescence is generated in direct proportion to the concentration of flaps generated in the first reaction.

mRNA immediately downstream of the primary probe. The nucleotide located at the 5′ end of the Invader oligonucleotide does not bind the mRNA target or the primary probe, but instead overlaps the junction of the primary probe and the target mRNA. This pushes on the 5′-most bond between the primary probe and the target mRNA to create a bulged secondary structure. The primary probe oligonucleotide is cleaved by a modified thermostable form of tTh polymerase termed Cleavase® to release the 5′ flap. The stacking oligonucleotide serves to modulate the Tm of the primary probe to 60°C, so that at the reaction temperature of 60°C, the primary probe oligonucleotide freely cycles on and off the target mRNA while the Invader oligonucleotide, which has a higher Tm, remains bound. Thus, multiple primary probes are cleaved per target, and the number of free flaps produced in this reaction is a product of the length of incubation and target abundance. In the secondary reaction, three additional oligonucleotides are added: an Arrestor oligonucleotide, a fluorescence resonance energy transfer (FRET) -labeled secondary probe oligonucleotide (containing both a fluorophore and a quencher molecule) and a secondary reaction template (SRT) oligonucleotide. The cleaved flap of the primary probe from the primary reaction acts as an Invader oligonucleotide in conjunction with the SRT and the FRET-labeled secondary probe oligonucleotides in another cleavage reaction in which the FRET-labeled secondary probe oligonucleotide is cleaved to separate the fluorophore on the flap from the quencher molecule. As in the primary reaction, the cleaved 5′ probe flap from the primary probe remains bound to the SRT whereas the FRET-labeled oligonucleotide cycles on and off, resulting in linear fluorescence signal generation. The Arrestor oligonucleotide binds uncleaved probe oligonucleotide from the primary reaction. This assay can be run in biplex format to allow the simultaneous detection of two targets from any one sample. This is helpful when using internal controls. Here, we describe a system based on the common reporter, human beta-globin mRNA, that takes advantage of the unique properties of the Invader RNA assay to study the polarity of mRNA decay.

2. Materials
2.1. In vitro Transcript Generation

1. pG-glo-PLEB-cDNA full-length pA14, a plasmid containing human beta-globin cDNA downstream of the SP6 promoter. After linearizing with SstII, transcription with SP6 polymerase (Ambion, Austin, TX, USA) generates full-length beta-globin mRNA with a 14-residue poly(A) tail.
2. SstII restriction enzyme (New England Biolabs, Ipswich, MA, USA) to linearize plasmid.
3. QIAquick PCR purification kit (Qiagen, Valencia, CA, USA).

4. Megascript kit (Ambion).
5. DNAse I (New England Biolabs).
6. Phenol/chloroform/isoamyl alcohol (25:24:1) (Sigma, St. Louis, MO, USA). Caution—this is corrosive and is a hazard to eyes, skin, upon ingestion, and inhalation. Avoid contact with this and wear gloves, lab coat, and protective eyewear. Light sensitive.
7. Isopropanol (Sigma).
8. Diethylpyrocarbonate (DEPC)-treated water. Caution, DEPC (Sigma) is irritating to eyes, skin, and mucus membranes. It is a suspected carcinogen. Light sensitive. Open DEPC only in fume hood, wear gloves (*see* **Note 1**).
9. Reagents for conventional agarose gel electrophoresis.
10. Ethidium bromide (Sigma). Caution, possible carcinogen. Avoid inhalation or skin contact. Light sensitive.

2.2. Cell Culture

1. LM(tk- tTA) cells, murine fibroblasts stably expressing the tTA protein for Tet off regulation of reporter gene.
2. pTet-CMV$_{min}$-glo-SPA, pTet-CMV$_{min}$-glo-ARE[*cfos*]-SPA, plasmids containing beta-globin reporter genes with or without the cfos ARE in the 3′-UTR, respectively, under control of tet-responsive promoter (*see* **Note 2**).
3. Transfection control plasmids, constitutively expressing a second reporter mRNA such as beta-galactosidase or luciferase.
4. Cell culture media. LM (tk- tTA) cells are grown in 90% Dulbecco's minimal essential medium (DMEM) supplemented with 10% fetal bovine serum (FBS) and 2 mM L-glutamine (Gibco, Carlsbad, CA, USA).
5. Transfection reagent, Superfect (Invitrogen) for plasmid transfection; Dreamfect (OZ Biosciences, Marseille, France) for siRNA transfection.
6. Tet or doxycycline (dox) (Sigma), at 1 mg/ml in 70% ethanol, stored at –20°C. Tet stored in this manner is stable 2 weeks and dox for 4 weeks. Both are light sensitive.
7. Cytoplasmic isolation buffer:
    0.14 M NaCl.
    1.5 mM MgCl$_2$.
    0.01 M Tris–HCl pH 8.0.
    0.0025% NP20.
    0.01 M dithiothreitol (DTT).
    RNAse OUT (Invitrogen).
    DEPC-treated water to desired volume.
    Add DTT and RNAse OUT just before use. Store buffer at 4°C.
8. Kit for RNA purification such as TRIzol (Invitrogen), Absolutely RNA RTPCR miniprep kit (Stratagene, La Jolla, CA, USA), RNA miniprep kit (Qiagen).
9. Phosphate-buffered saline (PBS) (Sigma).
10. Cell scrapers (Fisher, Waltham, MA, USA).

2.3. RNA Invader Assay

1. Invader RNA assay generic kit (Third Wave Technologies).
2. Primary oligonucleotide mix (*see* **Table 1** for specific sequences):
   Primary probe oligonucleotide, 40 µM.
   Invader oligonucleotide, 20 µM.
   Stacking oligonucleotide, 12 µM.
   Made in Te buffer (see below) to desired volume.
3. Secondary oligonucleotide mix (*see* **Table 1** for specific sequences):
   Arrestor oligonucleotide at 53.4 µM in Te buffer.
4. FRET-labeled oligonucleotide mix:
   SRT, 2.0 µM.
   Fluorescent oligonucleotide, 13.4 µM.
5. Te buffer (EDTA inhibits the Invader RNA assay): 10 mM Tris–HCl pH 8.0, 0.1 mM ethylenediamine tetraacetic acid (EDTA).
6. DEPC-treated water.
7. tRNA, diluted to 20 ng/ml in DEPC-treated water (Sigma).
8. Stop buffer (optional): 10 mM Tris–HCl pH 8.0, 10 mM EDTA.
9. 0.2 ml thin-walled strip tubes with individually attached caps (USA Scientific, Enfield, CT).
10. 96-well polypropylene round-bottom microplates.
11. Chill-Out 14 liquid wax (MJ Research, San Francisco, CA).
12. Thermal adhesive sealing film for PCR (Fisher).
13. 8-channel 0.1–10 µl multichannel pipettor and tips.
14. 0.1–100 µl electronic repeat pipettor and tips.
15. Hybridization oven or 96-well format thermocycler.
16. Fluorescence plate reader with appropriate filters attached to a computer with Magellan™ and Microsoft Excel™ software. We use the Tecan Genios fluorescence plate reader.

3. Methods

3.1. Primary Reaction Oligonucleotide Design, Synthesis, and Preparation

1. For a description of the functions of each oligonucleotide, *see* **Note 3**.
2. Choose the region for quantification on the target mRNA of interest. For this study, we developed probes targeting each of the 3 exons of human beta-globin mRNA (*see* **Table 1**).
3. Using a sequence of the top strand, in the 5′-3′ direction of the spliced and cleaved mRNA, pick a location where the cleavage site (the site of secondary structure) will be located (*see* **Note 4**).
4. Determine whether the surrounding approximately 50-nt region is likely to fall in a highly structured region of the mRNA (*see* **Note 5**).

Table 1
Oligonucleotides Used in the Invader RNA Assay

Probe set	Oligonucleotide	Sequence (5′–3′)
Exon 1	Probe	aacgaggcgcac**CTTCATCCAC**-NH2
	Invader oligo	CAGGGCCTCACCACCAAA
	Stacking oligo	GUUCACCUUGCC
	Arrestor oligo	GUGGAUGAAGGUGCGC
	FRET oligo	(6-FAM)-CAC-(EQ)-TGCTTCGTGG
	SRT	CCAGGAAGCAAGTGGTGCGCCTCGUUAA
Exon 2	Probe	aacgaggcgcac**CTAAAGGCACC**-NH2
	Invader oligo	GGTGAGCCAGGCCATCAA
	Stacking oligo	GAGCACUUUCUUGC
	Arrestor oligo	GGUGCCUUUAGGUGCGC
	FRET oligo	(6-FAM)-CAC-(EQ)-TGCTTCGTGG
	SRT	CCAGGAAGCAAGTGGTGCGCCTCGUUAA
Exon 3	Probe	aacgaggcgcac**ACCAGCACG**-NH2
	Invader oligo	GTGATGGGCCAGCACACAGC
	Stacking oligo	UUGCCCAGGAG
	Arrestor oligo	CGUGCUGGUGUGCGC
	FRET oligo	(6-FAM)-CAC-(EQ)-TGCTTCGTGG
	SRT	CCAGGAAGCAAGTGGTGCGCCTCGUUAA

SRT, secondary reaction template.

5. Determine the length of sequence each oligonucleotide needs to cover to get the appropriate Tm of 60°C or 80°C, as oligonucleotide sequences are strictly a function of Tm (*see* **Note 6**).
6. Check that the resulting oligonucleotides are free of common pitfalls:

   a. Length is too short to be sequence-specific. This can happen in GC-rich regions.
   b. Primer/dimer formation.
   c. Secondary structure within the oligonucleotide, e.g., runs of 4 or more Gs.
   d. 2 base invasion: the nucleotide upstream of the cleavage nucleotide in the probe (located in the flap region) must not bind to the target molecule.
   e. G as cleavage site in probe, as for some reason it is inefficiently cleaved. The most efficiently cleaved nucleotide is C (would be a G in the mRNA sequence).

7. Add a flap sequence to the 5′ end of the probe molecule (*see* **Note 7**).
8. Synthesize oligonucleotides at 1-µM scale, be sure to use high performance liquid chromatography (HPLC) purification (*see* **Note 8**).

9. Probe and Invader oligonucleotides are DNA; the stacking oligonucleotide is an RNA. The probe oligonucleotide has a 3′ amino modifier C7, the stacking oligonucleotide is 2′ O-methylated throughout (*see* **Note 9**).
10. To prepare oligonucleotide mixes, dissolve oligonucleotides in 300 µl Te and measure a 1:200 dilution at A_{260}. Dilute oligonucleotides in Te to make primary oligonucleotide mix:
    Primary probe oligonucleotide, 40 µM.
    Invader oligonucleotide, 20 µM.
    Stacking oligonucleotide, 12 µM.

3.2. Secondary Reaction Oligonucleotide Design, Synthesis, and Preparation

1. For a description of the functions of each oligonucleotide, *see* **Note 10**.
2. Design an arrestor oligonucleotide to be complementary to the uncleaved primary probe oligonucleotide. This is an RNA oligonucleotide and should be 2′ O-methylated throughout and purified by desalting.
3. Design a secondary probe oligonucleotide. For published secondary probe sequences see *refs. 3* and *5*. This is a DNA oligonucleotide containing a fluorescent tag on the 3′ end and a quencher molecule at the third position. This oligonucleotide should be HPLC purified.
4. Design the SRT. This oligonucleotide should hybridize with both the secondary probe oligonucleotide and the primary probe flap oligonucleotide to create the secondary structure recognized by Cleavase as a cleavage substrate. The 3′-most three to five nucleotides of this DNA oligonucleotide should be 2′ O-methylated.

3.3. In vitro Transcript Generation

1. For a description of the in vitro transcript, *see* **Note 11**.
2. Linearize plasmid template by restriction digestion.
3. Check linearization by running a portion of the product on a standard agarose gel and staining with ethidium bromide before UV visualization. Templates must re-cut or gel purified if more than one band is seen, as this indicates the restriction digest was not complete.
4. Purify the linearized plasmid. We use the QIAquick PCR purification kit (Qiagen) according to manufacturer's directions.
5. Perform transcription. We use the Megascript kit (Ambion). Each transcription reaction is composed of 5 µl plasmid (1 µg), 2 µl 10× RNA polymerase buffer, 2 µl each ATP, CTP, GTP, or UTP solution, 2 µl RNA polymerase and water to 20 µl. Incubate overnight at 37°C.
6. Remove the plasmid template from the in vitro transcript by adding 1 µl DNAse I and incubating for 15 min at 37°C.

7. Extract with phenol/chloroform/isoamyl alcohol. First, bring transcription reaction up to 100 µl volume with DEPC-treated water, then add 100 µl phenol/chloroform/isoamyl alcohol.
8. Precipitate with an equal volume of isopropanol.
9. Dissolve the pellet in 50 µl DEPC-treated water.
10. Check integrity of in vitro transcript by running on agarose gel and staining with ethidium bromide before UV visualization. If more than one band is observed or if there are shorter products detected the transcript must be gel purified before use. Extra bands suggest either transcription or RNAse degradation problems.
11. Quantify by measuring absorbance at A_{260} in a spectrophotometer.
12. Dilute the in vitro transcript to 100 amol/µl. Store aliquots at –80°C.

3.4. Transcription Pulse Assay

1. For general considerations of experimental design, *see* **Notes 12** and **13**.
2. On the first day, plate cells in normal growth medium plus 50–100 ng/ml tet or dox to repress reporter gene expression (*see* **Notes 14–16**).
3. To induce reporter gene expression, remove the growth medium and wash cells with PBS.
4. Replace with medium without tet or dox and grow cells for about 4 h (*see* **Note 17**).
5. To shut off reporter gene expression, add tet or dox back to the medium at 1000 ng/ml.
6. Cells may be removed from the incubator at desired timepoints following tet or dox readdition for protein or RNA isolation.

3.5. Transcription Turn Off Assay

1. For general considerations of experimental design, *see* **Notes 12** and **13**.
2. Plate cells in normal growth medium including antibiotics. Omit tet or dox to permit reporter gene expression to steady state.
3. At desired time, repress reporter gene expression by the addition of tet or dox to 1000 ng/ml.
4. Cells may be removed from the incubator at desired timepoints following tet or dox addition for protein or RNA isolation.

3.6. RNA Isolation

1. Remove growth medium from cells and wash plate twice with PBS.
2. Add 1 ml PBS to each plate and use a cell scraper to suspend the cells. The suspended cell solution is transferred to a 1.7-ml microcentrifuge tube.
3. Pellet cells by centrifuging for 1 min at $1000 \times g$, 4°C.
4. Remove the supernatant and resuspend the cells in 200 µl of cytoplasmic extraction buffer.
5. Incubate cells on ice for 10 min.

6. Remove the nuclei by centrifuging the lysed cell suspension for 5.5 min at 12,000 × g, 4°C.
7. Remove the supernatant containing cytoplasmic extract to a fresh tube. Cytoplasmic extract and the nuclear pellets can be stored frozen at –80°C until further use at this point.
8. Purify RNA from cytoplasmic or nuclear fractions using RNA isolation protocol of choice.
9. Quantify RNA yield by measuring A_{260} in a spectrophotometer.

3.7. Invader RNA Assay

1. *See* **Notes 18–20** for information on experimental design.
2. Dilute in vitro transcript in an aqueous solution of tRNA at 20 ng/ml for use as standard curve (*see* **Notes 21–23**). Be sure to thoroughly vortex and spin down each sample between dilutions.
3. Transfer each standard to one tube in a 8-well tube strip. Store on ice.
4. Prepare samples by diluting RNA isolated from transcription pulse- or transcription turn off-treated cells in water or in an aqueous solution of 20 ng/ml tRNA (*see* **Notes 24–26**).
5. Thoroughly vortex each sample, briefly centrifuge to collect the contents, and transfer each sample to a well of an 8-well tube strip. Store on ice.
6. Prepare primary reaction mix by combining (for each well) 4 µl RNA Primary Buffer (Third Wave Technologies), 0.25 µl beta-globin primary oligonucleotide mix, 0.25 µl glyceraldehyde 3-phosphate-dehydrogenase (GAPDH) primary oligonucleotide mix (if using, otherwise add Te), and 0.5 µl Cleavase enzyme (*see* **Note 27**). Add Cleavase enzyme last. This is then mixed on a vortex mixer, centrifuged, and placed on ice.
7. Prepare secondary reaction mix by combining (for each well) 2 µl RNA Secondary Buffer (Third Wave Technologies), 0.75 µl beta-globin secondary oligonucleotide, 0.75 µl carboxy-fluorescein (FAM) oligonucleotide and 1.5 µl GAPDH secondary oligonucleotide (if the experiment uses this as an internal control; otherwise add Te). Vortex the mixture, centrifuge to collect, and transfer into an 8-tube strip. Place on ice (*see* **Note 28**).
8. Load 96-well round-bottom plate with 5 µl primary reaction mix per well using the electronic repeat pipettor.
9. Add 5 µl standards or 5 µl sample RNA to the plate with 8-channel micropipettor.
10. Overlay each reaction with 10 µl Chill-Out wax using electronic repeat pipettor.
11. Cover plate with adhesive film.
12. Incubate at 60°C for 90 min in a thermocycler with 96-well plate format heating element or in a hybridization oven.
13. After 90 min, remove plate (if using hybridization oven), remove adhesive film, and add 5 µl secondary reaction mix to each sample with 8-channel micropipettor.

14. Incubate again 60°C for 60 min, or for up to 90 min, depending on signal strength desired.
15. Remove plate, remove adhesive film, and add 10 μl stop solution (optional).
16. Read plate in fluorescence plate reader immediately, using Magellan II software. For preliminary settings set: gain = 50, number of flashes = 10, lag time = 0 μs, integration time = 50 μs, plate definition file = NUN96ft.pdf (*see* **Note 29**).

3.8. Data Analysis, Standard Curves

1. Export data to Excel.
2. For each set of triplicates or quadruplicate points on the standard curve, calculate mean, standard deviation, and percentage coefficient of variation (*see* **Note 30**).
3. Subtract the mean value for the 0 amol in vitro transcript/well from each of the other standard curve means to generate net values.
4. For all standard samples >0 amol IVT/well, calculate FOLD > 0 by dividing the average gross signal for any particular sample by the average gross signal for 0 amol IVT/well.
5. For all standard samples >0 amol IVT/well, calculate SD > 0 by dividing the net signal for any particular sample by the standard deviation of the 0 amol IVT/well sample.
6. For all standard samples > 0 amol IVT/well, calculate a Student's *t* Test comparing that sample to zero. This gives the confidence that the value is different than zero. Use the TTest function in Excel and include all 4 zero signal values and all 4 sample signal values. Select 2 tails, type 1.
7. For all standard samples, calculate 95% confidence values. Use the Confidence function in Excel, choose alpha = 0.05, size as 4 (quadruplicates) or 3 (triplicates), and the worksheet cell containing the standard deviation of the sample with which you are working.
8. For all samples >0.01 amol, calculate a Student's *t* Test to compare it to the next lowest standard point. Use the TTest function in Excel as above, but select the 4 signal values for the sample being analyzed, and the 4 signal values for the sample below it. Other selections remain the same.
9. Plot the standard curve of net counts versus amol IVT/well using linear scale (not logarithmic). Examples of a standard curve showing a linear relationship between input RNA and output of relative light units is shown in **Fig. 2**. **Figures 3** and **4** provide examples of Invader sets that do not yield any light output or show a high background. Neither of these sets can be used for quantifying mRNA. Discard low points that have either TTest scores above 0.05 or FOLD > 0 scores <1.15. The lowest value that can be plotted is the limit of detection (LOD) of the assay. Discard high points that are plateaued. The linear range of the assay is the range of values between the lowest and highest plotted points. Fit with a linear trendline ($y = mx + b$) and calculate the R^2 value.

3.9. Data Analysis, mRNA Decay

1. Use only signal values that fall within the linear range of the standard curves run simultaneously with the experiment.

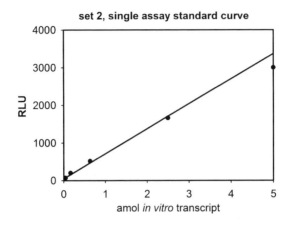

Fig. 2. Standard curve for an optimal Invader set. This is a standard curve for the Invader set directed against exon 1 of human beta-globin mRNA that is described in this chapter. Note that the line goes through zero and shows a linear increase in fluorescence output with increasing amount of in vitro transcript.

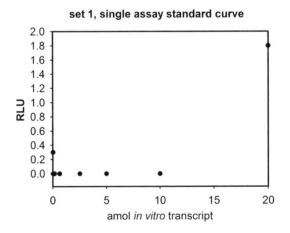

Fig. 3. Standard curve for an Invader set that shows no relationship to input RNA. The Invader set examined in this standard curve lies just upstream of the exon 1 set shown in **Fig. 2**. However, no change in fluorescence output is seen for this probe set when increasing amounts of the same RNA are added to the reaction.

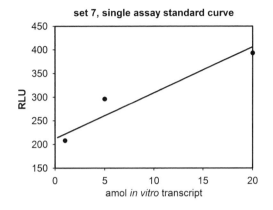

Fig. 4. Standard curve for an Invader set that does not go through zero. This is a standard curve for an Invader set in which fluorescence output is recorded with no input RNA. Although the amount of light-detected increases with increasing transcript this cannot be used.

2. Subtract the 0 amol IVT/well signal from the standard curve to calculate net signal values for all experimental values.
3. For biplex assays, divide the reporter gene signal in each well by the cotransfected control signal. This step is omitted when assaying a single mRNA in each reaction.
4. Calculate the mean and standard deviation for each set of quadruplicates or triplicates.
5. Using time = 0 h as 100%, calculate percent remaining. Percent remaining = [(signal of sample at time *x*)/(signal of sample at time 0)] × 100.
6. Plot as percent remaining (*y*-axis, logarithmic) versus time (*x*-axis, linear).
7. Fit the line and calculate the half-life.

3.10. Data Analysis, Calculating Amols Target Per Well

1. For each replicate in a set of triplicate or quadruplicate samples, plug the net RLU value into the equation describing the standard curve to get amol/well.
2. Calculate mean and standard deviation for each set of triplicates or quadruplicates.

4. Notes

1. All steps should be performed in a manner that is consistent with preventing RNAse contamination and protein degradation. If applicable, treat solutions with DEPC (autoclaving afterward to remove excess DEPC); wipe surfaces with RNAseAway (Invitrogen), wear gloves, use plasticware that is certified RNAse-free, and perform manipulations on ice. For all steps, use either double-distilled water or water that has been processed through a purification apparatus such as a Millipore system.

2. Alternately, one may co-transfect cells lacking tet repressor with a tet-regulated plasmid expressing the reporter mRNA and a plasmid expressing the tTA protein.

3. The primary oligonucleotide mix is composed of three oligonucleotides: a primary probe oligonucleotide, an Invader oligonucleotide, and a Stacking oligonucleotide. The primary probe oligonucleotide is composed of two regions, an assay specific region (ASR) and a flap region. The ASR is ≥10 bases in length to ensure specificity and anneals to the target mRNA. The flap region does not hybridize to the target mRNA and is a synthetic sequence designed only for use in the assay. The Stacking oligonucleotide binds the target mRNA at the 3′ end of the probe ASR and stacks coaxially with it. The function of the Stacking oligonucleotide is to improve assay performance by increasing the Tm of the primary probe and thus the assay reaction temperature. Together, the primary probe and Stacking oligonucleotide should have a Tm of 60°C. The Tm may be slightly higher, but a Tm of ≥65°C is detrimental to the assay performance. Because the Tm of the primary probe oligonucleotide is the temperature at which the assay is run, the primary probe oligonucleotide cycles on and off the target freely during the reaction. The Invader oligonucleotide binds the target mRNA immediately downstream (on the mRNA) from the primary probe oligonucleotide. The 3′ nucleotide of the Invader oligonucleotide does not hybridize with mRNA target or another oligomer, but instead through steric effects, pushes on the 5′-most bond between the primary probe oligonucleotide and the mRNA target. This creates a unique bulged secondary structure in the primary probe oligomer, which is recognized as a cleavage substrate by the Cleavase enzyme. The Cleavase enzyme then cuts the probe molecule to release the flap region from the ASR. The Tm of the Invader oligonucleotide is approximately 80°C to ensure that it is stably bound to the mRNA target throughout the assay. Because the difference in Tm between the stacking and probe oligomers and the Invader oligonucleotide is key to the success of the assay, the difference between them should always be approximately 20°C.

4. This is not typically a critical location but can be strategically chosen to quantify splice variants, spliced versus unspliced mRNAs or single nucleotide polymorphisms (SNPs) if desired.

5. Methods to determine mRNA secondary and tertiary structure include computer modeling and reverse transcription with random oligonucleotide libraries (RT-ROL) *(4)*, an experimental method. It might be necessary to design several possible sites for analysis and test each.

6. The Tm of the probe set is key to success with the Invader assay. A number of programs are available to determine the Tm of the hybrids, and this must be taken into account when determining the length or locations of particular oligonucleotides, or when either of these parameters are changed. Remember the combined Tm of the Stacking oligonucleotide and the primary probe oligonucleotide should be 60°C.

7. For published probe flap sequences, *see* **refs. 3** and **5**.

8. High purity oligonucleotides are essential for good results.

9. We have had success using oligonucleotides from Integrated DNA Technologies (IDT).

10. In the primary reaction, the Cleavase enzyme recognizes a distinct bulged secondary structure caused by the simultaneous hybridization of Stacking, primary probe, and Invader oligomers with the target mRNA. This recognition of the secondary structure leads to cleavage by Cleavase of the primary probe to release the flap portion. In the secondary reaction, the primary probe flap released in the first reaction becomes the Invader oligonucleotide in the formation of another secondary structure that serves to amplify the signal and couple cleavage with fluorescence production. The secondary reaction is performed by adding secondary oligonucleotides to the primary reaction. An Arrestor oligomer is added to bind up uncleaved primary probe oligonucleotide. A secondary probe molecule, containing a fluorophore molecule and a quencher molecule, is added to bind an SRT that simultaneously binds the primary probe flap to form the bulged structure. Cleavage of the secondary probe liberates the fluorophore from the quencher molecule, and the resulting free fluorophore can be determined by reading in a fluorescence plate reader.

11. The in vitro transcript serves as a control for the assay. It is used in optimizing the assay and in standard curves that must always be performed when running the assay. The in vitro transcript ideally spans the entire length of the mature mRNA and includes a poly(A) tail. The length of the tail may be short so that the molar concentration of the in vitro transcript may be accurately determined.

12. Transcriptional pulse assay versus transcriptional turn-off assay. The transcriptional pulse assay has the advantage of generating a homogeneous population of transcripts which should decay in a homogeneous fashion. However, expression of the reporter gene may be higher in cells subjected to only transcriptional turn-off. This may make it much easier to analyze decay in transcription turn off cells.

13. Transient versus stable cells. Of course, cells must be stably transfected with tTA, but whether to construct stable cells that also express the reporter gene is another question. Although transiently transfected cells typically give higher expression of the gene of interest, stably transfected cells have a more consistent level of expression because of a lower copy number per cell. This may make it easier to analyze the effect of overexpression of a regulatory protein on decay or to analyze the effect of knocking out a protein with siRNA on decay.

14. The level of tet or dox required to repress gene expression may be empirically determined through luciferase assay (using a tet-regulated luciferase reporter plasmid), northern blot, or RNase protection assay (perhaps using a specific tet-regulated reporter plasmid).

15. Tet has a half-life of 12 h in medium; dox has a 24-h half-life. If cells are grown for a few days, it will be necessary to add more antibiotic to maintain proper gene repression.

16. If using transiently transfected cells, care should be taken to determine the number of cells that will be needed to reach the desired degree of confluency at the times cells are harvested. On the day of cell harvest, the cells should nearly cover the plate but still be in log-phase growth.

17. The length of the pulse may also be empirically determined as above. If it is too long, a significant portion of the target mRNA may be undergoing decay, thus complicating interpretation of the data.

18. Assay design: A standard curve prepared from an in vitro transcript is required for each target mRNA. This is particularly important in establishing the linear range of the assay, which can vary between experiments—even for the same oligonucleotides and target.

19. To facilitate statistical analysis of the data, the assay should be run at least in triplicate and preferentially in quadruplicate.

20. When using cells transiently transfected with the reporter gene, be sure to include a cotransfection control as well as loading control. The loading control is accomplished by performing the Invader RNA assay in biplex format, that is, assaying for the presence of two RNA targets in each sample, the reporter gene and a housekeeping gene (beta-actin, GAPDH). Owing to the lack of availability of oligonucleotides for popular cotransfected controls such as luciferase, beta-galactosidase, or green fluorescent protein, cotransfected controls must often be quantitated by an alternative method such as northern blotting or RNase protection assay. It is not necessary to include a cotransfected control if using a cell line stably transfected with the reporter gene.

21. A typical standard curve includes in vitro transcript values of 0, 0.01, 0.02, 0.04, 0.16, 0.64, 2.5, 5.0, 10.0, and 20.0 amol of in vitro transcript per well. Because the volume of each diluted in vitro transcript added to each well is always 5 µl, the in vitro transcripts must be diluted to one-fifth of what will be in the well. For example, to add 5.0 amol in vitro transcript, the in vitro transcript must be diluted to 1.0 amol/µl. 5 µl of this is 5.0 amol in the well.

22. In vitro transcripts must be diluted in an aqueous solution of 20 ng/µl tRNA because the very low concentration of the transcripts leads to the risk of significant sample loss.

23. Each standard will be measured in triplicate or quadruplicate, so prepare at least 15–20 µl of each standard. Allow extra volume for pipetting error.

24. RNA samples should be diluted in water of the expression if the target gene is expected to be low such that a large amount of total cellular RNA will be added to each well (50–100 ng). RNA samples should be diluted in tRNA solution if the expression of the gene of interest is expected to be high such that a small amount of cellular RNA will be added to each well (≤50 ng). It is not advisable to use more than 100 ng cellular RNA per well.

25. The volume of cellular RNA that will be added to each well is always 5 µl. Therefore, cellular RNA must be diluted to a concentration one-fifth of the amount

that will be added to each well. For example, to add 30 ng of cellular RNA to a well, the cellular RNA will be diluted to a concentration of 6 ng/µl, such that the addition of 5 µl to a well gives 30 ng of sample RNA in the well.

26. Remember that each sample will be read in triplicate or quadruplicate, so prepare at least 15–20 µl of each sample. Allow extra volume for pipetting error.

27. The volume of primary buffer mix to prepare depends upon the number of wells in the assay plus extra for pipetting error. If preparing a small number of wells (≤20), use 20% extra mix volume. If preparing a large number of wells (~100 wells), it is sufficient to prepare enough for four extra wells. It is not usually necessary to prepare as large a volume as suggested in the assay manual.

28. This can be prepared before the assay or during the first incubation.

29. Adjust gain so that the "0 amol/well" sample of the standard curve has a value of approximately 100.

30. Percent coefficient of variation = [(standard deviation)/(average raw signal)] × 100. This value should be under 10%, preferably under 5%. This is a measure of the variation between replicates. The presence of an outlier can cause this value to be very high (≥10%).

Acknowledgments

This work was supported by PHS grant R01 GM38277 to D.R.S. E.L.M. was supported in part by PHS grant T32 CA09338, and support for core facilities was provided by PHS center grant P30 CA16058 from the National Cancer Institute to The Ohio State University Comprehensive Cancer Center. We thank Tsetka Takova for help with probe design, troubleshooting, and data analysis, Peggy Eis and David Fritz for help with design of the primary and secondary oligonucleotide probe sets, and Jim Dahlberg and members of the Schoenberg laboratory for helpful discussions. Invader® and Cleavase® are registered trademarks of Third Wave Technologies, Inc.

Referencess

1. Harrold, S., Genovese, C., Kobrin, B., Morrison, S.L., and Milcarek, C. (1991) A comparison of apparent mRNA half-life using kinetic labeling techniques vs decay following administration of transcriptional inhibitors. *Anal Biochem* 198, 19–29.

2. Loflin, P.T., Chen, C.-Y. A., Xu, N. and Shyu, A.-B. (1999) Transcriptional pulsing approaches for analysis of mRNA turnover in mammalian cells. *Methods* 17, 11–20.

3. Eis, P.S., Olson, M.C., Takova, T., Curtis, M.L., Olson, S.M., Vener, T.I., Ip, H.S., Vedvik, K.L., Bartholomay, C.T., Allawi, H.T., Ma, W.-P., Hall, J.G., Morin, M.D., Rushmore, T.H., Lyamichev, V.I., and Kwiatkowski, R.W. (2001) An invasive cleavage assay for direct quantitation of specific RNAs. *Nat Biotechnol* 19, 673–676.

4. Allawi, H.T., Dong, F., Ip, H.S., Neri, B.P., and Lyamichev, V.I. (2001) Mapping of RNA accessible sites by extension of random oligonucleotide libraries with reverse transcriptase. *RNA* 7, 314–327.

5. Wagner, E.J., Curtis, M.L., Robson, N.D., Baraniak, A.P., Eis, P.S. and Garcia-Blanco, M.A. (2003) Quantification of alternatively spliced FGFR2 RNAs using the RNA invasive cleavage assay. *RNA* 9, 1552–1561.

19

Development of an In Vitro mRNA Decay System in Insect Cells

Kevin Sokoloski, John R. Anderson, and Jeffrey Wilusz

Summary

Cytoplasmic extracts have proven to be a versatile system for assaying the mechanisms and interactions of RNA metabolism. Using *Aedes albopictus* (C6/36) cells adapted to suspension culture, we have been able to faithfully reproduce and manipulate all aspects of mRNA decay in vitro. Described in this chapter are the processes for both producing an active cytoplasmic extract and the subsequent applications of the extract with respect to mRNA decay. The following protocol for the production of cytoplasmic extracts from C6/36 cells can be altered to encompass a wide variety of cell types, including mammalian cell lines. In addition, a method for designing and implementing an in vitro transcription template to produce specific products are described in detail. Applications of the in vitro transcripts, specifically the deadenylation and exosome assays by which the decay of reporter transcripts is observed, are also examined in detail.

Key Words: mRNA decay; exosome; deadenylation; mosquito cells; cytoplasmic extracts.

1. Introduction

The level of an mRNA is determined by both its synthesis and its degradation. The contribution of mRNA decay to the control of gene expression has been highlighted by several microarray studies (*1*). These studies demonstrate that over 40% of observed changes in gene expression are due to regulation at the level of mRNA decay, making this post-transcriptional process a very major player in cellular expression. In addition to its influence on overall gene expression, mRNA decay also plays an important role in quality control. Processes such as nonsense mediated decay (NMD), non-stop decay,

From: *Methods in Molecular Biology, Vol. 419: Post-Transcriptional Gene Regulation*
Edited by: J. Wilusz © Humana Press, Totowa, NJ

and no-go decay ensure that only full-length, intact proteins are produced from mRNAs *(2–4)*. Through processes such as NMD and the interplay between stress granules and P-bodies *(5)*, mRNA decay is intricately associated with mRNA translation. Although the basic pathways of mRNA decay—deadenylation, decapping, 5′–3′ and 3′–5′ exonucleases, and endonucleases have been outlined, there are numerous questions in the field remaining to be addressed. These questions include the precise role of several factors that have been shown to play an integral role in decay pathways, the overall contributions of decay factors organized in P bodies, how premature termination codons are recognized by the decay machinery in conjunction with translation, the connections between the mRNA decay machinery and miRNAs, and a clear understanding of how the stability of mRNAs is regulated in response to extracellular stimuli.

To effectively approach questions such as these, it is very beneficial to have as large an experimental toolbox as possible. In vitro systems that faithfully reproduce in vivo phenomenon allow for the careful assessment of a plethora of biochemical parameters. Furthermore, in vitro assays provide a defined background that is often free of confounding side effects often observed in biological systems. In vitro systems have been instrumental in deciphering fundamental aspects of molecular processes such as transcription, translation, splicing, and polyadenylation. The purpose of this chapter is to present the methodologies we developed for a broadly applicable cell-free system to study regulated mRNA decay. The system that we derived reproduces many aspects of regulated deadenylation and 3′–5′ decay by the exosome complex. In addition, mRNA decapping, and 5′–3′ decay can also be studied in the system. Although originally described using HeLa cell cytoplasmic extracts *(6)*, our method for in vitro mRNA decay has been reproduced in numerous mammalian cell types as well as yeast *(7)* and trypanosomes *(8)*.

In this chapter, we provide a detailed protocol for our most recent success with in vitro mRNA decay—the establishment of the in vitro system that reproduces regulated mRNA decay in insect cell extracts, specifically mosquito (*Aedes albopictus*) C6/36 cells *(9)*. In addition to being interesting organisms, important laboratory model systems and/or disease vectors, insects possess several interesting and somewhat unique nuances in mRNA decay—most notably the presence of an endonuclease-mediated mechanism of NMD *(10)*. The technical approach outlined below creates a tool to help define many of the biochemical questions regarding mechanisms of insect mRNA decay and serves as a blueprint for the development of similar systems in other organisms.

2. Materials

2.1. Preparation of Cytoplasmic Extracts

1. *A. albopictus* (C6/36) spinner cells (ATCC# CRL-1660).
2. Phosphate-buffered saline (HyQ® Phosphate Buffered Saline, Hyclone, Logan, UT, USA).
3. SF-900 II SFM media (Gibco, Grand Island, NY, USA).
4. Penicillin/streptomycin solution (10,000 ug/ml, HyQ Penicillin-Streptomycin solution, Hyclone).
5. Buffer A: 10mM HEPES, pH 7.6 at room temperature, 1.5 mM $MgCl_2$, 10 mM KCl, 1 mM dithiothreitol (DTT) (added fresh before use), and 1mM phenylmethyl-sulfonate fluoride (PMSF) (added fresh before use). Store at 4°C.
6. Buffer B: 0.3 M HEPES, pH 7.6 at room temperature, 30 mM $MgCl_2$, 1.4 M KCl. Store at 4°C.
7. Glycerol.
8. Dounce homogenizer (Kontes, Vineland, NJ, USA).

2.2. Assembly and Transcription of a Plasmid-Based RNA Substrate

1. pGEM4 (or comparable in vitro expression vector) (Promega, Madison, WI, USA).
2. Tris–borate–EDTA (TBE) (10×): 890 mM Tris base, 890mM boric acid, 20mM EDTA pH 8.0. Store at room temperature.
3. Dialysis Tubing (Spectrapor Molecularporous membrane, Spectra).
4. Phenol : chloroform : isoamylalcohol (PCI) (25:24:1). Ensure that solution is mixed and saturated with 50mM Tris–HCl pH 8.0, 50 mM NaCl, and 5 mM EDTA pH 8.0.
5. 3 M sodium acetate, pH 5.0.
6. Ethanol. 100 and 80% solutions are required. Store at –20°C.
7. T4 DNA Ligase (New England Biolabs, Beverly, MA, USA).
8. T4 DNA Ligase Buffer (New England Biolabs).
9. Appropriate restriction enzymes and reaction buffers.
10. PureLink® HiPure Plasmid Filter Maxiprep Kit (Invitrogen, Carlsbad, CA, USA) or comparable alkaline lysis plasmid purification method/kit.
11. 10× RNA Polymerase Buffer (New England Biolabs).
12. rNTP Mixture: 5 mM rATP and rCTP; 0.5 mM rGTP and rUTP.
13. 5 mM $^{7me}G(5')ppp(5')G$ Cap (Amersham Biosciences, Piscataway, NJ, USA).
14. $\alpha^{32}P$-rUTP (800uCi/mmol; Perkin-Elmer, Shelton, CT, USA).
15. RNAse Inhibitor (New England Biolabs).
16. SP6 RNA Polymerase (New England Biolabs).
17. RNA loading buffer: 7 M urea, 1× TBE, 0.05% (w/v) bromphenol blue, 0.05% (w/v) xylene cyanol. Store at room temperature.
18. Autoradiography film.
19. High-salt column buffer (HSCB): 400 mM NaCl, 26 mM Tris–HCl, pH 7.6, 0.1% SDS. Store at room temperature.

20. Scintillation Fluid (ScintiSafe Econo 1, Fischer Scientific, Waltham, MA, USA).
21. 10 M Ammonium acetate.

2.3. In vitro RNA Decay Assays

1. Water, deionized and distilled water (ddH$_2$O). RNase-free. Used for the preparation of all solutions and buffers (*see* **Note 1**).
2. Polyadenylic acid (polyA), 500 ng/µL (Pharmacia Biotech, New York, NY, USA). Store in aliquots at $-80\,^\circ$C.
3. 10% (w/v) polyvinyl alcohol. Store at room temperature.
4. PC/ATP: 250 mM phosphocreatine and 12.5 mM adenosine triphosphate. The two are mixed and stored in aliquots at $-80\,^\circ$C.
5. C6/36 cytoplasmic cell extracts (*see* **Subheading 3.1.**).
6. ^{32}P-Labeled RNAs containing a 3′ 60 adenosine tail (polyadenylated) and non-polyadenylated RNA substrates (*see* **Subheading 3.2.**).
7. 40% (w/v) 38:1 acrylamide/bis-acrylamide. Solution is stirred until dissolved, vacuum filtered, and stored in brown bottles at 4°C.
8. TBE (10×).
9. Gel mix solution: 1× TBE, 7 M urea. Store at room temperature.
10. 10% ammonium persulfate (APS). Store at 4°C.
11. *N,N,N′,N′*-tetramethyl-ethylenediamine (TEMED) (Sigma, St. Louis, MO, USA).
12. RNA loading buffer.
13. HSCB.
14. tRNA (10 mg/mL). Store in aliquots at –20°C.
15. Phenol saturated with 1 M Tris–HCl, pH 8.0.
16. PCI (25:24:1).

3. Methods

3.1. Preparation of Cytoplasmic Extracts

1. In SF-900 II media supplemented with Pen/Strep (100U/ml), *A. albopictus* mosquito (C6/36) cells are cultured in spinner flasks at a concentration of 2–3 × 10^7 cells/ml. We generally use 3 L of cells per extract. As verified microscopically and through using Trypan blue analysis, cells should be healthy and actively growing in log phase to ensure high-quality, active extracts (*see* **Notes 2** and **3**).
2. All solutions used in this procedure should be pre-chilled and kept refrigerated or on ice throughout the protocol.
3. Harvest cells by centrifugation at 300 × *g*, for 5 min at 4°C. Media are removed, and cell pellet is gently resuspended in 30–50 mL of ice-cold PBS. Cells are transferred to a fresh 50-mL conical tube and are centrifuged at 300 × *g* for 5 min at 4°C. Discard the supernatant and record the packed cell volume (PCV).
4. Add 5× PCV of ice-cold Buffer A to the tube and gently shake to resuspend the cell pellet. Allow cells to swell in the Buffer A on ice for 10 min (*see* **Note 4**). Pellet

cells by centrifugation at 300 × g for 2 min at 4°C. The PCV should increase 1.5–2 times the original noted volume from cytoplasmic swelling because of osmosis.

5. Carefully remove and discard the supernatant, as the pellet of swelled cells will not be as firm as a standard cell pellet. Add 2× PCV ice-cold Buffer A to the swollen cell pellet, transfer mixture to appropriately sized dounce.

6. Gently dounce cells to rupture plasma membrane 10 times using an A pestle on ice. If desired, plasma membrane rupture can be verified by visualization under a microscope. Transfer the cell lysate from the dounce homogenizer to a fresh, pre-chilled 50-mL conical tube (*see* **Notes 5** and **6**).

7. Centrifuge the cell lysate at 900 × g for 5 min at 4°C. The pellet contains the cell nuclei, and the supernatant represents the cytoplasmic extract. Note the volume of the supernatant and remove it to fresh 50-mL conical tube kept on ice. Based on the volume of the supernatant, add 0.11× volume of ice-cold Buffer B to the cytoplasmic fraction.

8. Aliquot cytoplasmic fraction evenly into appropriate ultracentrifuge tubes and centrifuge at 100,000 × g for 1 h at 4°C in a pre-cooled rotor. Remove supernatant to a fresh tube (*see* **Note 7**), and add glycerol to a final concentration of 20%.

9. Aliquot the cytoplasmic extract into 1.5-mL tubes and freeze at –80°C (or quick freeze in liquid nitrogen and store at –80°C). Extracts have remained active in our experience for several years. However, it is a good idea to avoid repeated freeze–thaw cycles as activity will diminish with mistreatment (*see* **Note 8**).

3.2. Assembly and Transcription of a Plasmid-Based Template System for Preparation of RNA Substrates

3.2.1. Production of a Plasmid-Based Transcription Template

1. Digest vector containing desired sequence fragment and 1 ug pGEM4 with restriction enzymes compatible with subsequent ligation (*see* **Notes 9** and **10**). Visualize digestions using standard agarose gel electrophoresis to confirm that they are complete.

2. To gel purify desired DNA fragments, excise the correct molecular weight band from the gel. Gel slices are then loaded into dialysis tubing containing 200 µL of 1× TBE.

3. Place gel slices at 80 V for 5 min, reverse the polarity of the electrodes, and run for an additional minute. Carefully remove liquid from dialysis tubing and place into a fresh microcentrifuge tube.

4. Extract the sample with PCI. Add an equal volume PCI to the tube, vortex until the solution is homogenous, and spin for 3 min at full speed in a microcentrifuge. Remove the top aqueous phase to a second tube containing 1/10th initial volume 3 M sodium acetate. Add 2.5 vol 100% ethanol and let precipitate on dry ice for 10 min. Centrifuge samples for 10 min at full speed in a microcentrifuge. A pellet should be clearly visible. Remove the ethanol supernatant while not disturbing the

pellet. Wash the pellet by the addition of 200 μL 80% ethanol, remove the wash solution, and air dry pellets. Now the fragment and vector need to be resuspended at desirable concentrations. Resuspend the eluted DNA fragment in an appropriate volume based on your initial starting amount. Resuspend the digested vector to a final concentration of 50 ng/μL.

5. Insert the purified fragment into the digested vector using a standard DNA ligation reaction. Using a 3:1 ratio of insert to vector, assemble a 10-μL reaction consisting of an appropriate amount of fragment, 50 ng of digested vector, 1 μL T4 DNA ligase buffer, 1 μL T4 DNA Ligase, and ddH$_2$O to a final volume of 10 μL. Incubate at room temperature for 1 h.

6. Transform competent bacteria with the ligated plasmid product using standard techniques. Select for bacteria containing the plasmid of interest by using plating media supplemented with an appropriate antibiotic. Following overnight growth, select and culture positive colonies overnight in liquid culture. Harvest cells by centrifugation and extract DNA using a kit-based alkaline lysis method. We routinely sequence our purified plasmid ligation products for quality control purposes (*see* **Note 11**).

7. Digest 10 μg of the purified plasmid with a restriction enzyme that linearizes the vector just downstream of the insert (*see* **Note 12**). Proceed with a PCI extraction and ethanol precipitation as described in **step 4** of this procedure, resuspending the digested DNA in a final concentration of 1 μg/μl.

3.2.2. In vitro Transcription Using a Linearized Plasmid Transcription Template

1. Assemble the following reaction in a fresh tube: 1 ug plasmid template, 1μL 10× RNA polymerase buffer, 1 μL rNTP mixture, 1 μL 5 mM Cap, 4.5 μL radiolabeled rUTP, 0.5 μL RNAse Inhibitor, and 1 μL SP6 RNA polymerase. Mix gently and incubate for 1–6 h at 37°C (*see* **Notes 13–15**).

2. Adjust the volume to 100 uL and process the sample immediately with PCI extraction and ethanol precipitation as described in **Subheading 3.2.1., step 4**, but using one-third of initial volume 10 M ammonium acetate as the precipitation salt instead of sodium acetate (*see* **Notes 16** and **17**). Resuspend in 10 μL RNA loading buffer.

3. Analyze transcription products by electrophoresis on a 5% denaturing acrylamide gel. Following electrophoresis, expose the gel to autoradiography film, excise desired transcription product, and place the gel slice into a fresh microfuge tube containing 400 μL HSCB. Elute overnight at room temperature. Remember to dispose of all waste using methods conforming to institutional standards.

4. Remove the supernatant to a fresh microfuge tube, leaving the gel slice behind. Proceed with PCI extraction and ethanol precipitation of the sample. Resuspend in 21 μL dH$_2$O. Add 1 μL of resuspended transcript to 3 mL scintillation fluid and assess the amount of radioactivity using a liquid scintillation counter. In our hands,

a typical 10-μL transcription should give a total yield of at least 1.5×10^7 counts per minute (cpm).

5. Dilute the transcript to desired concentration. For the assays described in **Subheading 3.3.**, 1×10^5 cpm is sufficient.

3.3. In Vitro RNA Deadenylation and Exosome-mediated Decay Assays

The following method describes the basic reaction for an in vitro deadenylation assay. Conditions for our assay to assess exosome activity are almost identical. Differences between the deadenylation and exosome assay are noted in the method. The order of the steps is presented in a way to maximize time efficiency while performing the assay.

1. The following gel preparation assumes using glass plates of 20×21 cm; however, any gel size can be used depending on the RNA resolution that is needed for the particular RNA substrate.
2. Extensively clean two glass plates. Clamp the two glass plates together with 0.75-mm spacers on each side and the bottom, leaving the top space open to pour in the gel and place the comb.
4. Prepare a 5% polyacrylamide gel by combining 3.75 mL of 40% 38:1 polyacrylamide solution, 26.25 mL gel mix solution, 300 μL 10% APS, and 30 μL TEMED. Pour the gel solution into the space between the glass plates. Insert a comb that can generate wells wide enough to hold 20–30 μL of sample. Let gel solidify.
5. Thaw an appropriate ^{32}P-labeled RNA substrate (polyadenylated with a 60 base adenosine tail for deadenylation assays or non-polyadenylated for exosome assays), poly(A), C6/36 cytoplasmic cell extract, and PC/ATP on ice. Combine 6.5 μL 10% (PVA) Polyvinyl Alcohol, 2.0 μL RNA substrate (100,000 cpm/μL), 2.0 μL 500ng/μL poly A, 2.0 μL PC/ATP, and 16.0 μL cell extract (it is important to note that the cell extract should be added last) in a 1.5-mL microfuge tube. Exosome assays are set up in a similar fashion, with the exception being the omission of poly(A) from the reaction conditions. Immediately after addition of cell extract, mix gently and place the tube at 28°C (*see* **Notes 18** and **19**).
6. For each time point (the zero time point should be taken immediately after the cell extract is added), remove 6 μL of the reaction mixture to a new microfuge tube containing 400 μL of HSCB with 1 μL of tRNA. This procedure stops deadenylation, giving a snapshot of the process at any particular certain time point (*see* **Notes 20** and **21**).
7. Add 400 μL of PCI, vortex for 10 s and spin at $16,100 \times$ g in a microcentrifuge for 2 min. Remove as much of the aqueous phase as possible, taking care not to remove any phenol or protein at the interface, and place in a new 1.5-mL tube containing 1 mL of 100% ethanol. Thoroughly mix and place at –80°C or on dry ice for a minimum of 10 min.

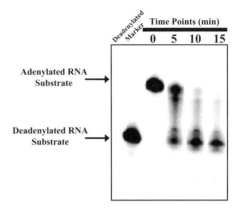

Fig. 1. A Gem4-based substrate RNA containing a 3′ tail of 60 adenosines is effectively deadenylated in C6/36 cytoplasmic cell extract. Polyadenylated RNA was incubated with C6/36 cell extract for the indicated amount of time. Reactions products were run on a 5% acylamide gel containing 7 M urea and analyzed using phosphorimaging. The deadenylated marker, which has the same sequence as the adenylated RNA substrate minus the 3′ poly(A) tail is shown in the rightmost lane.

8. Centrifuge samples at 16,100 × g in a microcentrifuge for 10 min. Remove the supernatant and appropriately discard it. Wash the RNA pellet with 200 μL 80% ethanol and spin at 16,100 × g for 5 min. Remove the supernatant and discard it.
9. Vacuum dry the pellets for 1 min (or until dry) in a Speed-vac or equivalent apparatus.
10. Add 8 μL RNA loading dye and vortex aggressively for 20 s to resuspend pellets. Heat the resuspended pellets for 30 s at 90°C and immediately place on ice for 30 s.
11. Load samples onto a pre-run 5% denaturing polyacrylamide gel and run at 600 V for approximately 30 min. Note that running time can be adjusted for optimum resolution between the adenylated and deadenylated products (*see* **Note 22**).

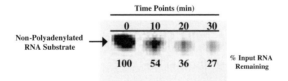

Fig. 2. A Gem4-based substrate RNA lacking a poly(A) tail is effectively degraded in the 3′–5′ direction in C6/36 cytoplasmic cell extract. The RNA substrate was incubated with C6/36 cell extract for the indicated amount of time. Reactions products were run on a 5% acylamide gel containing 7 M urea and analyzed using phosphorimaging. The percentage of the input RNA remaining at each time point is indicated at the bottom.

12. The gel is then removed from the glass plates and placed on thick chromatography paper (Fisher Scientific, Waltham, MA, USA) cut to be just larger than the gel. Plastic wrap is placed over the gel, and the gel is then dried in a slab gel dryer (Savant or equivalent) for 30 min.

13. The dried gel is exposed to a storage phosphor screen (Molecular Dynamics, Piscataway, NJ, USA) and viewed using a molecular imager (Bio-Rad, Herculus, CA, USA). An example of the deadenylation assay is shown in **Fig. 1** (*see* **Note 23**) and an exosome assay is depicted in **Fig. 2**.

4. Notes

1. We have found that collecting deionized water directly from a still into clear (never used for anything else) bottles and autoclaving them results in a consistent and reproducible source of RNAse-free water. We do not use DEPC-treatment or purchase certified RNAse-free water for our studies.

2. As previously discussed, a major versatility of this procedure lies in the fact that extracts may be produced from a multitude of cell types. The methods presented are applicable to the vast majority of tissue culture cells that we have tested.

3. In our experience, cells that are actively growing make the best extracts. Cells that are overgrown or stressed generally make poor or inactive extracts. Therefore, one of the key steps in the success of preparing a cell extract lies in the care and attention given to the cell culture.

4. One of the most common problems resulting in inactive cytoplasmic extracts is the failure to keep all solutions and equipment chilled on ice for the duration of the extract preparation; this cannot be stressed enough and is another important key to producing active cytoplasmic extracts.

5. We have observed that with certain cell types the nuclei tend to lyse along with the plasma membrane. Premature lysis can occur if the tonicity of the swelling buffer is not compatible with the osmolarity of the cell. We recommend altering the concentration of KCl in the Buffer A to find the best possible conditions for cell swelling without loss because of lysis. The pH of the solutions can also cause premature lysis, and buffers should have a pH of 7.6 at room temperature and should have a pH of 7.9 at 4°C. Adding $MgCl_2$ may also help stabilize the nuclear membrane.

6. Avoid rupturing the pelleted cell nuclei when removing the cytoplasmic fraction; doing so will result in contamination with nuclear proteins and extracts will exhibit increased non-specific nuclease activity, particularly from 3′–5′ exonucleases, as a result.

7. After centrifugation at 100,000 × g, a white film will appear on the surface of the cytoplasmic extracts. This film consists of cellular lipid and should generally be avoided to maintain extract consistency. However, we do note that recovering some of this film along with the soluble cytoplasm does not appear to appreciably affect the activity of an extract.

8. Traditionally, an active extract yields 4–7 mg/mL of protein when assayed using the BioRad assay or similar method for detection of protein concentration. Cell

extracts should be aliquoted in small volumes (enough for 1–2 uses) and stored at −80°C. Repeated freeze/thawing can sometimes lead to a decreased efficiency of decay assays.

9. Any commercially available vector containing an RNA polymerase initiation site (for example SP6 or T7) can be used as a transcription vector. Restriction enzyme selection will depend on the sites unique to both the vector and the desired insert.

10. Modifications can be introduced to the transcription vector using a similar method as the ligation of the insert. For example, we routinely use a variant we constructed called pGEM4-A60 *(9)* which contains an encoded poly(A) tail of 60 residues after the end of the multiple cloning site.

11. Occasionally, the insertion of sequences into a plasmid vector can be unusually inefficient. As always, a vector control should be included with the transformations to readily detect the level of background colonies in the procedure. If the background is high, treatment of the vector with 1 µL calf intestinal alkaline phosphatase (New England Biolabs) before PCI extraction is recommended. Additionally, following ligation, the residual vector can be re-linearized through digestion with a restriction enzyme unique to the vector (e.g., located in the middle of the sequence excised and replaced by the fragment of interest). This will greatly reduce the appearance of background, vector-only colonies.

12. Experimental evidence suggests that 3′ overhangs unfavorably influence the efficiency of bacteriophage RNA polymerases.

13. If template quality is poor or transcriptional efficiency is weak, the amount of template per reaction can be increased to 2 ug.

14. The reaction described is designed to produce a capped transcript internally radiolabeled at U residues. To use other labels, simply reduce the corresponding rNTP to 0.5-mM stock concentration. Alternatively, to produce uncapped transcripts, increase the concentration of GTP to 5mM and omit cap analog from the reaction. In addition, cold transcriptions can be performed by increasing all rNTPs to a stock concentration of 5 mM. Always remember when working with radiation, use proper shielding and follow all institutional policies regarding the use, storage, and disposal of hazardous materials.

15. The described reaction will produce a majority of capped transcripts; however, a small population of uncapped transcripts will unavoidably also be produced.

16. We have found that precipitating transcription reactions using ammonium acetate is an effective way of removing unincorporated nucleotides and cap analog from your RNA without having to use gel exclusion chromatography. Ammonium ions can inhibit some enzymatic reactions, however, so wash RNA pellets well before using RNA precipitated in this fashion.

17. Because occasionally it is difficult to observe the RNA pellet after ethanol precipitation, 1 ug of tRNA (or other suitable carrier) can be added to the reaction to enhance the visibility of the pellet.

18. The assay incubation temperature (28°C for C6/36) corresponds to the growth temperature for the cells used to make the cell extract. For many other tissue culture cells, we run the assay at 30°C.

19. If more reaction volume is needed (e.g., to do an extended time course), increase the amount of the overall reaction accordingly, keeping the proportions consistent.

20. With the reaction set up as above, time points of 0, 5, 10, and 15 min are usually adequate to view clear deadenylation of an RNA substrate containing a poly(A) tail of 60 residues in C6/36 mosquito cell extract (*see* **Fig. 1**). However, different cell extracts may require either shorter or longer time points depending on overall deadenylase concentration and activity. For example, HeLa cytoplasmic cell extracts generally require time points of 0, 10, 20, and 30 min to view complete deadenylation.

21. Exosome assays in C6/36 cytoplasmic cell extracts require time points of 0, 10, 20, and 30 min to see approximately 75% degradation of the input RNA (*see* **Fig. 2**). Again, longer or shorter time points may be necessary to see significant exonucleolytic degradation in other cell extracts. HeLa cell extracts, for example, often require up to 45 min for greater than 75% exosomal degradation of the input RNA to occur.

22. A non-polyadenylated form of the RNA substrate used can be loaded onto the gel and used as a marker for the fully deadenylated RNA substrate (*see* **Fig. 1**). For the marker to give you an optimal image, use cpm of the marker RNA equal to the cpm in the RNA pellets of **step 8** in **Subheading 3.3**.

23. Free methylated cap (7Me-GpppG) (200 µM final concentration) can be added to the deadenylation reaction to determine whether the active deadenylase is (PARN) Poly A Ribo Nuclease. As PARN is a cap-dependent deadenylase and readily binds free cap, it becomes sequestered when an abundance of cap analog is added to the reaction.

Acknowledgments

We thank all members, past and present, of the Wilusz laboratory who have contributed to the development of the in vitro mRNA decay system in a multitude of cell lines and organisms. This work was supported by NIH grants GM072481 and AI 063434 to J. W.

References

1. Cheadle, C., Fan, J, Cho-Chung, Y.S., Werner, T., Ray, J., Do, L., Gorospe, M., and Becker, K.G. (2005) Stability regulation of mRNA and the control of gene expression. *Ann N Y Acad Sci* **1058**:196–204.

2. Amrani, N., Sachs, M.S., and Jacobson, A. Early nonsense: mRNA decay solves a translational problem. (2006) *Nat Rev Mol Cell Biol* **7**:415–425.

3. van Hoof, A., Frischmeyer, P.A., Dietz, H.C., and Parker, R. (2002) Exosome-mediated recognition and degradation of mRNAs lacking a termination codon. *Science* **295**:2262–2264.

4. Doma, M.K. and Parker, R. (2006) Endonucleolytic cleavage of eukaryotic mRNAs with stalls in translation elongation. *Nature* **440**:561–564.

5. Anderson, P. and Kedersha, N. (2006) RNA granules. *J Cell Biol* **172**:803–808.

6. Ford, L.P., Watson, J., Keene, J.D., and Wilusz, J. (1999) ELAV proteins stabilize deadenylated intermediates in a novel *in vitro* mRNA deadenylation/degradation system. *Genes Dev* **13**:188–201.

7. Wilusz, C.J., Gao, M., Wilusz, J., and Peltz, S.W. (2001) Poly(A)-binding proteins regulate both mRNA deadenylation and decapping in yeast cytoplasmic extracts. *RNA* **7**:1416–1424.

8. Milone, J., Wilusz, J., and Bellofatto, V. (2004) Characterization of deadenylation in trypanosome extracts and its inhibition by poly(A)-binding protein Pab1p. *RNA* **10**:448–457.

9. Opyrchal, M., Anderson, J.R., Sokoloski, K.J., Wilusz, C.J., and Wilusz, J. (2005) A cell-free mRNA stability assay reveals conservation of the enzymes and mechanisms of mRNA decay between mosquito and mammalian cell lines. *Insect Biochem Mol Biol* **35**:1321–1334.

10. Conti, E. and Izaurralde, E. (2005) Nonsense-mediated mRNA decay: molecular insights and mechanistic variations across species. *Curr Opin Cell Biol* **17**: 316–325.

20

Using Synthetic Precursor and Inhibitor miRNAs to Understand miRNA Function

Lance P. Ford and Angie Cheng

Summary

Although the majority of gene function studies center themselves around protein-encoding RNAs, the study of non-protein-encoding RNAs is becoming more widespread because of the discovery of hundreds of small RNA termed micro (mi) RNA that have regulator functions within cells. Currently, over 470 human miRNA genes are predicted to exist and are annotated within the "miRBase" public miRNA database (http://microrna.sanger.ac.uk/). There is no denying that short interfering (si) and short hairpin (sh) RNAs have revolutionized how scientists approach understanding gene function; however, si and shRNAs are not effective for analyzing the function of miRNAs given that miRNAs are typically short (17–24 bases). In turn, new sets of agents that allow for the expression of miRNA above endogenous levels and inhibition of miRNAs have become a valuable technology for the study of these small regulatory RNAs. In this chapter, we provide step-by-step methods on how to utilize synthetic precursor and antisense inhibitor molecules for understanding miRNA function.

Key Words: miRNA function; miRNA inhibition; miRNA expression.

1. Introduction

Micro (mi) RNAs were originally thought to be restricted to expression and function in worms *(1–3)*. It is now known that miRNAs are a widely conserved, abundant, and diverse gene family that are found in many organisms. This has changed how scientists think about regulation of gene expression and the human genetic makeup. It turns out that the number of miRNA that exist corresponds to overwhelmingly greater than 1% of all protein-encoding genes. To expand

From: *Methods in Molecular Biology, Vol. 419: Post-Transcriptional Gene Regulation*
Edited by: J. Wilusz © Humana Press, Totowa, NJ

our interests further, it turns out that each individual miRNA has the potential to target several hundred to possibly even a thousand targets (determined by bioinformatic predictions). Based on these predictions, it has been suggested that miRNAs may regulate 20–80% of all genes.

To help identify miRNA-binding targets in mRNAs, several groups have developed predictive algorithms that take advantage of the observation that each of the known miRNA target sites in animal mRNA are complementary to positions 2 through 8 of the miRNA *(4)* and are normally restricted to the 3′ untranslated region (UTR) of an mRNA. Another commonly used criterion is that the predicted target site must be conserved between genomes of different organisms. These restrictions help narrow the number of putative targets and increase the chances that the predictions are correct. In a comprehensive study, Lewis et al. predicted mRNA targets for human miRNAs and used a reporter construct with the 3′ UTRs of sixteen target genes to confirm target sites for eight miRNAs. The system of transfecting miRNA or miRNA expression vectors and reporter constructs containing predicted miRNA-binding sites in its 3′ UTR offers a straightforward method for confirming target sites and is frequently used in miRNA studies.

Although miRNAs appear to regulate gene expression by impacting translation of mRNAs that contain sequences that are at least partially complementary to the miRNAs *(1,2)*, evidence also suggests that miRNAs can target mRNA for degradation and chromatin for transcriptional silencing *(5–7)*. For example, a recent study has found that mRNAs have an inverse expression relationship with the miRNAs that are predicted to target them *(8)*. In addition, two miRNAs have well-defined roles in mRNA degradation, where miR-196 targets HOXB8 mRNA for degradation and miR-16 influences A/U rich element-mediated mRNA turnover *(9)*. This indicates that miRNAs can have a direct impact on mRNA stability. Thus, at least three mechanisms of regulation of gene expression are known for miRNAs (i.e., at the level of mRNA degradation, mRNA transcription, and translation).

Ultimately, miRNAs impact the expression of proteins that impact critical biological processes such as development, differentiation, and disease *(10–18)*. Whereas a growing number of miRNAs are being implicated in important cellular functions, miR-181a has a clear involvement in cellular differentiation. Chen et al. *(19)* found miR-181a expressed specifically in the thymus, primary lymphoid organ, brain, and lung. They also found that in bone-marrow-derived undifferentiated progenitor (Lin-) cells, miR-181a is expressed at significantly lower levels than in differentiated B-lymphocytes. Ectopic expression of miR-181a using a mammalian expression vector increased the fraction of B lineage cells *(19)*. In addition, a study by Guimaraes-Sternberg demonstrated that miR-181 is involved in cellular differentiation in megakaryocytic lineage cells; as synthetic miRNA oligonucleotide mimicking the miR-181a sequence was administered, differentiation of Meg-01 cells was blocked and the physiological function of the mimicked miRNA was found to increase nitric oxide production indicative of macrophage activation *(20,21)*. Lastly, Naguibneva et al. *(14)* demonstrated that miR-181 targets Hox-A11 mRNA and modulates mammalian myoblast differentiation. Together, these studies indicate not only that miR-181 has significant biological roles in the development of blood and muscle cell lineages but also suggest that modulation of miRNA expression has the potential for tissue engineering of stem cells.

Although miRNAs can be expressed from mammalian polymerase II promoters *(22–25)*, adding about 30–100 bases of miRNA precursor flanking sequence has been found to be essential to get proper miRNA processing and expression *(24,26)*. In turn, when designing vectors to express miRNAs, it is suggested that the miRNA precursor sequence along with flanking genomic sequence be included in its construction *(17)*. These expression constructs are very useful for long-term analysis of miRNA mis-expression. Synthetic miRNA precursor-like molecules have also been developed and shown to function in short-term studies in cells *(27)*. In addition to expressing miRNAs in cells using vectors or synthetic miRNA precursor-like molecules, a clear need was identified and reagents were developed to inhibit miRNA function. Typically, miRNA inhibition is achieved using antisense oligonucleotides that bind to the mature miRNA sequence and prevent this mature miRNA from associating with its endogenous or exogenous targets *(28–30)*. Antisense molecules are effective at inhibiting the ability of miRNAs to function on exogenous reporters containing miRNA target sites in human cells and have been used to analyze miRNA activity in biological assays *(29,31)*. Additionally, in *Caenorhabditis elegans*, let-7 antisense molecules induced a let-7 loss-of-function phenocopy *(30)*.

These miRNA inhibitors, precursors, and expression vectors have begun to speed up the process of understanding the rich biological roles miRNAs play. In addition to being used for identifying miRNA activity in cellular processes and specific disease states, the technologies listed above and their extensions also have the potential to be used in therapeutic intervention *(18,32)*.

In studying miRNA function in mammalian cells, two main areas are important. First, the assay being used to understand the process needs to be functioning as expected. Second, proper controls and molecules need to be in place and function as expected in the assay. We will focus on the use of miRNAs to influence the expression of a reporter construct containing miRNA-binding sites located in its 3′ UTR as this is a key step in the understanding of the miRNA targets under evaluation as well as utilizing a library of miRNA precursor molecules on cell cycle progression.

It is important that transfection be optimized to ensure that the small nucleic acids are getting into the cells and capable of having biological activity. Using positive and negative control precursors or inhibitor molecules identify conditions where maximal activity of the molecules is detected with minimal amounts of cellular toxicity. Non-targeting control miRNAs should be used at the same concentration as positive control and experimental miRNAs because nucleic acid concentrations within cells can impact the activity and specificity. The miRNA precursor and inhibitor delivery optimization process relies on non-targeting control miRNAs to provide a baseline reference for ensuring that the transfection process itself is not impacting the process under evaluation or that the conditions being used are not leading to toxic conditions for the cells. This is important because the miRNA delivery process itself can induce changes in gene expression profiles, making a non-transfected or non-electroporated control meaningless for evaluation of silencing in delivery optimization

miRNA	Target(s)	Reference
miR-1	PTK9	(5)
miR-10a	HOXA1	(32a)
Let7	Ras	(10)
miR-122	CAT-1	(11)
miR-122	Cholesterol biosynthesis genes	(18)
miR-129 and miR-32	β-actin	(12)
miR-132	CREB	(13)
miR-181	Hox-A11	(14)
miR-196	Hoxb8	(32b)
miR-223	NFI-A	(15)
miR-221 and miR-223	kit	(16)
miR-327 and miR-323	LAST2	(17)

Fig. 1. Select miRNAs and targets. These miRNAs and their targets can be used as controls for analyzing and confirming the activity of miRNA precursor and inhibitor molecules following transfections into cell lines and tissues.

experiments. Non-targeting control miRNAs will cause the same non-specific effects on gene expression in the cell as the experimental miRNAs being intro-duced (regardless of the sequence) and therefore create the basis on which to evaluate the impact on the specific cellular process or gene being analyzed.

Choose positive controls to analyze activity of precursor and inhibitor miRNA molecules. Many miRNAs with defined targets can be used as positive controls in the optimization assays; for example, based on Vatolin et al., the β-actin gene is targeted by miR-129 and miR-32. Optimizations can be done by analyzing the impact of the precursor or inhibitor miRNAs on the endogenous target RNA or protein levels. Alternatively, the impact of miRNA inhibition or expression can be analyzed with a reporter vector that contains the 3′ UTR miRNA-binding sites from a validated target. Depending on the cell lines being used, an appropriate control may be chosen from the list of published miRNA targets as outlined in **Fig. 1**.

2. Materials

2.1. Cell Culture

1. Dulbecco's Modified Eagle's Medium (DMEM) (Gibco/BRL, Carlsbad, CA; cat. no. 10569-010) supplemented with 10% fetal bovine serum (heat-Inactivated) (Mediatech, Herndon, VA Inc.; cat. no. 35-011-CV).
2. Trypsin–ethylenediaminetetraacetic acid (EDTA) 0.05% (Gibco; cat. no. 25300-062).
3. 96- or 24-well cell culture plates (Nunc, Rochester, NY; cat. no. 167008 or no. 142475).
4. 1× Phosphate-buffered saline (PBS) is made from Ambion, Austin, TX; 10× PBS cat. no. 9625).

2.2. Transfection

1. siPORT™ *NeoFX*™ transfection agent (Ambion; cat. no. 4510).
2. miRNA precursors and inhibitors (Ambion; cat. no. 17100 and 17000).
3. Negative control precursors and inhibitors (Ambion; cat. no. 17110, 17111, and 17010).
4. DMEM (Gibco/BRL; cat. no. 10569-010) supplemented with 10% fetal bovine serum (heat-Inactivated) (Mediatech, Inc., cat. no. 35-011-CV).
5. Opti-MEM® reduced-serum medium (Gibco/BRL; cat. no. 31985-070).
6. 96- or 24-well cell culture plates (Nunc; cat. no. 167008 or no. 142475).

2.3. miRNA Precursor and Inhibitor Molecules

1. Anti-miR™ miRNA inhibitor (Ambion; cat. no. 17000).
2. Anti-miR™ miRNA inhibitor negative control (Ambion; cat. no. 17010).
3. Pre-miR™ miRNA Precursor molecule (Ambion; cat. no. 17100).
4. Pre-miR™ miRNA Precursor molecule negative control 1 (Ambion; cat. no. 17110).
5. Anti-miR™ miRNA inhibitor library (Ambion; cat. no. 17200V2).

2.4. Reporter Vector Construction

1. DNA oligonucleotides for either amplifying miRNA targets from cDNA or hybridization of short sequences as used in this chapter.
2. pMIR-REPORT™ miRNA vector (Ambion; cat. no. 5795).
3. Restriction enzymes.
4. Competent cells.
5. DNA ligase (Ambion; cat. no. 2134).
6. 1× Oligonucleotide hybridization solution: 100 mM potassium acetate, 30 mM HEPES-KOH, pH 7.4, and 2 mM magnesium acetate.
7. Nuclease-free water (Ambion; cat. no. 9937).
8. QIAEX II Gel Extraction kit (Qiagen, Germantown, MD; cat. no. 20021).
9. QIAprep Spin Miniprep Kit (Qiagen; cat. no. 27104).
10. TE Buffer: 10mM Tris–Hcl, pH 8.0, 1 mM EDTA.
11. Annealing Solution: 100 mM potassium acetate, 30 mM HEPES-KOH, pH 7.4 and 2 mM magnesium acetate.

2.5. Analysis of miRNA Function with a Reporter Vector

1. Applied Biosystems Dual-light® System (cat. no. BD100LP).
2. POLARstar OPTIMA microplate-based multi-detection reader.
3. 96-well *U*-bottom microtiter black plates (ABgene, United Kingdom cat. no. AB-0796/K)

2.6. Cell Cycle Analysis Assay

1. Acumen Explorer™ instrument from TTP LabTech.
2. 1× PBS is made from Ambion 10× PBS cat. no. 9625.
3. 2% Paraformaldyhyde solution (Sigma, St. Louis, MO; cat. no. P6148) (*see* **Note 1**).
4. Propidium Iodide (Molecular Probes, Carlsbad, CA; cat. no. P3566).

3. Methods
3.1. miRNA Reporter Vector Cloning (see Fig. 2)

1. Dilute the template oligonucleotides in approximately 100 µl of nuclease-free water.
2. Dilute 1 µl of each oligonucleotide 1:100 to 1:1000 in TE and determine the absorbance at 260 nm. Calculate the concentration (in µg/ml) of the oligonucleotides by multiplying the A260 by the dilution factor and then by the extinction coefficient (~33 µg/ml).
3. Dilute the oligonucleotides to approximately 1 µg/µl using nuclease-free water.
4. Anneal the sense and antisense strands of the oligonucleotides by adding 2 µg of each oligonucleotide to 46 µl of annealing solution and incubate at 90°C for 3 min and then at 37°C for 1 h.
5. Digest 5 µg of the annealed oligonucleotides and pMIR-Report™ luciferase vector with *Hin*dIII and *Spe*I and gel purify from a 2% agarose gel (*see* **Note 2**).

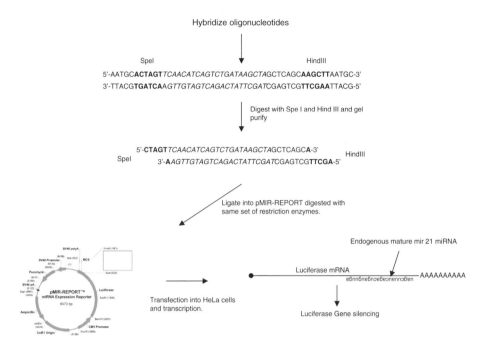

Fig. 2. pMIR-REPORT™ miRNA Expression Reporter Vector cloning. This figure depicts the cloning strategy for the reporter vectors including the design of the oligonucleotides, restriction sites, and map of the vector.

6. Mix approximately 5 ng of gel-purified hybridized oligonucleotide, 50 ng of linearized gel-purified pMIR-Report™ luciferase vector, 1 µl of 10× DNA ligase buffer and 1 µl of DNA ligase and water to bring the volume to a total of 10 µl, and incubate at room temperature for 1–3 h (*see* **Note 3**).

7. Transform an aliquot of competent cells with the plus-ligation products and transform a second aliquot with the minus-ligation products. (1) Use an appropriate amount of ligation product according to how the competent cells were prepared and the transformation method. (For chemically competent cells, we routinely transform with 3 µl of the ligation reaction.) (2) Plate the transformed cells on LB plates containing 50–200 µg/ml ampicillin or carbenicillin and grow overnight at 37°C. Generally, it is a good idea to plate 2–3 different amounts of transformed cells so that at least one of the plates will have distinct colonies. Always include a non-transformed competent cell control: this negative control is a culture of your competent cells plated at the same density as your transformed cells. (3) Examine each plate and evaluate the number of colonies promptly after overnight growth at 37°C (or store the plates at 4°C until they are evaluated) (*see* **Note 4**).

8. Evaluate for positive clones by growing colonies, isolating DNA, digesting the DNA with *Hin*dIII and *Spe*I, and analyzing them on a 10% native acrylamide gel. A positive clone should have an insert of approximately 35 bases that is released when digested with the restriction enzymes (*see* **Note 5**).

3.2. Transfections of HeLa Cells in 24-well Plates for miRNA Functional Analysis With a Reporter Vector Bearing miRNA-Binding Sites (see Fig. 3)

1. Pre-plate 50–60 K HeLa cells 24 h before transfection in 24-well tissue culture plates.
2. The next day, approximately 200 ng of each vector (control and experimental vector) and 10–30 pmole inhibitors are diluted into 50 µl Opti-MEM® in round-bottom polystyrene tubes.
3. Next, 3 µl of Lipofectamine™ 2000 is diluted into 50 µl Opti-MEM® and incubated at room temperature for 5 min.
4. The diluted DNA/miRNA inhibitor is next added into the transfection agent complex and incubated at room temperature for 20 min.
5. The growth media is changed and 100 µl of the complex is added to the cells.
6. Twenty-four hours following transfection, reporter activity is measured as described in **Subheading 3.3.**

3.3. Analysis of miRNA Function on a Reporter Vector Bearing a miRNA-Binding Site Located in its 3´ UTR

1. Rinse the cells twice with 1× PBS.
2. Add 100 µl of lysis solution (for a 24-well plate) and incubate for 10 min on ice.
3. Transfer 5–10 µl of the cell lysate to a 96-well *U*-bottom microtiter plate.
4. Add 25 µl of buffer A to each well of the 96-well plate that contains the lysed cells.
5. Within 10 min, inject 100 µl of buffer B (supplied in kit) and after a 1- to 2-s delay, read the luciferase signal for 0.1–1 s per well using POLARstar OPTIMA microplate-based multi-detection reader.
6. After the luciferase read, incubate the plate with the samples for 30–60 min at room temperature.
7. Add 100 µl of Accelerator II (supplied in kit). After 1–2 s, read the ß-galactosidase signal for 0.1–1 s per well using a POLARstar OPTIMA microplate-based multi-detection reader (*see* **Note 6**).

3.4. Transfections of HeLa cells in 96-Well Plates for miRNA Functional Analysis Screening, i.e., Cell Cycle Regulation

1. Dilute 3 pmole precursor miRNA molecules into 10 µl Opti-MEM® into each well of a 96-well plate.

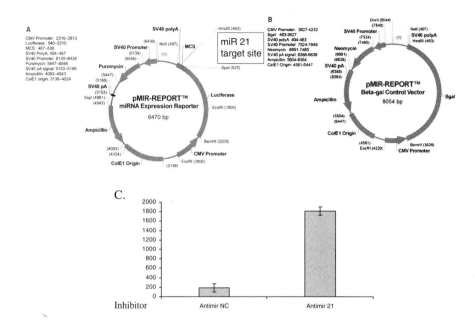

Endogenous expression of mir-21 in Hela cells as determined using array analysis: +++

Fig. 3. (**A** and **B**) miRNA reporter vectors used to analyze miRNA activity. The restriction sites for cloning the miRNA-binding sites are located in the 3′ UTR of the luciferase gene in the pMIR-REPORT™ luciferase vector. A ß-gal vector is used as a control for transfection efficiency—once cloned, the miRNA-binding site vector, pMIR-REPORT™ luciferase vector, is co-transfected with the pMIR-REPORT™ ß-gal vector. (**C**) Enhanced expression of a miRNA-regulated reporter vector with the use of a miRNA inhibitor. HeLa cells were plated at 50,000 cells/well in a 24-well plate. Cells were transfected using Lipofectamine™ 2000 in duplicate with pMIR-REPORT™ ß-gal vector, pMIR-REPORT™ luciferase vector that contained one target site for miR-21, and either an inhibitor for miR-21 or a negative control 2′-OMe-RNA (NC). Twenty-four hours post-transfection, cells were assayed for luciferase and ß-gal expression. ß-Gal is used to normalize for differences in transfection efficiency.

2. Dilute 0.3 µl siPORT™*NeoFX*™ transfection agent (Ambion, Inc.) into 10 µl Opti-MEM® for each sample.
3. Incubate the Opti-MEM/*NeoFX* mixture for 10 min at room temperature.
4. Add 10.3 µl of the Opti-MEM/*NeoFX* mixture to wells that contain the miRNA inhibitors or precursor miRNAs.
5. Incubate for 10 min at room temperature.

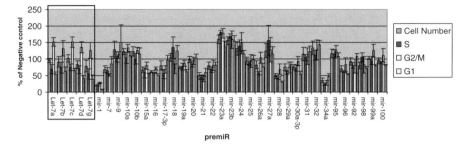

Fig. 4. Analysis of miRNA activity of cell cycle. After 72 h, the percent of cells in different cell cycles was measured. Different miRNAs affect the number of cells in a particular cell cycle.

6. Add 100 μl of diluted cell suspension mix containing 8000 cells/well to the Opti-MEM/*NeoFX* mixture.
7. After 24 h, the media is changed, and the samples are assayed after 72 h for cell cycle analysis as described in **Subheading 3.5** (*see* **Note 7**).

3.5. Analysis of miRNAs on Cell Cycle Regulation (Fig. 4)

1. 72-h post-transfection, media from three of the "cells alone" wells in each plate are replaced with media + vinblastine sulfate (3 μM) and incubated overnight at 37°C.
2. Cells are then fixed with paraformaldehyde (2% final concentration).
3. Cells are washed with 1× PBS and stained with 3 μM propidium iodide for >1 h at room temperature.
4. Individually scan plates with the Acumen Explorer instrument.
5. Analyze the percent of cells in G1, S, and G2/M versus other phases of the cell cycle as compared with negative control transfected sample.

4. Notes

1. When making 4% paraformaldehyde, add small amounts of 5 M NaOH and heat for 5 min at 37°C to get the paraformaldehyde to go into solution. Mix the paraformaldehyde powder in 1× PBS. Use the paraformaldehyde at a 2% final concentration.
2. To avoid digesting the inserts with restriction enzymes, it is also possible to use oligonucleotides that contain the correct overhang sequence for direct cloning into linearized and gel-purified pMIR-Report™ luciferase vector.
3. Important parameters for successful ligation: oligonucleotides should be at least 70% full length and accurately quantitated; this makes it possible to anneal equimolar quantities of sense and antisense hairpin siRNA oligonucleotides to make full-length, complementary hairpin siRNA-encoding inserts for ligation. Always include a minus-insert ligation control: colonies that grow after transformation with the

minus-insert ligation represent "background." The number of background colonies compared with the number of colonies from the plus-insert ligation indicates whether the ligation worked well or not. Successful plus-insert ligations will generate at least three times as many colonies as the minus-insert ligations.

4. Tips for transformation of *Escherichia coli*: Always include a non-transformed competent cell control: this negative control is a culture of your competent cells plated at the same dilutions as your transformed cells. No colonies should be seen on this plate, indicating that the ampicillin or carbenicillin in the culture medium is effective at inhibiting the growth of *E. coli* that does not contain the p*Silencer*™ plasmid. Use an appropriate amount of ligation product for the transformation. This amount will vary depending on how the competent cells were prepared and on the transformation method. Be sure that the bacterial culture medium has cooled adequately before adding 50–200 μg/ml ampicillin or carbenicillin. Store antibiotic-containing plates at 4°C and use them before the antibiotic loses it potency. Dilute the transformed cells in LB (the appropriate dilution will depend on the transformation method) and plate 2–3 dilutions. Incubate the transformants overnight and evaluate the results promptly (or store the plates at 4°C until they are evaluated).

5. Evaluating *E. coli* transformants: Examine each plate. The non-transformed control culture should yield no colonies (indicating that the ampicillin or carbenicillin in the culture medium is effective at inhibiting the growth of *E. coli* that does not contain the p*Silencer* plasmid). Identify the dilution of plus- and minus-insert ligation transformations that yields a reasonable number of well-spaced colonies. The minus-insert ligation will probably result in some ampicillin-resistant colonies, but the plus-insert ligation should yield 2- to 10-fold more colonies than the minus-insert ligation. (Remember to take the dilution into account when calculating the percent background.)

6. Normalize the luciferase readings by dividing by the β-gal readings to control for variation in transfections.

7. All transfections should be done in at least triplicate to ensure statistical significance of the data and if possible done on different days to ensure day-to-day reproducibility. This procedure is for reverse transfection; however, forward transfection or optimal conditions for different cells need to be determined empirically.

References

1. Lee, R.C., R.L. Feinbaum, and V. Ambros. The C. elegans heterochronic gene lin-4 encodes small RNAs with antisense complementarity to lin-14. *Cell* 1993; **75**(5): p. 843–54.

2. Wightman, B., I. Ha, and G. Ruvkun. Posttranscriptional regulation of the heterochronic gene lin-14 by lin-4 mediates temporal pattern formation in *C. elegans*. *Cell* 1993; **75**(5): p. 855–62.

3. Ha, I., B. Wightman, and G. Ruvkun. A bulged lin-4/lin-14 RNA duplex is sufficient for Caenorhabditis elegans lin-14 temporal gradient formation. *Genes Dev* 1996; **10**(23): p. 3041–50.

4. Lewis, B.P., C.B. Burge, and D.P. Bartel. Conserved seed pairing, often flanked by adenosines, indicates that thousands of human genes are microRNA targets. *Cell* 2005; **120**(1): p. 15–20.

5. Lim, L.P., et al. Microarray analysis shows that some microRNAs downregulate large numbers of target mRNAs. *Nature* 2005; **433**(7027): p. 769–73.

6. Yekta, S., I.H. Shih, and D.P. Bartel. MicroRNA-directed cleavage of HOXB8 mRNA. *Science* 2004; **304**(5670): p. 594–6.

7. Schramke, V., et al. RNA-interference-directed chromatin modification coupled to RNA polymerase II transcription. *Nature* 2005; **435**(7046): p. 1275–9.

8. Bagga, S., et al. Regulation by let-7 and lin-4 miRNAs results in target mRNA degradation. *Cell* 2005; **122**(4): p. 553–63.

9. Jing, Q., et al. Involvement of microRNA in AU-rich element-mediated mRNA instability. *Cell* 2005; **120**(5): p. 623–34.

10. Johnson, S.M., et al. RAS is regulated by the let-7 microRNA family. *Cell* 2005; **120**(5): p. 635–47.

11. Bhattacharyya, S.N., et al. Relief of microRNA-mediated translational repression in human cells subjected to stress. *Cell* 2006; **125**(6): p. 1111–24.

12. Vatolin, S., K. Navaratne, and R.J. Weil. A novel method to detect functional microRNA targets. *J Mol Biol* 2006; **358**(4): p. 983–96.

13. Vo, N., et al. A cAMP-response element binding protein-induced microRNA regulates neuronal morphogenesis. *Proc Natl Acad Sci USA* 2005; **102**(45): p. 16426–31.

14. Naguibneva, I., et al. The microRNA miR-181 targets the homeobox protein Hox-A11 during mammalian myoblast differentiation. *Nat Cell Biol* 2006; **8**(3): p. 278–84.

15. Fazi, F., et al. A minicircuitry comprised of microRNA-223 and transcription factors NFI-A and C/EBPalpha regulates human granulopoiesis. *Cell* 2005; **123**(5): p. 819–31.

16. Felli, N., et al. MicroRNAs 221 and 222 inhibit normal erythropoiesis and erythroleukemic cell growth via kit receptor down-modulation. *Proc Natl Acad Sci USA* 2005; **102**(50): p. 18081–6.

17. Voorhoeve, P.M., et al. A genetic screen implicates miRNA-372 and miRNA-373 as oncogenes in testicular germ cell tumors. *Cell* 2006; **124**(6): p. 1169–81.

18. Krutzfeldt, J., et al. Silencing of microRNAs in vivo with 'antagomirs'. *Nature* 2005.

19. Chen, C.Z., et al. MicroRNAs modulate hematopoietic lineage differentiation. *Science* 2004; **303**(5654): p. 83–6.

20. Guimaraes-Sternberg, C., et al. MicroRNA modulation of megakaryoblast fate involves cholinergic signaling. *Leuk Res* 2006; **30**(5): p. 583–95.

21. Guimaraes-Sternberg, C., et al. MicroRNA modulation of megakaryoblast fate involves cholinergic signaling. *Leuk Res* 2005.

22. Zeng, Y., X. Cai, and B.R. Cullen. Use of RNA polymerase II to transcribe artificial microRNAs. *Methods Enzymol* 2005; **392**: p. 371–80.

23. Zeng, Y. and B.R. Cullen. Sequence requirements for micro RNA processing and function in human cells. *RNA* 2003; **9**(1): p. 112–23.

24. Zeng, Y. and B.R. Cullen. Efficient processing of primary microRNA hairpins by Drosha requires flanking nonstructured RNA sequences. *J Biol Chem* 2005; **280**(30): p. 27595–603.

25. Zeng, Y., E.J. Wagner, and B.R. Cullen. Both natural and designed micro RNAs can inhibit the expression of cognate mRNAs when expressed in human cells. *Mol Cell* 2002; **9**(6): p. 1327–33.

26. Chen, C.Z. and H.F. Lodish. MicroRNAs as regulators of mammalian hematopoiesis. *Semin Immunol* 2005; **17**(2): p. 155–65.

27. McManus, M.T., et al. Gene silencing using micro-RNA designed hairpins. *RNA* 2002; **8**(6): p. 842–50.

28. Boutla, A., C. Delidakis, and M. Tabler. Developmental defects by antisense-mediated inactivation of micro-RNAs 2 and 13 in Drosophila and the identification of putative target genes. Nucleic *Acids Res* 2003; **31**(17): p. 4973–80.

29. Meister, G., et al. Sequence-specific inhibition of microRNA- and siRNA-induced RNA silencing. *RNA* 2004; **10**(3): p. 544–50.

30. Hutvagner, G., et al. Sequence-specific inhibition of small RNA function. *PLoS Biol* 2004; **2**(4): p. E98.

31. Cheng, A.M., et al. Antisense inhibition of human miRNAs and indications for an involvement of miRNA in cell growth and apoptosis. *Nucleic Acids Res* 2005; **33**(4): p. 1290–7.

32. Ford, L.P. Using synthetic miRNA mimics for diverting cell fate: a possibility of miRNA-based therapeutics? *Leuk Res* 2006; **30**(5): p. 511–3.

32a. Garzon, R., et al. MicroRNA fingerprints during human megakaryocytopoiesis. *Proc Natl Acad Sci U S A* 2006; **103**(13): p. 5078–83.

32b. Hornstein, E., et al. The microRNA miR-196 acts upstream of Hoxb8 and Shh in limb development. *Nature* 2005; **438**(7068): p. 671–4.

21

A Step-by-Step Procedure to Analyze the Efficacy of siRNA Using Real-Time PCR

Angie Cheng, Charles L. Johnson, and Lance P. Ford

Summary

Knockdown of cellular RNA using short interfering RNA has enabled researchers to perform loss-of-function (LOF) experiments in a wide variety of cell types and model systems. RNA interference techniques and reagents have made possible experiments that test everything from the analysis of function of single genes to screening for genes that are involved in critical biological pathways on a genome-wide scale. Although siRNA experiments are generally common practice in research laboratories, it is still important to keep in mind that many factors can influence efficacy of knockdown. A properly designed siRNA, optimized protocols of siRNA delivery, and an appropriate and well-optimized readout are all critical parameters for ensuring the success of your experiment. In this chapter, we provide step-by-step procedures for performing an siRNA knockdown experiment from cell culture to analysis of knockdown using quantitative real-time PCR.

Key Words: siRNA; transfection; knockdown; RNA isolations; reverse transcription; real-time PCR; TaqMan® gene expression assays.

1. Introduction

RNA interference (RNAi) is a mechanism by which short double-stranded RNA (siRNA) silences the expression of complementary target RNAs by inducing RNA cleavage and subsequent reduction in protein expression levels *(1–3)*. RNAi has changed how researchers study gene function in many diverse model systems and has tangible applications for tissue/cell engineering and human therapeutics *(4–9)*. Because siRNAs degrade their target mRNA, assessing siRNA efficacy is most often performed by quantifying target mRNA

From: *Methods in Molecular Biology, Vol. 419: Post-Transcriptional Gene Regulation*
Edited by: J. Wilusz © Humana Press, Totowa, NJ

levels using real-time polymerase chain reaction (PCR). Analyzing mRNA reduction is typically quantitative and provides a means to rapidly determine efficacy of the siRNA under different conditions and cell backgrounds. Following analysis of mRNA, experimentation to determine protein levels and phenotypic changes is often utilized to gain an understanding of how mRNA correlates with protein reduction (assuming an antibody-based assay is available for your gene of interest) and induction of a cellular phenotypic effect. Although not fully characterized, different genes have different threshold requirements of knockdown to obtain a biological phenotype. For example, one gene may require a 99% reduction of its protein to induce an observable phenotype whereas another gene may require only a 10% reduction. In turn, a negative result in a phenotypic activity assay does not necessarily mean the gene is not involved in the process under evaluation.

In addition to considering the threshold effects of genes, many other factors influence the success of a siRNA experiment. **Figure 1** depicts a general workflow of an siRNA experiment and lists many of the choices that need to be made before starting out. For example, some of the choices that need to be made are: What cell lines and growth conditions are going to be used? Should an siRNA or short hairpin (sh)RNA be used? What is the target or targets that are going to be analyzed? Should one or more than one siRNA be used? At what time or times should knockdown be analyzed? What method should be used to analyze the extent of the knockdown of the target gene? What agents should be used to deliver the siRNA into the cells? What controls (positive and negative) should be used? What concentration of siRNA should be used? Typically, control siRNAs are utilized for optimizing transfection conditions for the cell type and growth conditions that are chosen. Several types of positive and negative controls (NCs) are offered from commercial suppliers of siRNAs and can be easily obtained. To detect siRNA-mediated gene reduction, real-time polymerase chain reaction (PCR) is typically the most quantitative, easily accessible, and most frequently utilized.

Real-time PCR is a method to quantify the amount of a specific mRNA in a sample using a fluorescent readout during the progression of the PCR *(10,11)*. As the PCR reaction occurs, an increase in fluorescence is detected because of the increase in PCR product formed. The starting amount of the target can be determined based on the number of cycles the fluorescent signal takes to reach a given threshold relative to a standard. The standard is used to calculate the relative quantity of a product in a given sample. In our case, the standard is NC (an siRNA with no identity to endogenous genes) transfected cells. Targets in greater abundance will reach the threshold in fewer cycles; conversely, targets in lower abundance will take more cycles to reach the

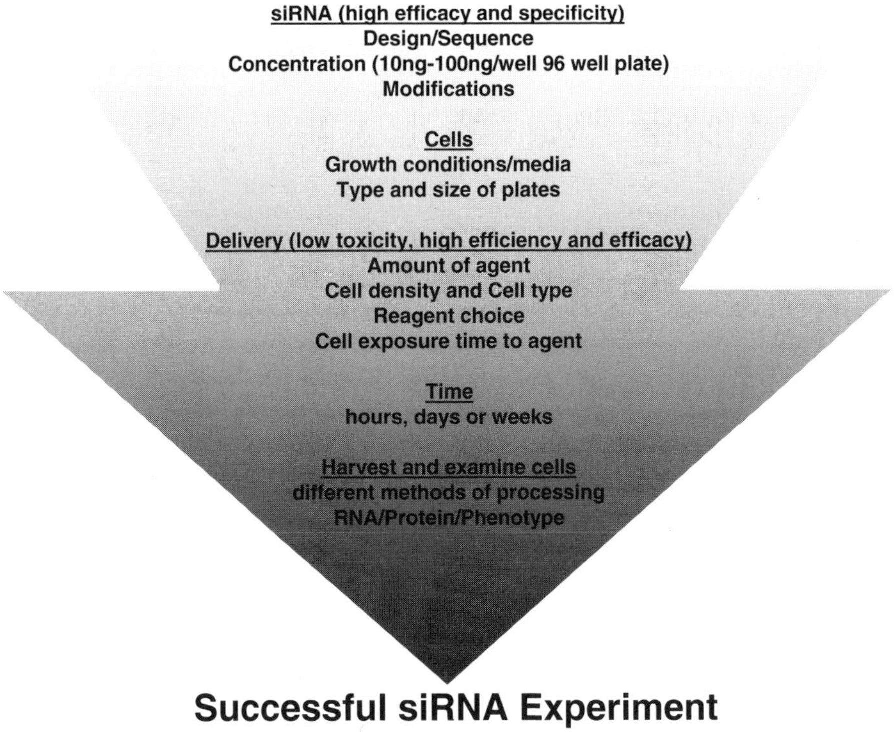

siRNA (high efficacy and specificity)
Design/Sequence
Concentration (10ng-100ng/well 96 well plate)
Modifications

Cells
Growth conditions/media
Type and size of plates

Delivery (low toxicity, high efficiency and efficacy)
Amount of agent
Cell density and Cell type
Reagent choice
Cell exposure time to agent

Time
hours, days or weeks

Harvest and examine cells
different methods of processing
RNA/Protein/Phenotype

Successful siRNA Experiment

Fig. 1. Diagram of the siRNA experimental workflow and the variables that can be optimized to achieve effective siRNA-mediated knockdown.

same threshold. There are two commonly used strategies for determining gene expression levels depending on the experimental strategy employed: absolute quantification and relative quantification. Absolute quantification compares the product amplification in a sample to a given set of standards, allowing for accurate relative levels using a standard curve to compare samples or accurate absolute levels that allows for copy number determination.

Experiments designed to look at the knockdown of a gene with siRNA are well suited for relative quantification and data analysis using the Comparative Ct or $\Delta\Delta$Ct method. This is the method we will describe in this chapter. This method requires that four measurements be taken, the Target, the Calibrator (NC-treated sample), and two Endogenous Controls. The Target is the nucleic acid sequence you are trying to knockdown using the gene-specific siRNA. The Calibrator is the nucleic acid sequence used to normalize for fluctuations in relative cell numbers. The Endogenous Control/calibrator is a gene with consistent expression in each sample measured (one measurement is taken from both the treated sample and the control sample). Typically, housekeeping genes such as β-actin, GAPDH, and ribosomal RNAs such as 18S are used as the endogenous controls. In turn, we will describe the materials and methods used to determine knockdown activity using siRNA transfection, RNA purification, reverse transcription, and real-time PCR techniques in this chapter.

2. Materials

2.1. Cell Culture and Transfection

1. Dulbecco's Modified Eagle Medium (DMEM) high glucose (Invitrogen, Carlsbad, CA) supplemented with 10% fetal bovine serum (FBS) (MediaTech, Inc.) and 1% penicillin (5000 Units)-streptomycin (5000 μg) (Pen-Strep) (Invitrogen). Store at 4°C (*see* **Note 1**).
2. Opti-MEM® I Reduced-Serum Medium (Invitrogen). Store in aliquots at 4°C.
3. PBS 10×, pH 7.4 (Ambion, Inc.). Store at room temperature (RT).
4. Nuclease-Free water (Ambion, Inc.). Store at RT.
5. 1000 mL 0.2 μM filter unit (Nalgene Labware, Rochester, NY).
6. Trypsin–EDTA (0.05% Trypsin with EDTA 4NA) 1× (Invitrogen). Store in aliquots at –20°C. Working aliquot can be stored at 4°C.
7. siPORT™ NeoFX™ Transfection Agent (Ambion, Inc.) Store at 4°C (*see* **Note 2**).
8. *Silencer®* Pre-designed siRNAs or *Silencer®* Validated siRNAs (Ambion Inc., Austin, TX). Store at –20°C or long term at –80°C.

2.2. siRNA

1. GAPDH control siRNA (Ambion, Inc.).
2. Neg1 control siRNA (Ambion, Inc.).
3. *Silencer®* CellReady™ siRNA Transfection Optimization Kit (Ambion, Inc.).

2.3. RNA Isolations

1. MagMAX™-96 Total RNA Isolation Kit (Ambion, Inc.).
2. 96-Well *U*-bottom plates and lids (Evergreen, Los Angeles, CA).
3. 96-Well magnetic-ring stand (Ambion, Inc.) (*see* **Note 3**).
4. 96-Well plate shaker (VWR).
5. 96-Well plate seals (Axygen Scientific).

2.4. Reverse Transcription

1. M-MLV Reverse Transcription Enzyme (100 U/μL) (Ambion, Inc.).
2. 10× buffer accompanied with M-MLV (Ambion, Inc.).
3. Random Decamers (50 μM) (Ambion, Inc.).
4. Nuclease-free Water (Ambion, Inc.).
5. RNase Inhibitor (40 U/μL) (Ambion, Inc.).

6. dNTP mixture (2.5 mM each) (Ambion, Inc.).
7. PCR plates or tubes, specific to PCR instrument.
8. PCR instrument (Applied Biosystems).

2.5. Real-Time PCR

1. Real-Time PCR Instrument—Applied Biosystems 7900HT Fast Real-Time PCR System (Applied Biosystems).
2. Real-Time PCR Plate (optically clear) specific to real-time PCR instrument.
3. TaqMan® Universal Master Mix, No AmpErase UNG (Applied Biosystems).
4. Nuclease-free Water (Ambion).
5. TaqMan® Gene Expression Assay targeted to Gene of Interest (GOI) (Applied Biosystems).
6. TaqMan® Gene Expression Assay targeted to 18S and GAPDH (Applied Biosystems).

2.6. Tissue Culture

1. Cell Growth Medium: DMEM + 10% FBS + 1% P/S. Thaw (1) 50 mL aliquot of FBS and (1) 5 mL aliquot of Pen/Strep (P/S) in a 37°C water bath, at RT or overnight at 4°C. When thawed, remove 55 mL from a new DMEM bottle and dispose. Then add the 50 ml of FBS and 5 ml of Pen/Strep, and invert the bottle 3 times to mix.
2. 1× PBS: Mix 100 ml of 10× PBS and 900 ml of nuclease-free water in a 1000-mL graduated cylinder. Seal the top with parafilm and invert three times to mix. Then filter sterilize it with a 1000 mL 0.2 μM Filter Unit using a vacuum pump.
3. Suspend siRNAs: siRNAs are typically shipped lyophilized or suspended at a specific concentration (usually 50 μM). If necessary, resuspend the siRNAs with nuclease-free water to a desirable concentration (usually 20–50 μM) (*see* **Note 4**).
4. Warm-up Reagents Before Use:
   a. Warm the Opti-MEM® I to RT before use.
   b. Warm cell growth media and trypsin in a 37°C water bath or incubator before use.
   c. Warm the siPORT™ *NeoFX*™ to RT before use.

3. Methods

As off-target effects are typically sequence-dependent, we recommend using two functional siRNA against one gene to ensure that the siRNA results are valid in the particular biological assay being analyzed *(12–17)*. In some cases, the best transfection method may be electroporation or may only be achieved using viral infections. General optimization of transfection is needed, which include testing the transfection agent type, amount, cell number, siRNA concentration, and assay time while balancing cellular toxicity and maximizing gene knockdown. In our case, we are using the GAPDH-validated siRNA and NeoFx pre-optimized conditions for knockdown and viability. See **Fig. 2** for overview of the procedure.

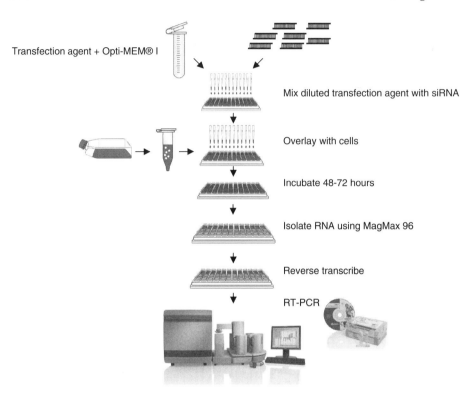

Transfection agent + Opti-MEM® I

Mix diluted transfection agent with siRNA

Overlay with cells

Incubate 48-72 hours

Isolate RNA using MagMax 96

Reverse transcribe

RT-PCR

Fig. 2. Real-time PCR amplification plot and knockdown of GAPDH using real time PCR generated using the GAPDH gene expression taqman assay and 7900 instrument. **(A)** Shows a picture of the GAPDH amplification plot that contains the cycle number on the X axis and the change in fluorescent value on the *y*-axis. The lines are depicted for the GAPDH siRNA treated sample (18S and GAPDH analysis) and for the negative control (NC) transfected sample (18S and GAPDH analysis). **(B)** The graph of the information generated using the ΔΔ CT method with the replicates as shown in **Note 15**.

3.1. Preparation of Cells for Transfections

1. In a standard T 75-cm^2 tissue culture flask, you can get 4–8 million cells depending on how confluent the cells are.
2. Make sure cells in the flask are healthy by observing for media color, cell morphology, media/cell contamination, and cell confluence.
3. After cells are determined to be healthy, trypsinize cells according to standard tissue culture protocols.
4. Collect the cells in a 15-mL polypropylene tube (*see* **Note 5**).
5. Count the cells with a hemocytometer or cell-counting instrument according to standard tissue culture protocols.

6. Dilute cells to the appropriate cell density, which is 4 K cells/well for HeLa cells in a 96-well plate, with cell growth media in a total of 80 µl per well. For the example used in our studies, we plated 10 wells worth of cells (1 GAPDH control siRNA in triplicate, triplicate Neg1 wells, triplicate non-transfected wells, and 1 extra well). Place 96-well cell suspensions in a 37°C tissue culture incubator or water bath until ready for use (*see* **Note 6**).

3.2. Preparation of siRNAs for Transfections in a 96-Well Plate

1. Dilute 0.15 ul of a 20-uM siRNA stock solution in 10 µl Opti-MEM® I per well in a sterile round bottom polystyrene tissue culture plate or polystyrene 12 × 75-mm tubes. Make master mixes when applicable for replicates to minimize variability. For example, make enough for 4 wells (40 µl) (*see* **Note 7**).
2. Dispense the 10-µl diluted siRNA into a 96-well plate in the desired format (*see* **Note 8**).

3.3. Reverse Transfection in a 96-Well Plate-Preparation of Transfection Complexes

1. Dilute the optimized amount of siPORT™ *NeoFX*™ transfection agent (which is 0.3 µl/well for HeLa cells) in Opti-MEM® I for a total volume of 10 µl in a polystyrene 12 × 75-mm tube. Make a master mix for 10 wells by mixing 3 µl siPORT™ *NeoFX*™ transfection agent + 97 µl Opti-MEM® I. Mix by gently flicking the tube but not so hard that liquid splashes onto the wall of the tube. It may be hard to collect all the liquid on the bottom. Incubate this mixture at RT for 10 min (*see* **Note 9**).
2. Combine 10 µl of the *NeoFX*™ mixture per 10 µl of diluted siRNA for a total of 20 µl with a multichannel pipettor or repeater. Mix by tapping the four corners of the plate against the bench to ensure complete coverage of the whole well. Incubate this mixture for 10 min at RT.
3. Next add 80 µl of the diluted cell suspension using a multichannel or repeater pipette to each well to bring the total volume to 100 µl. Be sure to mix the cells before pipetting to suspend cells that may have settled to the bottom. After the cells have been added to the plate, rock the plate back and forth to mix—but do not swirl.
4. Place the plate in a 37°C incubator under normal cell culture conditions. Remove and assay at the desired time point. Maximal knockdown is usually observed 48 h after transfection (*see* **Note 10**).

3.4. RNA Isolations with MagMAX™-96 Total RNA Isolation Kit

1. Empirically determine the maximum shaker setting that maintains efficient mixing without splashing. We use a Lab-line shaker instrument model no. 4625 with a setting speed between 5 and 8.5.
2. Prepare the reagents in the kit according to the manufacturer's protocol (*see* **Note 11**).
3. After 48 h following the transfection, remove the media from the cells, wash 2× with 1× PBS and add 140 µl of the Lysis/Binding solution to the wells. Shake plate for at least 1 min to lyse (*see* **Note 12**).

4. Transfer the lysate into a new 96-well *U*-bottom plate.
5. Add 20 μl of the freshly prepared bead mix and shake for 5 min. After shaking, beads should be completely mixed to a uniform brown color.
6. Move the plate to the magnetic stand to pellet the beads. Capture time is approximately 1–3 min.
7. Aspirate the supernatant without disturbing the beads (*see* **Note 13**).
8. Add 150 μl Wash Solution 1 and shake for 1 min. After shaking, beads should be completely mixed to a uniform brown color.
9. Move the plate to the magnetic stand to pellet the beads. Capture time is approximately 1–3 min.
10. Aspirate the supernatant without disturbing the beads.
11. Add 150 μl Wash Solution 2 and shake for 1 min. After shaking, beads should look like dark specs in a semi-clear liquid.
12. Prepare Diluted TURBO DNase™ (*see* **Note 14**).
13. Move the plate to the magnetic stand to pellet the beads. Capture time is approximately 1–3 min.
14. Aspirate the supernatant without disturbing the beads.
15. Add 50 μl of the Diluted TURBO DNase™ and shake for 15 min. After shaking, beads should be completely mixed to a uniform brown color. If not, pipet up and down to effectively break up the beads.
16. Do not place the plate on the magnetic stand. Directly add 100 μl of RNA Rebinding Solution and shake for 3 min.
17. Move the plate to the magnetic stand to pellet the beads. Capture time is approximately 1–3 min.
18. Aspirate the supernatant without disturbing the beads.
19. Add 150 μl Wash Solution 2 and shake for 1 min. Beads should look like darks specs in a semi-clear liquid.
20. Move the plate to the magnetic stand to pellet the beads. Capture time is approximately 1–3 min.
21. Aspirate the supernatant without disturbing the beads.
22. Repeat **steps 19–21**.
23. Dry the beads by shaking the plate for 2 min. Do not let the beads overdry.
24. Add 50 μl of the Elution Solution and shake for 3 min. After shaking, beads should be completely mixed to a uniform brown color.
25. Move the plate to the magnetic stand to pellet the beads. Capture time is approximately 1–3 min.
26. Transfer the purified RNA supernatant into a new 96-well *U*-bottom plate and seal. The RNA can be stored at –20°C or at –80°C for long-term storage.

3.5. Reverse Transcription

1. Combine the following in the appropriate wells of a 96-well PCR plate or tubes that will work with the PCR machine you will use:

1. 1 µL purified RNA (20–100 ng/µL); 1 µL random decamers.
2. Seal the plate or tube with a suitable PCR plate cover or cap and heat the plate to 72°C for 3 min and then cool to 4°C for at least 3 min. Spin plate or tube after cooling.
3. Prepare the following mixture (numbers listed are for a single well, but a master mix can be prepared for multiple samples plus 10% overage). In this example, make enough for 10 wells.

> Nuclease-Free Water 3.8 µL
> 10× Buffer 2.0 µL
> 2.5 mM dNTP Mixture 2.0 µL
> RNase Inhibitor (40 U/µL) 0.1 µL
> M-MLV RT Enzyme 0.1 µL

4. Add 8 µL of the above mixture to each well or tube and then mix, and spin the plate or tube for a total of 20 µL. Incubate at 42°C for 1 h and then at 92°C for 10 min. The cDNA may be stored on ice or frozen until real-time PCR is performed.

3.6. Real-Time PCR

1. Mix the following components for each gene being amplified (numbers listed are for a single reaction, but a master mix can be prepared for multiple samples plus 10% overage). Real-time PCR should be performed in triplicate for each sample for the Target (GAPDH), Calibrator (Neg1), and Endogenous Controls (18S). In our example, there will be 60 total wells.

> Nuclease Free Water 2.5 µL
> TaqMan® Universal Master Mix 5.0 µL
> TaqMan® Gene Expression Assay 0.5 µL

2. For each sample being examined, plate 2.0 µL of cDNA in 6 wells of a real-time PCR plate for GAPDH detection and 18S detection in triplicate.
3. Add 8 µL of the GOI (GAPDH) mixture to 30 of the wells and the Endogenous Control (18S) mixture to 30 wells.
4. Seal the plate with an optical cover and spin the plate.
5. The thermocycling conditions using TaqMan® Universal Master Mix and a TaqMan® Gene Expression Assay are 95°C for 10 min followed by 40 cycles of 95°C for 15 s and 60°C for 1 min.

3.7. Data Analysis

1. Once the PCR is complete, there are two settings that need to be adjusted for each gene, the threshold and the baseline. Baseline—This separates the background that occurs before the exponential phase of the amplification takes place. Set the upper limit just before the exponential phase begins. Set the lower limit such that it minimizes background. Threshold—This is the amount of fluorescence at which

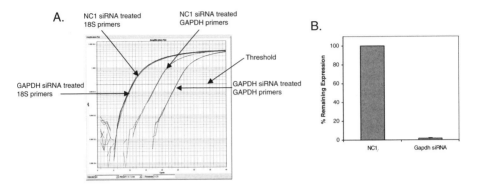

Fig. 3. Overview of the siRNA knockdown experiment. Diluted transfection agent is added to siRNA, incubated for 15 minutes and cells are added. Typically the media is changed 24 h following the transfection with fresh media to reduce toxic effects due to prolonged exposure to the transfection agent. At about 48–72 h following transfection, the RNA is isolated using MagMax, reverse transcribed, and used in PCR reactions using gene specific gene-expression assays on a 7900 instrument (depicted).

a Ct will be determined. This should be set such that it falls on the exponential amplification portion of the amplification plot for all of the samples for which that gene will be compared (see the threshold line in **Fig. 3**).

2. Once your baseline and threshold are set, export the file such that it can be edited in the spreadsheet program of your choice.

3. Calculate the average Ct and the standard deviation for the PCR triplicates for each sample and each gene (for each comparison there should be four averages).

4. Calculate the ΔCt for each sample: Subtract the Ct value of the endogenous control for the Target sample from the Target Ct. Then, subtract the Ct value of the endogenous control for the Calibrator sample from the Calibrator Ct.

5. Calculate the $\Delta\Delta$Ct for the two samples: Subtract the ΔCt for the Target sample from the ΔCt of the Calibrator sample. This value represents the fold difference between the levels of gene expression in the two samples.

6. To calculate the percent remaining use the following formula: Percent remaining = $100 \times 2^{-\Delta\Delta Ct}$ (*see* **Note 15**).

4. Notes

1. It is not necessary to supplement the DMEM bottle with 1% Pen/Strep antibiotics. Transfections work with or without antibiotics. With new bottles, make 50 ml aliquots of FBS and 5 ml aliquots of Pen/Strep upon arrival and store at –20°C.

2. siPORT™ *NeoFX*™ is a lipid-based transfection agent specifically designed for reverse transfections.

3. Ambion, Inc., sells two kinds of magnetic stands. Both isolate RNA well, so choice is determined by user preference. The ring stand is slightly easier to use if isolating by hand because the pellet is looser and so easier to suspend during washes, but be careful when removing the supernatant because you need to go in the middle of the ring pellet. If isolating by hand with a vacuum pump aspirator, the pedestal stand may be easier.

4. Resuspension of siRNA: Briefly centrifuge tubes to ensure that the dried oligonucleotide is at the bottom of the tube. Resuspend siRNA at the desired concentration in the nuclease-free water. For instance, resuspend 5 nmol of siRNA with 100 µl nuclease-free water to obtain a 50-µM solution. Ambion provides an online calculator for suspension of dry oligonucleotides on its Web Site at http://www.ambion.com/techlib/append/oligo_dilution.html. Once reconstituted in nuclease-free water, the siRNA is ready to transfect and can be used at your choice of final concentration (e.g., 1–100 nM). Store the reconstituted siRNA precursor at −20°C in a non-frost-free freezer.

5. A polypropylene tube is necessary so that the cells do not stick to the wall of the tube.

6. Here is a sample calculation to determine how many cells are needed:
Cell count = 400,000 cells/mL
Desired concentration = 4000 cells/well
Number of wells + overages = 10 wells
volume per well = 80 µl
Total volume needed = number of wells × volume per well = (10) (80 µl) = 800 µl
Total amount of cells needed = (number of wells) × (desired concentration) = (10 wells) (4000 cells/well) = 40,000 cells
Volume of cells needed = (amount of cells needed)/(cell count) = (40,000 cells)/(400,000 cells/mL) = 0.1 mL Mix the 0.1 mL cells with 0.7 mL growth media to give the total volume needed of 800µl.

7. Here is a sample calculation to determine how much siRNA you need: $M_1 V_1 = M_2 V_2$
Stock siRNA concentration = 20,000 nM
Desired siRNA concentration = 30 nM
Final transfection volume = 100 µl
(20,000 nM)(x) = (30 nM) (100µl) x = 0.15 µl/well
Triplicate wells + 1 well overage = 0.15 µl × 4 = 0.60 µl

8. A repeater pipettor may be useful for your replicates. Pipet 30 µl of your master mix and multi-dispense 10 µl to your triplicate wells. There should be approximately 10 µl left in your master mix tube because you made enough for 4 wells. To save time and if there are multiple siRNAs to screen at once on the actual transfection day, pre-plating siRNAs is an option. In a deep-well 96 dish, dilute siRNAs to the appropriate concentration in water for 20 µl/well and pre-plate the 20 µl in 96-well plates using a multichannel pipet. For example, for a final siRNA concentration of 30 nM in a 100-µl total transfection volume, dilute 3 pmoles in 20 µl water. Then, make a master mix for how many replicates and plates. Freeze plates at −20°C if required.

9. If the siPORT™ *NeoFX*™ transfection agent/ Opti-MEM® I mixture does not fit in a polystyrene 12 × 75-mm tube, the *NeoFX*™ can be diluted in a polypropylene tube such as a 15-mL conical tube.

10. After 8–24 h, the media may be changed to reduce cellular toxicity because of transfections. Aspirate media and replace with 100 µl fresh growth media.

11. Add 44 mL of 100% EtOH to the Wash 2 Solution. Add 6 mL of 100% isopropanol to the Wash 1 Solution and RNA-Rebinding Solution. Lysis/Binding solution composition depends on cell and tissue type. For mammalian cells in culture, make the Lysis/Binding Solution for 10 wells by mixing 770 µl Lysis Binding Solution Concentrate and 630 µl 100% isopropanol. Prepare the bead mix fresh on the day it will be used. For 10 wells, mix 100 µl beads + 100 µl water. Lysis/Binding Enhancer is included with the kit to resuspend the beads, but if you do not remove it completely during washes, it may interfere with downstream applications. If dealing with low cell number, water can be substituted. Prepared solutions with the ethanol and isopropanol are stable at RT for months.

12. After plates are lysed, they can be frozen at –80°C until ready for isolations.

13. A vacuum pump aspirator with an 8-channel plastic adaptor for pipet tips (without filters) can be used to remove the supernatant cleaner and faster.

14. Prepare Diluted TURBO DNase™ nuclease for 10 wells: 490 µl of MagMAX™ TURBO DNase Buffer + 10 µl TURBO DNase™ nuclease.

15. Here is a sample data analysis calculation using a transfected GAPDH control siRNA and a Neg1 siRNA. Triplicate wells were set up for real-time PCR and analyzed as follows:

a. GAPDH target: GAPDH siRNA

Ct	dCT	ddCT	% RE	Avgerage % RE	SD
24.63711	13.333993	5.6995324	1.9242868	1.884353	0.3224205
24.724567	13.42145	5.7869894	1.8111007		
24.348963	13.045846	5.4113854	2.3496906		
24.538855	13.235738	5.6012774	2.0599064		
24.48225	13.179133	5.5446724	2.1423346		
24.502605	13.199488	5.5650274	2.1123206		
25.017992	13.714875	6.0804144	1.4777905		
25.125948	13.822831	6.1883704	1.3712445		
24.807014	13.503897	5.8694364	1.710502		

dCT = Ct – Avg 18S target for GAPDH siRNA ddCT = dCT – Avg GAPDH target for Neg1 %RE = 100 × $2^{\wedge - ddCt}$

b. GAPDH target: Neg1 siRNA

Neg1 siRNA	Ct	dCT
18.574251	7.7142828	7.6344609
18.432425	7.5724568 Avg. Neg1	
18.506578	7.6466098	
18.50076	7.6407918	
18.463566	7.6035978	
18.455627	7.5956588	
18.583641	7.7236728	
18.505386	7.6454178	
18.427628	7.5676598	

dCT = Ct – Avg 18S target for Neg1

c. 18S target for Neg1

Ct	Avg.
10.873851	10.859968
10.844239	
10.861815	

d. 18S target for GAPDH siRNA

Ct	Avg.
11.446755	11.303117
11.466453	
10.996142	

References

1. Elbashir, S.M., et al. (2001) Duplexes of 21-nucleotide RNAs mediate RNA interference in cultured mammalian cells. *Nature* **411**(6836): p. 494–8.
2. Yang, D., H. Lu, and J.W. Erickson (2000) Evidence that processed small dsRNAs may mediate sequence-specific mRNA degradation during RNAi in Drosophila embryos. *Curr Biol* **10**(19): p. 1191–200.
3. Tuschl, T., et al. (1999) Targeted mRNA degradation by double-stranded RNA in vitro. *Genes Dev* **13**(24): p. 3191–7.
4. Mocellin, S., R. Costa, and D. Nitti (2006) RNA interference: ready to silence cancer? *J Mol Med* **84**(1): p. 4–15.

5. Saulnier, A., et al. (2006) Complete cure of persistent virus infections by antiviral siRNAs. *Mol Ther* **13**(1): p. 142–50.

6. Campochiaro, P.A. (2006) Potential applications for RNAi to probe pathogenesis and develop new treatments for ocular disorders. *Gene Ther* **13**(6): p. 559–62.

7. Mallanna, S.K., et al. (2006) Inhibition of Anatid Herpes Virus-1 replication by small interfering RNAs in cell culture system. *Virus Res* **115**(2): p. 192–7.

8. Manfredsson, F.P., A.S. Lewin, and R.J. Mandel (2006) RNA knockdown as a potential therapeutic strategy in Parkinson's disease. *Gene Ther* **13**(6): p. 517–24.

9. Rodriguez-Lebron, E. and H.L. Paulson (2006) Allele-specific RNA interference for neurological disease. *Gene Ther* **13**(6): p. 576–81.

10. Bustin, S.A (2002) Quantification of mRNA using real-time reverse transcription PCR (RT-PCR): trends and problems. *J Mol Endocrinol* **29**(1): p. 23–39.

11. Peters, I.R., et al. (2004) Real-time RT-PCR: considerations for efficient and sensitive assay design. *J Immunol Methods* **286**(1–2): p. 203–17.

12. Jackson, A.L. and P.S. Linsley (2004) Noise amidst the silence: off-target effects of siRNAs? *Trends Genet* **20**(11): p. 521–4.

13. Snove, O., Jr. and T. Holen (2004) Many commonly used siRNAs risk off-target activity. *Biochem Biophys Res Commun* **319**(1): p. 256–63.

14. Saxena, S., Z.O. Jonsson, and A. Dutta (2003) Small RNAs with imperfect match to endogenous mRNA repress translation. Implications for off-target activity of small inhibitory RNA in mammalian cells. *J Biol Chem* **278**(45): p. 44312–9.

15. Jackson, A.L., et al. (2003) Expression profiling reveals off-target gene regulation by RNAi. *Nat Biotechnol* **21**(6): p. 635–7.

16. Qiu, S., C.M. Adema, and T. Lane (2005) A computational study of off-target effects of RNA interference. *Nucleic Acids Res* **33**(6): p. 1834–47.

17. Birmingham, A., et al. (2006) 3´ UTR seed matches, but not overall identity, are associated with RNAi off-targets. *Nat Methods* **3**(3): p. 199–204.

Index

Printed in the United States of America.